Histone Deacetylase Inhibitors in Combinatorial Anticancer Therapy

Shabir Ahmad Ganai

Histone Deacetylase Inhibitors in Combinatorial Anticancer Therapy

Springer

Shabir Ahmad Ganai
Division of Basic Sciences and Humanities
Faculty of Agriculture
Wadura Sopore, SKUAST-Kashmir, India

ISBN 978-981-15-8178-6 ISBN 978-981-15-8179-3 (eBook)
https://doi.org/10.1007/978-981-15-8179-3

This Springer imprint is published by the registered company Springer Nature Singapore Pte Ltd.
The registered company address is: 152 Beach Road, #21-01/04 Gateway East, Singapore 189721, Singapore

I dedicate this book to my wife Fatima Akhter, to my son Mohammad Faiq Ganai, and to my daughter Hafizah Shabir. While the writing of this book was ongoing, I was not able to give them company.

Foreword

Apart from the recent novel developments in cancer therapy such as immune checkpoint inhibitors or targeted agents focusing on altered cancer signaling pathways, the important role of epigenetic changes for tumorigenesis and tumor development as well as the potential of targeting these epigenetic changes as an alternative antineoplastic approach is increasingly accepted. In this context, the term epigenetics does not only refer to concepts that can explain how very different cellular phenotypes can arise without changes in the sequence or amount of inherited DNA, paradigmatic for the development of an organism but also to mechanisms in the cell nucleus that establish stable states of gene expression. Though it is increasingly well understood how mechanisms interact with each other to produce complex phenomena such as cell differentiation, gene dose compensation, or genomic imprinting, the question of how epigenetic changes interact in the development of malignant diseases and how we can target these changes remains a scientifically challenging one.

In this context, Dr. Shabir Ahmad Ganai aggregates the current knowledge on one of the most promising epigenetic targets, namely histone deacetylases (HDACs), in his book.

Dr. Shabir Ahmad Ganai has formerly remained the Principal Investigator/Young Scientist in the Department of Biotechnology, University of Kashmir. Following a masters degree from Karnataka and earning his Ph.D from Tamil Nadu, he dedicated his scientific efforts on understanding the modulations of the nuclear geometry and post-translational modifications during treatment with structurally distinct HDAC inhibitors. Further, he focuses on designing target-selective HDAC inhibitors using various Bioinformatics tools. So, given his scientific focus and the research efforts of our group and the use of isoenzyme-specific HDAC inhibitors in the antineoplastic treatment of urothelial carcinoma, it was just a matter of time that we stumbled upon each other's work.

In his book "Histone Deacetylase Inhibitors in Combinatorial Anticancer Therapy," Dr. Shabir Ahmad Ganai now explores the role of tumor epigenetics and antineoplastic epigenetic therapy strategies with a special focus on histone deacetylases (HDACs) and HDAC inhibitors.

Following a concise overview of the concept of epigenetics in health and disease, the epigenetic "key players," with a special focus on HDACs, important for human cancer biology are introduced to the interested reader. Further, from a biochemical point of view, current anticancer strategies including the use of epigenetic modulators in general and HDAC inhibitors in particular are outlined. Equipped with this fundamental knowledge, the potential use of these inhibitors as novel therapeutic agents both as singlet agent and as combination partners for other anticancer agents is discussed. For the interested reader, also novel solutions regarding the design of isozyme-selective inhibitors to improve both therapeutic efficacy and toxicity profile of HDAC inhibitors are discussed.

Both the basic researcher as the clinician-scientist (as myself) will find in this interesting book a valuable source of knowledge regarding tumor epigenetics in general and HDAC inhibitors in cancer medicine in particular bridging the gap between "bench" and "bedside."

Günter Niegisch

Department of Urology, Division of
Conservative Uro-Oncology,
Medical Faculty of the Heinrich-Heine-University
Düsseldorf, Germany

Preface

During my doctorate, I started working on small molecule inhibitors of histone deacetylases (HDACs). These inhibitors known as histone deacetylase inhibitors (HDACi) modulate gene expression programs following the epigenetic route. Through wet-lab experiments, I explored how structurally different HDACi sodium butyrate and entinostat alter the nuclear architecture and how they influence site-specific methylation in HeLa cells (cervical cancer cells) stably or transiently expressing H2B-EGFP and H3-EGFP and H3K9R-EGFP. These questions were addressed using **cell culture techniques, transient transfection, therapeutic intervention, mutagenesis (H3K9 to H3R9 conversion), immunofluorescence, and laser scanning confocal microscopy**. Following this, the second phase of my Ph.D. started with drug designing. In this phase, I worked on how minute structural differences at the active sites of Class II HDACs can be exploited for designing isozyme selective inhibitors. I used a variety of methods including **extra-precision molecular docking, molecular mechanics generalized born surface area (MMGBSA), and energetically optimized pharmacophores approach.** These methods were available in **Schrödinger Suite** which the university had purchased for the research purpose. In the third phase of my Ph.D., I worked on how HDAC inhibitor valproic acid in the human cell model alters site-specific methylation marks. A variety of techniques like **cell culture, pharmacological intervention, immunofluorescence, nuclear staining, quantitative PCR, confocal imaging, and molecular docking** were used for reaching a logical conclusion. As a **Principal Investigator/Young Scientist** at the **University of Kashmir (Department of Biotechnology)**, I worked on finding isozyme selective inhibitors of HDAC2. Further, I investigated the binding inclination and interaction mechanism of two plant-derived inhibitors apigenin and luteolin against Class I HDAC isozymes. Importantly, through hardcore Molecular Biology and Biochemistry approach we explored an entirely novel Plant based inhibitor against epigenetic therapeutic target from certain plant sources of Kashmir Valley. Different techniques like HEK-293T, HL-60 cell cultures, Pharmacological Intervention, Liquid chromatography-tandem mass spectrometry (LC-MS/MS), Western Blots, Molecular Docking/ MMGBSA, Real time PCR, Chromatin Immunoprecipitation (ChIP) followed by PCR, MTT assays,

DNA laddering assay and ChIP-sequencing were used to arrive at logical conclusion.

During this time that is from 2009 to the present day, I came across a lot of research, review papers, and books on HDACi. This made me quite aware of the current lacunae in the field. Thus, first I wrote a book on histone deacetylase inhibitors in the context of neurological disorders entitled **"Histone Deacetylase Inhibitors—Epidrugs for Neurological Disorders."** However, I was knowing that no single book is available that has thoroughly discussed histone deacetylase inhibitors in combined form with other conventional and molecular targeted agents, shortcomings of singlet therapy of HDACi and conventional therapies, toxicities associated with pan-HDAC inhibitors, and escalating need of isozyme selective inhibitors. These unavoidable facts resulted in the genesis of this book. Due to COVID-19 pandemic, the movement became restricted and I fruitfully spent this time in writing this precious book at home.

Wadura Sopore, SKUAST-Kashmir, India
June 28, 2020

Shabir Ahmad Ganai

Acknowledgment

Dr. Shabir Ahmad Ganai whole-heartedly thanks **Science and Engineering Research Board** (India) for helping financially in the form of a big grant (Start-Up Grant for Young Scientists). The file number of this project is **YSS/2015/001267**. Moreover, Dr. Ganai thanks referees for their valuable suggestions towards the improvement of this book. Further, the author sincerely thanks his family members for keeping me away from home-related activities.

Contents

About the Author

Dr. Shabir Ahmad Ganai has formerly worked as Principal Investigator/Young Scientist in the University of Kashmir. He has over 9 years of working experience in the field of histone deacetylases and histone deacetylase inhibitors.

He has published more than 24 articles in highly reputed international journals, 3 international book chapters and has authored a book.

He is currently serving as a referee for various international journals like Scientific Reports, PLOS ONE, Medicinal Chemistry Research, Current Drug Targets, Current Topics in Medicinal Chemistry, and Journal of Agricultural and Food Chemistry (American Chemical Society Publications). He is a member of two international societies, including the Epigenetics Society.

1 Overview of Epigenetic Signatures and Their Regulation by Epigenetic Modification Enzymes

It is well endorsed in higher organisms that development and differentiation are strongly reliant on multiplex interactions between the genome and the surroundings (environment) (Aguilera et al. 2010). At the beginning the term epigenetics was used to refer these interactions. The word epigenetics was coined by Conrad Hall Waddington, a polymath scientist in 1942 (Waddington 1942). According to his opinion preformation and epigenesis are not independent but rather interdependent (Handy et al. 2011; Waddington 1942, 1968, 2012). All biologists are familiar that development begins from fertilized egg (zygote) containing certain "preformed" characters. These characters must interconnect with each other in processes of "epigenesis" prior to adulthood. Study restricted to these "preformed" characters, currently, is termed as genetics. The word "epigenetics" has been proposed for studying those processes composing epigenesis. From this discussion, it becomes crystal clear that the term "epigenetics" is nothing but reminiscent of "epigenesis" and has been derived from union of two terms "epigenesis" and "genetics" (Waddington 1956). According to Waddingtonian equation, epigenetics = epigenesis + genetics (Van Speybroeck 2002). Waddington's meaning for development was the same what we call today differential gene expression and regulation. He introduced a metaphor for how gene regulation modifies development and that was "epigenetic landscape". Actually he introduced this idea to explain the procedure of "decision making" in the course of development (Baedke 2013; Burbano 2006; Tronick and Hunter 2016; Waddington 1940). Almost 2 years after "epigenetic landscape" metaphor (1940), Waddington presented the term epigenetics (1942) as a refined version of this landscape (Tronick and Hunter 2016; Waddington 1942). About 15 years later in David Nanney's article, the term epigenetic systems was used. According to him expression of genetically decided potentialities are controlled by epigenetic systems (Nanney 1958). Hall has defined epigenetics/epigenetic control as the sum total of genetic and non-genetic factors that act upon cells resulting in choosy regulation of gene expression generating elevated phenotypic complexity in the course of development (Burbano 2006; Wagner 1993). Robust technological refinement and in-depth studies on molecular biology gave

S. A. Ganai, *Histone Deacetylase Inhibitors in Combinatorial Anticancer Therapy*,
https://doi.org/10.1007/978-981-15-8179-3_1

timely amendments to the epigenetics. For instance the term heritable added to the usual Waddingtonian definition was a remarkable improvement (Holliday 2006; Holliday and Pugh 1975). Some use the term epigenetics for heritable alterations that are not due to change in nucleotide sequence of DNA (Wu and Morris 2001). Literally speaking, the term epigenetics means above or on top of genetics and thus designates external modifications (DNA methylation and histone post-translational modifications) turning genes on or off (Burbano 2006; Jablonka and Lamb 2002). Thus it is quite understandable that epigenetic modifications do not change actual DNA sequence but affect the reading of genes by cells. In other words, epigenetics may be defined as a regulatory system controlling gene expression without affecting their original makeup. Among the advanced definitions of epigenetics, one thing that usual DNA sequence is not altered stands firmly (Felsenfeld 2014; Ganai 2019; Goldberg et al. 2007; Tronick and Hunter 2016).

1.1 Introduction to Epigenetic Modifications

Regulatory mechanisms of gene transcription determine several processes including protein production and differentiation and contribute substantially to the pathogenesis of several diseases. Epigenetic regulation of gene expression involves the selective and reversible modification(s) of overlying DNA and underlying histone proteins. These modifications regulate the conformational transition between open (transcriptionally active) and closed (transcriptionally inactive) chromatin states. Thus epigenetic modifications cover both post-translational modifications of nucleosomal histones and DNA methylation.

Chromatin, a highly dynamic nucleoprotein structure, responds to different exogenous and endogenous stimuli (Bannister and Kouzarides 2011). The underlying histones have unstructured tails that protrude through nucleosome (own) making contact with adjacent ones (nucleosomes) (Luger et al. 1997). This modulates interaction in between nucleosomes and consequently the overall chromatin structure (Bannister and Kouzarides 2011; Ganai 2015; Shaytan et al. 2016). Post-translational modifications predominantly on the unstructured ends of histone proteins play a central role in epigenetic regulation of gene expression. These tails undergo several modifications including acetylation, methylation, phosphorylation etc. Histone substrates undergo eight types of post-translational/chemical modifications (Kouzarides 2007; Strahl and Allis 2000). These modifications are deposited by epigenetic players termed as writers and function in a combinatorial pattern referred to as "histone code". The main crux of histone code hypothesis is that post-translational modifications recruit other specific proteins and modulate chromatin architecture rather than by simply influencing the histone–DNA interactions (Jenuwein and Allis 2001; Kouzarides 2007). Importantly, post-translational modifications recruit enzymes having the potential to write or erase or to read these modifications (Torres and Fujimori 2015). Almost 150 such enzymes have been found in human systems. Certain modifications like acetylation and phosphorylation modify the overall charge on histone substrates interfering histone–

DNA interactions and consequently the nucleosome stability (Fenley et al. 2018). For better understanding of these modifications, I will explain them individually with strong emphasis on the enzymes regulating these modifications, which is the prime focus of this chapter.

1.1.1 Histone Acetylation and Its Enzymatic Regulation

One of the dynamic post-translational modifications concentrated on the lysine residues of histone tails is acetylation. Histone acetylation mediated by histone acetyltransferases (HATs) or lysine acetyl transferases (KATs) (Allis et al. 2007; Kurdistani and Grunstein 2003) has great impact on conformational transition of chromatin. These enzymes catalyse the transfer of acetyl group from cofactor acetyl-CoA to ε-amino group of histone lysines (Ganai 2015; Ganai et al. 2016). The weight of histone protein escalates approximately by 42 Da on the deposition of single acetyl moiety. Addition of acetyl moiety to nucleosomal histones powers electrostatic repulsion between histones and polyanioinic DNA (Barnes et al. 2019). This causes passive chromatin remodelling as no energy is invested for this purpose. Acetylated histones result in chromatin decondensation, which is further promoted by active remodelling complexes ultimately making the conditions supportive for RNA polymerase to participate in transcription.

1.1.1.1 Different Types of HATs and Their Target Residues on Histones

HATs, the workaholics of the epigenome, have two main categories based on their subcellular confinement. While type A HATs are nuclear residents and acetylate nucleosomal histones, type B HATs are restricted to cytoplasm and acetylate newly formed histones to promote their assemblage into nucleosomes. Moreover, type B HATs have importance in repairing DNA double strand breaks.

HATs have five distinct families based on domain structure and the sequence resemblance. These include p300/CBP family having p300 and CBP as its members (Chan and La Thangue 2001). The MYST family has nine representatives including KAT6B/MORF, MOF and TIP60 within its boundary (Avvakumov and Côté 2007). While GNAT (Gcn5-related HAT) superfamily includes GCN5 and PCAF (Dyda et al. 2000), the NCoA (nuclear receptor coactivator) family has ten members including NCoA1 and NCoA2 under its umbrella. Certain HATs like TAF1 and GTF3C2/TFIIIC110 come under the transcription-related HATs (Sheikh 2014). Only few years before, a new family of active HATs has been identified. These HATs known as camello proteins have developmental significance in zebrafish. Apart from this they are perinuclear in localization and deposit acetyl tags to histone H4 (Ganai et al. 2016; Karmodiya et al. 2014).

Some HATs acetylate a single histone at different lysine residues or even lysine residues of different histone, while others are specific for a particular site of single histone. For instance, circadian locomoter output cycles protein kaput (CLOCK), a histone acetyl transferase regulating circadian rhythms, catalyses the acetylation of histone H3 at lysine 14 (H3K14ac), while elongator complex protein 3, a HAT belonging to GNAT family, performs H3K9 and H3K18 acetylation (Close et al. 2006; Doi et al. 2006; Igarashi et al. 2007; Kim et al. 2002). CREB-binding protein acetylates both histone H3 and H4 at different sites (H3K18, H3K27, H3K56, H4K16, H4K12, H4K5 and H4K8) and its activity has direct involvement in transcriptional activation (Bedford et al. 2010; Martinez-Balbás et al. 1998).

Further, nuclear cytoplasmic O-GlcNAcase and acetyltransferase (NCOAT), a bifunctional enzyme acquainted with both glycoside hydrolase and HAT activity, acts as writer for H3K14 and H4K8 acetylation (Toleman et al. 2004, 2006). MYST3/MOZ (monocytic leukaemia zinc-finger protein) specifically acetylates histone H3 at lysine 14 (H3K14), whereas MYST3/Hbo1 (histone acetyl transferase binding to ORC1) acetylates histone H4 at three different lysine spots (H4K12, H4K8 and H4K5) (Iizuka and Stillman 1999; Katsumoto et al. 2008; Kitabayashi et al. 2001; Miotto and Struhl 2010). Certain HATs such as CYDL (chromodomain Y-like) protein, *N*-acetyltransferase 10 (NAT10), nuclear receptor coactivator 1 (NCoA-1), nuclear receptor coactivator 3 (NCoA-3) and testis-specific chromodomain protein Y 1 (CDY1B) have not been yet assigned any acetylation spot (Lahn et al. 2002; Liu et al. 2007; Lv et al. 2003).

HATs have strong cross-talk with distinct processes such as gene derepression, DNA repair (Guo et al. 2018; Sun et al. 2009), autophagy (Lin et al. 2012), cell cycle progression (Alomer et al. 2017) and cell signal transduction (Sun et al. 2005). Acetylation further facilitates gene expression programs by creating the sites favourable for attachment of bromodomain-containing proteins (Ganai et al. 2016; Zeng and Zhou 2002).

1.1.1.2 Histone Deacetylases

The histones with acetyl tags are deacetylated by antagonistic family of enzymes termed as histone deacetylases (HDACs) or lysine deacetylases (KDACs). These enzymes have multiple classes, which will be discussed rigorously in Chap. 3.

1.1.2 Histone Methylation and Its Regulation by Functionally Antagonistic Enzymes

This modification occurs on the basic amino acid residues of histones such as lysine, arginine and to some extent on histidine (Byvoet et al. 1972; Fischle et al. 2008; Kim et al. 2014; Ng et al. 2009). Histone methylation like acetylation falls within the most thoroughly studied post-translational modifications. The effect of histone methylation on gene expression depends not only on site but also on the extent this modification. Whereas the methylation on lysine 4 and 36 and 79 of histone H3 promotes transcription, the same signature on lysine 9 and 27 pampers transcriptional silencing. Moreover, methylation of lysine 20 of histone H4 (H4K20) also supports transcriptional silencing (Gu and Lee 2013; Sims 3rd et al. 2003).

1.1.2.1 Lysine Methylation and Its Impact on Transcriptional Events

Methylation occurring on lysine residues of nucleosomal histones is termed as lysine methylation. The degree of methylation of lysine residues ranges from mono- to trimethylation (Bannister et al. 2002; Kim et al. 2014). The enzymes depositing methylation tags on lysine residues are known as lysine methyl transferases (KMTs). They are widely studied enzymes and contain [Su (var) 3–9, Enhancer of zeste, Trithorax] domain briefly known as SET domain (Krajewski and Reese 2010; Qian and Zhou 2006). However, H3K79-specific methyltransferase namely disruptor of telomeric silencing 1-like (DOT1L) is atypical and lacks SET domain (Feng et al. 2002; Min et al. 2003; Wong et al. 2015; Wood et al. 2018).

Lysine methyl transferases follow SN-2 mechanism in transferring methyl group from *S*-adenosyl methionine to ε-amino group of specific lysine residue (Zhang and Bruice 2008). At the end of the reaction, this methyl group donor gets converted into *S*-adeno-L-homocysteine (Dillon et al. 2005). Unlike histone acetylation that alters histone–DNA interactions, histone methylation does not affect these interactions but provides a platform for the recruitment of methylation reader proteins that have the potential to modulate gene expression (Bannister and Kouzarides 2011; Gu and Lee 2013; Sims 3rd et al. 2003). Histone protein nearly gains 14 Da on deposition of a single methyl moiety. The SET domain of lysine methyltransferases possesses enzymatic activity leading to lysine methylation. Histone H3 lysine 9 trimethylation (H3K79me3) is carried out by SUV39H1 and SUV39H2 (Frontelo et al. 2004; Lachner et al. 2001). This site-specific methylation results in the recruitment of heterochromatin protein 1 (HP1) facilitating heterochromatinization.

Certain methyltransferases like MLL1, MML3 and MLL4 mediate H3K4 trimethylation (H3K4me3) (Ansari and Mandal 2010; Guenther et al. 2005; Wang et al. 2011). Another meiosis-specific methyltransferase mediating H3K4me3 namely PR domain-containing protein 9 (PRDM9) has been identified (Baudat et al. 2010; McVean and Myers 2010). While enhancer of zeste homolog 1 (EZH1) is writer for H3K27 mono- and dimethylation (H3K27me1 and H3K27me2), enhancer of zeste homolog 2 (EZH2) performs trimethylation as well. They are the components of Polycomb repressor complex and maintain chromatin in repressive state through distinct mechanisms (Abel et al. 1996; Cao and Zhang 2004; Margueron et al. 2008).

Though the notion that histone methylation is stable and static modification hold ground for many years, a death blow came to this concept when demethylation reactions received certification from experimental evidences (Bannister et al. 2002).

1.1.2.2 Lysine Demethylation and the Associated Enzymes

Gene regulatory mechanisms turn the genes on or off depending on the requirement of a particular gene product. Lysine demethylases, antagonistic enzymes to lysine methyltransferases, come under erasers as they remove methyl moieties from histone lysines (Upadhyay and Cheng 2011). These enzymes function downstream to certain clues acquired by the cell and may facilitate or hamper gene expression (De Santa et al. 2007). These enzymes participate in configuring chromatin landscape, which in turn regulates the transcriptional output (García et al. 2016; Okada et al. 2007).

On the basis of structure and mechanism, lysine demethylases come under the confines of two major families. The lysine-specific demethylases (LSD) family includes the demethylases showing resemblance with flavin-containing monoamine oxidases. These enzymes use flavin adenine dinucleotide (FAD) as co-factor for oxidizing methylated lysines to demethylated lysine and formaldehyde through corresponding imine intermediate (Heightman 2011; Shi and Tsukada 2013; Yang et al. 2007). These enzymes are not able to demethylate trimethyllysine residues as the formation of required imine intermediate is not possible from quaternary ammonium group because it lacks protonated nitrogen (Shi and Tsukada 2013). The first lysine-specific demethylase 1A (encoding gene KDM1A) was discovered unexpectedly. This enzyme demethylates mono- and dimethyl states of histone H3 at K4 and K9 (H3K4me1, H3K4me2, H3K9me1 and H3K9me2) (Metzger et al. 2005; Shi et al. 2004). LSD1 forms the part of a complex known as RE1-silencing transcription factor (REST), and its recruitment by the corepressor of REST (CoREST) capacitates it to demethylate nucleosomal histones at defined positions. This demethylase activity of LSD1 culminates in repression of REST target genes (Lee et al. 2005c). Another demethylase of the same family LSD 1B encoded by KDM1B gene has also been reported. This enzyme demethylates demethylated histone H3 at lysine 9 (H3K9me2). Unlike LSD1, this enzyme causes transcriptional activation (Fang et al. 2010).

It has been already discussed that LSD family of demethylases lack the potential to demethylate trimethyllysines. This was enough clue for the existence of other demethylases that are compatible to demethylate lysine residues in trimethylated state. Thus another family of histone demethylases namely Jumonji C (JmjC) family was discovered. These enzymes also known as Jumonji histone demethylases form the major group of demethylases (Mosammaparast and Shi 2010). Among the 30 identified proteins possessing JmJC domain, 20 have the ability to demethylate specific lysine residues of histones (Hojfeldt et al. 2013; Rotili and Mai 2011). These enzymes are 2-oxoglutarate-dependent dioxygenases and need Fe^{2+} and oxygen to demethylate their substrates. This family of demethylases can remove trimethylation as well (Cloos et al. 2008). Several different subfamilies of these demethylases have been discovered ranging from KDM2 to KDM6 and others (D'Oto et al. 2016).

The first subfamily KDM2 includes two members namely KDM2A and KDM2B encoded by lysine (K)-specific demethylase 2A (KDM2A) and lysine (K)-specific demethylase 2B (KDM2B) genes, respectively. While the former demethylase specifically demethylates the dimethylated lysine residue of histone H3 at position 36, the latter demethylates trimethylated lysine residue of histone H3 at position 4 (H3K4me3) as well (Frescas et al. 2007; He et al. 2008; Tsukada et al. 2006). Lysine (K)-specific demethylase 2A has cross-talk with HP1 and is associated with heterochromatic spots of the genome (Frescas et al. 2008). This demethylase promotes the silencing of non-coding RNAs and thus makes the conditions conducive for attachment of HP1 to centromeric regions (Frescas et al. 2008). Like KDM2 subfamily, KDM3 also includes two members namely lysine (K)-specific demethylase 3A (KDM3A) and lysine (K)-specific demethylase 3B (KDM3B). They are encoded by KDM3A and KDM3B genes, respectively. Both these enzymes remove the methyl groups from histone H3 lysine 9 (H3K9) when the lysine is in mono- or dimethylated form (Yamane et al. 2006).

KDM4 subfamily has four members within its boundary. This subfamily includes KDM4A, KDM4B, KDM4C and KDM4C. While KDM4A is encoded by lysine (K)-specific demethylase 4A (KDM4A) gene, the rest are encoded by KDM4B, KDM4C and KDM4D genes, respectively. While KDM4A demethylates H3K79me3 and H3K36me3, KDM4B is specific for former mark only (Whetstine et al. 2006). KDM4C performs the demethylation of H3K9me2 and H3K9me3 wherein KDM4D has H3K9me2, H3K9me3 and H1K25me1 as its substrates (Cloos et al. 2006; Labbé et al. 2013; Lee et al. 2020; Whetstine et al. 2006). KDM4C promotes euchromatinization and has direct impact on NOTCH1 signalling (Cloos et al. 2006; D'Oto et al. 2016; Liu et al. 2009).

The subfamily KDM5 like KDM4 is composed of four members (KDM5A–KDM5D). KDM5A is known by alternative names as Jumonji/ARID domain-containing protein 1A/JARID1A or retinoblastoma-binding protein 2 (RBBP-2) (Christensen et al. 2007). KDM5B and KDM5C demethylate only H3K4me3, while the remaining members also target H3K4me2 (Iwase et al. 2007; Yamane et al. 2007). Expression of KDM5B is more predominant in testes and regulates transcription in association with androgen receptor (D'Oto et al. 2016; Xiang et al. 2007b).

Subfamily KDM6 also consists of two members only like KDM2 and KDM3 subfamilies. These members including KDM6A (histone demethylase UTX) and KDM6B (JmjC domain-containing protein 3) have lysine (K)-specific demethylase 6A and lysine (K)-specific demethylase 6B as their encoding genes. In addition to H3K27me3 demethylated by KDM6A, KDM6B demethylates H3K9me2 also (Agger et al. 2007; Cribbs et al. 2020; D'Oto et al. 2016; Xiang et al. 2007a).

Other demethylases like lysine-specific demethylase 7 (JmjC domain-containing histone demethylation protein 1D/KDM7A) and lysine-specific demethylase 8 (JmjC domain-containing protein 5/KDM8), lysine-specific demethylase NO66 (nucleolar protein 66) and histone lysine demethylase PHF8 (PHD finger protein 8) also possess JmJC domain (Horton et al. 2010; Hsia et al. 2010; Sinha et al. 2010). While KDM7A removes methyl groups from H3K9me2 and H3K27me2, KDM8 demethylates H3K36me2 (Horton et al. 2010; Hsia et al. 2010). Similarly NO66 demethylates H3K4me3 and H3K36me3, the transcriptional activation marks. PHF8 acts as eraser for H3K4me3, H3K9me2, H3K27me2 and H4K20me1 (Sinha et al. 2010; Zhu et al. 2010).

1.1.2.3 Arginine Methylation and Its Dynamic Control

Methylation occurring on the specific arginine residues of histone proteins is termed as arginine methylation. This methylation is function of enzymes known as protein arginine methyl transferases (PRMTs). Although nine such enzymes have been reported in human systems, only seven have the methylation ability (Di Lorenzo and Bedford 2011). These enzymes catalyse mono- or dimethylation of the basic arginine residues. Making use of *S*-adenosyl-methionine as methyl donor, these enzymes transfer the methyl moiety to the guanidinium side chain of basic arginine residue (Tewary et al. 2019). PRMTs have two types on the basis of position of methyl moiety addition. While type 1 PRMTs form the largest group composed of seven members, type II PRMTs consist of only two members. Moreover, type I PRMTs methylate arginine residue asymmetrically (*N,N*-dimethylation), whereas symmetric methylation (*N,N′*-dimethylation) is catalysed by type II PRMTs (Fulton et al. 2019). PRMTs facilitate transcription, but some like PRMT6 show the opposite trend (Stein et al. 2012). In order to regulate transcriptional events various PRMTs function in close coordination. First I will discuss the functions of type I PRMTs following which the type II ones will also be discussed.

Type I PRMTs

Histone-arginine methyltransferase CARM1 methylates histone H3 at various sites H3R17, H3R2, H3R26 and H3R17. At the first three sites described monomethylation occurs while at the last site asymmetrical dimethylation occurs (Bauer et al. 2002; Miao et al. 2006). Protein arginine *N*-methyltransferase 1 (PRMT1) through mono- and asymmetric dimethylation of histone H4 at arginine 3 site (H3R3me1 and H3R3me2) favours transcription (Wang et al. 2001).

Protein arginine *N*-methyltransferase 2 (PRMT2) has been reported to methylate histone H4 under in vitro conditions (Lakowski and Frankel 2009). Its effect on the expression of NF-kB target genes is opposite to that of PRMT4 and is thus antagonistic to the latter. Further, this arginine methyltransferase facilitates apoptosis by obstructing NF-kB function (Ganesh et al. 2006).

PRMT6 (protein arginine *N*-methyltransferase 6), a full nuclear resident, catalyses the methylation of Tat (HIV protein) hampering the HIV replication (Boulanger et al. 2005; Frankel et al. 2002). This enzyme by methylating DNA polymerase beta plays a regulatory role in base excision repair (El-Andaloussi et al. 2006).

Protein arginine *N*-methyltransferase 8 that is encoded by PRMT8 gene like PRMT4 transferase has been reported to transfer methyl group/groups to histone H4 under the conditions of in vitro. This methyltransferase not only mediates monomethylation but performs asymmetric dimethylation as well. Being tissue-restricted, is expressed in the brain and differs from other PRMT enzymes in being the cell membrane bound (Lee et al. 2005a). The amino terminal tail of this methyltransferase is acquainted with autoregulatory activity (Dillon et al. 2013; Sayegh et al. 2007). In ageing motoneurons this enzyme has been reported to neutralize cellular stress. PRMT8 knock-out male mice models with advancing age showed alleviated muscle strength due to early impairment of neuromuscular junctions (Simandi et al. 2018).

Type II PRMTs

As already mentioned these PRMTs methylate arginine symmetrically and have PRMT5 and PRMT7 under its umbrella. Protein arginine *N*-methyltransferase 5/Jak-binding protein 1/JBP1 with inclination towards histone H2A and H4 mediates both monomethylation and dimethylation (symmetric) of arginine residues (Branscombe et al. 2001; Pollack et al. 1999; Shailesh et al. 2018). This enzyme acts as writer for post-translational modifications including H4R3me2 and H3R8me2. This transferase contributes to gene repression by methylating H4R3 and through the recruitment of DNA (cytosine-5)-methyltransferase 3A/DNMT3A (Zhao et al. 2009). Protein arginine *N*-methyltransferase 7 (PRMT7) under in vitro conditions has been reported to methylate histone H2A and H4. Being a type II PRMT results in the formation of symmetric dimethyl arginines (Jain and Clarke 2019; Lee et al. 2005b). This kinase by methylating the arginine residue 70 of p38MAPK has been found to facilitate differentiation of myoblasts (Jeong et al. 2020).

1.1.2.4 Arginine Demethylation and the Relevant Enzymes

A panel of proteins participating in distinct cellular processes undergoes arginine methylation. In humans a single enzyme has been reported to have arginine demethylase activity. This enzyme namely bifunctional arginine demethylase and lysyl-hydroxylase JMJD6 has JMJD6 as the encoding gene. JMJD6 demethylase reported to be iron- and 2-oxoglutarate-dependent dioxygenases mediates the demethylation of monomethyl and dimethyl arginine residues (H4R3me2, H4R3me1 and H3R2me2). These findings have strong support from both cell-based and biochemical assays (Chang et al. 2007). Studies have shown that stress granule formation (SG) provides cytoprotection against environmental stress, and it has also been reported that methylation of G3BP1, a SG-nucleating protein hampers SG formation. JMDJ6 demethylates G3BP1 by direct or indirect mechanisms, thereby facilitating the formation of stress granules (Tsai et al. 2017).

1.1.3 Histone Phosphorylation/Dephosphorylation and Its Dynamic Enzymatic Regulation

The phosphorylation of histone substrates is highly dynamic like histone acetylation. It occurs mainly on serine, threonine and tyrosine residues of amino terminal tails of histones though not exclusively (Bannister and Kouzarides 2011). Phosphorylation has been reported to occur on the tails of all the four nucleosomal histones. Turnover of histone phosphorylation is regulated by functionally antagonistic enzymes known as kinases and phosphatases. Kinases and phosphatases write and erase this modification, respectively (Oki et al. 2007). This modification serves as a key intermediate step in various nuclear events including cell division, transcriptional regulation, chromosome condensation and DNA damage repair (Kschonsak and Haering 2015; Rossetto et al. 2012). Phosphorylation of histone H3 on two serine residues (H3S10 and H3S28) and core histone H2A on threonine 120 (H2AT20) plays a critical role in chromatin condensation besides regulating its structure and function in mitosis (Hsu et al. 2000). Recruitment of DNA damage repair proteins is facilitated by H2AX phosphorylation at serine 139 position (S139) (Lowndes and Toh 2005; Turinetto and Giachino 2015). This phosphorylation is among the earliest events following the DNA double strand breaks (Lowndes and Toh 2005; Turinetto and Giachino 2015). Although H2B phosphorylation is relatively less studied, it has been reported to promote apoptosis-induced chromatin compaction, fragmentation of DNA and cell death (Fullgrabe et al. 2010).

A lone aurora kinases namely Ark1 has been related to histone H3 serine 10 phosphorylation in *Schizosaccharomyces pombe* (Petersen et al. 2001). Two aurora kinases (aurora A and aurora B) have been reported in certain multicellular organisms like *Caenorhabditis elegans* and *Drosophila melanogaster* (Glover et al. 1995; Prigent and Dimitrov 2003; Reich et al. 1999; Schumacher et al. 1998). However, in human's kinase aurora C dedicated to spermatogenesis has also been reported (Adams et al. 2001a; Prigent and Dimitrov 2003). RNA interference-based studies in nematode and fruit fly cells have revealed that aurora A is immaterial but aurora B a crucial kinase for Ser10 phosphorylation (Adams et al. 2001b; Giet and Glover 2001; Hsu et al. 2000). The better kinase activity of aurora B has been attributed to its subcellular localizations. Unlike aurora A, which is a centrosomal protein, aurora B is a passenger protein (Gopalan et al. 1997; Terada et al. 1998). The identity of mitotic Ser 10 phosphatase namely protein phosphatase 1 (PP1) is highly clear unlike the mitotic kinase meant for this spot (Ganai 2018; Hsu et al. 2000). Both Ser 10 kinase and PP1 are in connection with mitotic chromosomes in vertebrates. This phosphatase not only dephosphorylates Ser10 of histone H3 but also inactivates the kinase responsible for this phosphorylation (aurora B) (Murnion et al. 2001). Evidences suggest that aurora B also phosphorylates H3S28 during mitosis and thus acts as kinase for both H3S10 and H3S28 (Goto et al. 2002).

Haspin, a mitotically essential and unusual serine/threonine protein kinase composed of 798 amino acid residues in humans, phosphorylates histone H3 at threonine 3 position (H3T3). Its kinase domain responsible for catalytic activity lies towards C-terminal region from residue number 470–798 (Eswaran et al. 2009; Feizbakhsh et al. 2017; Villa et al. 2009). Another kinase Bub1 has been found to mediate the phosphorylation of core histone H2A at threonine 120 (H2A120). At kinetochores this phosphorylation facilitates the centromeric sister-chromatid togetherness. Defect in this phosphorylation at centromeric/pericentromeric H2A causes problems in segregation of chromosomes (Gil and Vagnarelli 2019; Lin et al. 2014; Maeda et al. 2018).

Studies have shown that during mitotic exit Repo-man/PP1 complex dephosphorylates all the mitotic phosphosites of core histone H3 (S10, S28 and T3). CDK1 activation at mitotic entry phosphorylates Repo-man at various sites, thereby lowering its affinity towards PP1 and chromatin (Gil and Vagnarelli 2019; Qian et al. 2015; Senthil Kumar et al. 2016; Vagnarelli et al. 2011). Mitosis-specific phosphorylation of histone H3 at threonine 11 (H3T11) has also been reported. This phosphorylation has been ascribed to death-associated protein (DAP)-like kinase (Dlk) and has significance in kinetochore assembly (Preuss et al. 2003). Phosphorylation of linker histone H1 markedly increases during mitosis and the middle phase of interphase. The phosphosites identified for this histone result in chromatin decompaction (Gil and Vagnarelli 2019; Roth and Allis 1992).

1.1.4 Histone Ubiquitinataion/Deubiquitination and Its Regulation by Epigenetic Players

Ubiquitination, a bulky modification to small histones, provides signals necessary for transcription regulation. However, this modification usually sends proteins for degradation by proteasomes. Among histone proteins, H2A has peculiarity of being the first protein that was identified to be modified by ubiquitinataion (Goldknopf et al. 1975). Inside the nuclear confines histone H2A and H2B are the most profuse ubiquitinated proteins. It has been estimated that vertebrate cells contain 5–15% ubiquitin conjugated H2A and only 1–2% H2B, while in yeast cells the percentage of ubiquitinated H2B is about 10% (Goldknopf et al. 1975; Matsui et al. 1979). In humans ubiquitination occurs at lysine 119 of histone H2A and at lysine 120 of H2B. Ubiquitination results in the formation of isopeptide bond between the glycine of ubiquitin at carboxyl-terminal and the lysine residue located at the carboxyl-end of histones. Ubiquitination involves three sequential steps: activation, conjugation and ligation. The first two steps are performed by ubiquitin-activating enzyme and ubiquitin-conjugating enzymes, respectively. Following this, the ubiquitin ligase installs ubiquitin on specific lysine residues of histone substrates through isopeptide linkage (Scheffner et al. 1995). H2B ubiquitination has cross-talk with histone H3 methylation and has been reported to promote its methylation at lysine residues at 4 and 79 positions (Sun and Allis 2002).

Ubiquitination of histone proteins like histone acetylation and methylation is dynamically regulated. This is reversed by deubiquitinating enzymes also known as deubiquitinases (DUBs) (Huang and Cochran 2013). Deubiquitinases fall under two major classes and may be zinc metalloproteases (JAMM motif zinc metalloproteases) or cysteine proteases (Heideker and Wertz 2015). Ninety deubiquitinases have been discovered in human system, and the functional role of many is yet to be elucidated. Certain deubiquitinases are H2A specific; some are H2B restricted while others are dual ones. As an example USP16, 2A-DUB, BAP1 (BRCA1-associated protein-1 [BAP1]) and USP21 are H2A-specific deubiquitinases, whereas UBP10 and UBP8 are restricted to H2B. Deubiquitinases like USP12, USP3, USP46 and USP22 show double specificity and thus have inclination towards H2A and H2B (Cao and Yan 2012; Chen et al. 2015; Gu et al. 2016). Only recently a novel deubiquitinase, USP36, specific for H2B has been discovered. This deubiquitinase has been proved to deubiquitinate monoubiquitinated H2B in cells as well as under in vitro conditions (DeVine et al. 2018). Experimental evidences suggest that ubiquitinated H2A restrains transcription initiation by obstructing H3K4 di- and trimethylation. This indicates that trans-histone cross-talk has potential to modulate transcriptional events (Nakagawa et al. 2008).

1.1.5 Histone SUMOylation/deSUMOylation and the Respective Enzymes

Another post-translational modification SUMOylation is more or less similar to ubiquitination. SUMOylation involves the formation of covalent linkage between the lysine residue of a target protein and small ubiquitin-like modifier (SUMO) (Johnson 2004; Melchior 2000). Over 1000 proteins have been recognized as potential targets of this modification (Hochstrasser 2009). SUMOylation like ubiquitination is a heavy modification as the size of SUMO and ubiquitin is 11 kDa and 9 kDa, respectively. Unlike ubiquitination, SUMO has no role in protein degradation (Nathan et al. 2003). This modification was first reported in the context of histone H4 in 2003 by Shiio and Eisenman. They related this modification with gene suppression as SUMOylation recruits the molecular players well known for chromatin condensation and heterochromatinization (HDACs and HP1) (Shiio and Eisenman 2003). In yeast model it has been demonstrated that all core histones can undergo sumoylation. The sites identified were K6/K7 and to meagre extent K16/K17 of core histone H2B while for histone H2A the putative sumoylation site identified was K126. Regarding histone H4 all the lysine residues present in the amino terminal tail were considered as SUMO installation sites (Nathan et al. 2006).

In mammalian cells three SUMOs namely SUMO-1, SUMO-2 and SUMO-3 have been reported. The last two SUMOs (SUMO-2 and SUMO-3) that show nearly 95% identity cannot be differentiated in most contexts and are thus collectively designated as SUMO2–2/3. While certain targets of SUMO are SUMO-1 restricted, some to SUMO-2/3, other targets show dual behaviour and thus can be conjugated to all SUMOs (Vertegaal et al. 2006). Further, the overall cellular concentrations of SUMO-1 are relatively lesser than the SUMO-2/3 (Saitoh and Hinchey 2000) and as such majority of SUMOylation involves SUMO2/3 (Ayaydin and Dasso 2004; Saitoh and Hinchey 2000). Photobleaching-based studies have shown that other SUMOs are comparatively more dynamic than SUMO-1 (Ayaydin and Dasso 2004; Saitoh and Hinchey 2000). It has been established that all mammalian SUMOs have the common activating (E1) and conjugating enzymes (E2), which in turn have structural resemblance with corresponding enzymes of ubiquitin (Hochstrasser 2009). SUMO E1 is heterodimer unlike E1 meant for ubiquitin activation and is composed of Aos1p and Uba2p (Sae1 and Sae2 in vertebrates) (Johnson et al. 1997). An AMP-intermediate is formed by SUMO E1, the latter then through thiol-transfer reaction (intermolecular) passes SUMO to cysteine residue located at the active site of Ubc9, a conjugating enzyme (E2). Although two ligase-independent mechanisms of SUMO transfer from Ubc9 to target protein have been reported, bulk of the SUMOylation under physiological conditions is performed by SUMO ligases (E3 enzymes) (Bernier-Villamor et al. 2002; Meulmeester et al. 2008; Yunus and Lima 2006). In mammals PIAS1 and in yeast, Siz1 and Siz2 have been identified as SUMO ligases. These ligases possess Siz/PIAS RING-finger-like domains (SP-RING domains) crucial for ligase activity (Hochstrasser 2001). The PIAS protein family comes under SP-RING ligases and apart from SP-RING possess N-terminal domain composed of 400 amino acid residues. In mammalian system five PIAS (protein inhibitor of activated STAT) proteins have been reported namely PIAS1, PIAS2 having two splice variants (PIASxα, PIASxβ), PIAS3 and PIASy. Another SUMO ligase namely MMS21 has been reported in yeast as well as in human system (Stephan et al. 2011). These PIAS proteins have strong implications in various cellular processes including genome maintenance, signal transduction and in transcriptional events (Niu et al. 2018; Palvimo 2007; Potts and Yu 2005; Shuai 2006; Shuai and Liu 2005; Wu and Zou 2016). RanBP2, a nuclear pore protein localized towards the cytoplasmic side of the pore, has been reported to facilitate SUMOylation. Fragments of RanBP2 having internal repeat domain have shown SUMO ligase function under in vitro set-up (Pichler et al. 2002; Reverter and Lima 2005; Saitoh et al. 1998). Additionally other proteins have been proved to have potential SUMO ligase activity. These proteins include histone deacetylase 4 (HDAC4), Pc2 (polycomb group protein), Topors (RING-finger protein) and KRAB-associated protein 1 (Peng and Wysocka 2008; Weger et al. 2005; Wotton and Merrill 2007; Zeng et al. 2008). Expression of HDAC4 escalates the myocyte-specific enhancer factor 2 (MEF2) SUMOylation (Zhao et al. 2005). This HDAC has the capacity to bind Ubc9, and it has also been suggested that HDAC4 facilitates SUMOylation by augmenting the target protein phosphorylation (Geiss-Friedlander and Melchior 2007; Zhao et al. 2005).

For deconjugating SUMOylated species, Ulps (budding yeast)/SENPs (humans) have been reported to be involved (Hay 2007; Mukhopadhyay and Dasso 2007). Six mammalian SENPs have been identified ranging from SENP1 to SENP3 and then from SENP5 to SENP7 (Mukhopadhyay and Dasso 2007). Among these enzymes, SENP1 and SENP2 possess processing and deconjugation function both for SUMO-1 and SUMO-2/3. Contrastingly the remaining SENPs have more inclination towards SUMO-2/3. Processing and deconjugation of SUMO-2/3 are performed by SENP3 and SENP5 having nucleolar localization (Di Bacco et al. 2006; Gong and Yeh 2006; Wang and Dasso 2009; Yang et al. 2017).

1.1.6 Histone ADP-Ribosylation, Its Dynamics and Enzyme-Related Regulation

ADP-ribosylation, one of the covalent post-translational modifications, is mediated by ADP-ribosyltransferases. This modification has significant contribution in DNA damage response, gene expression and cell cycle regulation (Hottiger 2011; Martinez-Zamudio and Ha 2012; Messner and Hottiger 2011). ADP-ribosyltransferases covalently links ADP-ribose to target proteins on certain residues using nicotinamide adenine dinucleotide (NAD+) as cofactor. Core histones undergo poly-ADP-ribosylation mainly at their positively charged N-terminus. These N-terminal tails protrude from the fundamental unit of chromatin, the nucleosome. Among the ADP-ribose acceptor sites, specific glutamic acid residues of H2B and H1 have been reported (Ogata et al. 1980a, b). Apart from this C-terminal lysine of histone H1 serves as installation site for ADP-ribose (Ogata et al. 1980b). However, these reports require authentication from mass spectrometry.

Electron-transfer dissociation and mass spectrometry-based study has revealed lysine 13 of histone H2A (H2AK13) (Messner et al. 2010), lysine 30 of H2B (H2BK30) (Messner et al. 2010), lysine 27 and 37 of H3 (H3K27 and H3K37) (Messner et al. 2010) and lysine 16 of H4 (H4K16) (Boulikas 1990) as the potential sites for ADP-ribosyltransferase diphtheria toxin-like 1 (ARTD1).

This modification (ADP-ribosylation) can be either mono- or poly-ADP-ribosylation (Lüscher et al. 2018). ADP-ribosylation like other post-translational modifications imparts allosteric effects on enzymatic proteins and as such regulates their enzymatic function. As this modification is readable by various protein motifs and domains, it gives rise to protein–protein interactions (Verheugd et al. 2016).

On the whole, less than 1% of all histone proteins undergo this modification (Boulikas 1989; Nolan et al. 1980; Stone et al. 1977). Studies have shown that whole of the core histones and linker histone H1 can be modified by ADP-ribosylation (Burzio et al. 1979; Giri et al. 1978; Jump et al. 1979). This pattern of this modification varies according to chromatin composition. For instance, in native chromatin histone H1 undergoes ADP-ribosylation most heavily, whereas in H1 depleted condition H2B dominates (Huletsky et al. 1989).

The protein family of ADP-ribosyltransferases has 22 members in humans and all possess ADP-ribosyltransferase domain (Hottiger et al. 2010). ADP-ribosyltransferase diphtheria toxin-like 1 (ARTD1) is the best studied ADP-ribosyltransferase of diphtheria toxin-like subclass. In vitro studies have shown that this transferase (ARTD1) can ribosylate H2A, H2B, H3, H4 and H1 in their free state, whereas ARTD2 (previously known as PARP2) has no capacity to modify histones singly (Caplan et al. 1979; Ferro and Olivera 1982; Messner et al. 2010; Okazaki et al. 1980; Poirier et al. 1982). Although it has been observed that ARTD3 (formerly PARP-3) interacts with histones (H2B and H3), whether it ribosylates them or not is unclear (Ewing et al. 2007). Another ADP-ribosyltransferase, ARTD3, has been linked to ribosylation of H1.2, the most abundant variant of linker histone H1 (Rulten et al. 2011). ARTD10 was formerly termed as PARP-10, the resident of cytoplasm and nucleus, mono-ADP-ribosylates core histone substrates (Chou et al. 2006; Hottiger 2011; Hottiger et al. 2010; Kleine et al. 2008). However, extensive studies are required for remaining members of ARDT family to arrive at the logical conclusion. Sirtuins, the Class III HDACs, have also been reported to mediate mono-ADP-ribosylation (Saunders and Verdin 2007).

This post-translational signature is removed by hydrolases acquainted with the potential to break the glycosidic bonds. These bonds may be either between ADP-ribose units or between ADP-ribose (protein proximal) and amino acid (side chain) (Lüscher et al. 2018). Turnover of ADP-ribose polymers is of strong importance. In mouse and fly model, deletion of crucial poly-ADP-ribose glycohydrolase (PARG) culminated in embryonic lethality (Gagné et al. 2006; Koh et al. 2004). As of now only single PARG and three ADP-ribosyl hydrolases (ARHs) have been identified from human system (Glowacki et al. 2002; Hottiger 2011; Koch-Nolte et al. 2008; Oka et al. 2006). PARG and ADP-ribosylhydrolase 3 (ARH3) cleave glycosidic bonds in between two ADP-ribose units and thus can degrade poly-ADP-ribose polymer (Oka et al. 2006). Among the ARH family, ARH1 is certified mono-ADP-ribosyl-arginine hydrolase. This enzyme cleaves the *N*-glycosidic bond between the ADP-ribose and guanidino group of arginine liberating ADP-ribose (Mashimo et al. 2014; Moss et al. 1992).

1.1.7 Histone Biotinylation and Its Writers and Erasers

Biotinylation of histone proteins has been identified as another post-translational modification. This epigenetic signature is enriched in transcriptionally inactive regions of chromatin and is likely to be involved in gene silencing. Biotinylation of histones is predominantly catalysed by holocarboxylase synthetase and biotinidase (Hymes et al. 1995; Hymes and Wolf 1999). In the first step of biotinylation, biotin is adenylylated (AMP addition) to form biotinyl-5′-AMP. Following this, the biotin is transferred to specific residue of histone proteins and AMP is set free (Kothapalli et al. 2005). In addition to biocytin hydrolase activity, serum biotinidase has showed biotinyl-transferase activity (Hymes et al. 1995; Hymes and Wolf 1999). This enzyme belongs to nitrilase superfamily and has been characterized even at molecular level (Brenner 2002; Knight et al. 1998). Biotinidase enzyme is ubiquitously present in mammalian cells, and 26% of biotinidase (cellular) activity has been attributed to nuclear fraction (Zempleni et al. 2011). Certain studies have demonstrated the presence of biotinylated histones in human lymphocytes (Stanley et al. 2001), lymphoma cells (Manthey et al. 2002), choriocarcinoma cells (Crisp et al. 2004), small cell lung cancer cells (Scheerger and Zempleni 2003) and in chicken erythrocytes (Peters et al. 2002) ruling out biotinidase as the sole enzyme responsible for histone biotinylation. At the end, holocarboxylase synthetase was identified as another enzymatic player that may mediate the histone biotinylation (Narang et al. 2004).

Several biotinylation installation sites have been confirmed on human histones. These include K9, K13, K125, K127 and K129 for H2A, whereas for H3 lysine 9 and lysine 18 (K9 and K18) (Bao et al. 2011; Hassan and Zempleni 2006) are the defined sites. The authenticated biotinylation sites for histone H4 are H4K8 and H4K12 (Camporeale et al. 2004). Histone H4 biotinylation is concentrated in pericentromeric heterochromatin and has considerable role in transcriptional silencing, chromatin condensation (mitotic) and DNA damage response (Hassan and Zempleni 2006). Biotinylation of histone H3 at mentioned spots is mediated by holocarboxylase synthetase, whereas in vitro biotinylation of H2A at K9 and K13 is done by biotinidase.

Biotinylation occurring on the specific lysine residues of histone proteins is a dynamic modification. Experimental evidences suggest that in debiotinylation of histones, biotinidase is involved at least partially (Ballard et al. 2002; Chew et al. 2007). This suggestion is in alignment with the findings where this enzyme has been reported to have biotin-ε-lysine hydrolase activity (Wolf 2005; Zempleni et al. 2011).

1.1.8 DNA Methylation and Its Regulation by Functionally Antagonistic Enzymes

Till now, methylation of DNA is the only known epigenetic modification impacting gene expression. DNA methylation involves covalent addition of methyl group to the C5 position of base cytosine resulting in the formation of 5-methylcytosine (Moore et al. 2013). This modification regulates gene expression either by recruiting transcriptional corepressors (Jones et al. 1998) or by hampering transcription factor–DNA binding (Watt and Molloy 1988).

DNA methylation is catalysed by DNA methyl transferases, the writers of epigenome, and the methyl group transferred comes from *S*-adenosyl-L-methionine (Jin and Robertson 2013; Robertson 2005). Among the members of DNMT family, only DNMT3L lacks congenital enzymatic activity but accentuates the activities of DNMT3A and DNMT3B significantly (Gowher et al. 2005; Kareta et al. 2006). From these findings it is clear that DNMT3L serves as an accessory protein for these de novo DNMTs (Bourc'his et al. 2001). It is special to mention that DNMT3L is not expressed in somatic cells but noticeably in germ and embryonic stem cells (Chedin 2011). While DNMT3A and DNMT3B are de novo methyltransferases, DNMT1, the primary methyltransferase is maintenance DNMT (Ganai 2019; Jin and Robertson 2013; Liang et al. 2002; Okano et al. 1999). DNMT2, an atypical DNMT lacking the regulatory domain and methylating aspartic acid transfer RNA instead of usual DNA, has also been reported. In other words this enzyme functions as tRNA methyltransferase rather than DNA methyltransferase (Goll et al. 2006; Gujar et al. 2019). Recently another DNMT (DNMT3C) has been discovered that safeguards male fertility by methylating the young retrotransposon promoters (Barau et al. 2016).

DNA demethylation involving the reversal of DNA methylation has a potential role in development and differentiation of mammals (Dvoriantchikova et al. 2019; Greenberg and Bourc'his 2019). Demethylation has positive correlation with transcriptional activation. While global DNA methylation has been reported in progressing zygote and primordial germ cells, locus-specific demethylation has been identified in certain somatic cells (Chen and Riggs 2011; Hill et al. 2014). DNA methylation has been linked to cellular differentiation, and this conclusion has been extracted from the study where DNA demethylation proved to be requisite for differentiation of monocytes into other cell types (monocytes and dendritic cells) (Vento-Tormo et al. 2016). DNA demethylation, a complicated process, is fulfilled by either passive or active mechanism. Although the easiest way seems to be the direct demethylation of DNA, this is not thermodynamically feasible (Wu and Zhang 2014).

In passive mechanism the newly formed DNA strand following replication is not methylated (Chen and Riggs 2011). Thus this mechanism seems to dilute 5-methyl cytosine in replication-dependent manner and thus is the failure to sustain DNA methylation patterns following rounds of replication. Probably passive demethylation mechanism is operative during mammalian development (Ganai 2019; Mayer et al. 2000). Moreover passive DNA demethylation has been attributed to downregulation of molecular players meant for DNA methylation or the machinery becomes cytoplasm restricted. Deposition of 5-hydroxymethylcytosine has also been suggested as the possible factor for passive demethylation as DNMT1 activity is drastically lowered (maximum 60-fold) on a DNA substrate having this modification (Ganai 2019; Hashimoto et al. 2012; Rasmussen and Helin 2016; Valinluck and Sowers 2007). DNA methylation till long was considered an irreversible modification, and DNA replication was thought to be the only mode of alleviation. This notion received a death blow with the discovery of the ten-eleven translocation protein 1 (TET1) having the innate tendency to modify methylcytosine and excise DNA methylation (Tahiliani et al. 2009). This protein is one of the three representatives of ten-eleven translocation (TET) family, others being TET2 and TET3. These enzymes mediate the successive oxidation of 5-methylcytosine to 5-carboxylcytosine through intermediates 5-hydroxymethylcytosine and 5-formylcytosine (Hahn et al. 2014; Ito et al. 2011; Tahiliani et al. 2009). After this thymine-DNA glycosylase comes into play excising 5-carboxylcytosine from the DNA. From here base excision repair mechanism comes into action and regenerates unmodified cytosine (Fig. 1.1) (Hahn et al. 2014; He et al. 2011; Hu et al. 2014; Kohli and Zhang 2013; Shen et al. 2013).

1.2 Brief Introduction of "Writers" and "Erasers" of Epigenetic Tags

Here it is worthy to mention that enzymes involved in depositing or removing the various epigenetic signatures of histones or DNA are termed as epigenetic players or epigenetic enzymes. These enzymes may be "writers" installing epigenetic signatures or "erasers" uninstalling such signatures or tags (Biswas and Rao 2018). While certain "writers" are ubiquitously expressed across cell types, the expression of some writers is tissue restricted (Fischle et al. 1999; Ganai 2019; Ganai et al. 2015). About 100 unique "writers" belonging to 8 distinct categories have been identified in various model organisms. Enzymes like HATs (KATs), HMTs (KMTs), ubiquitinases, kinases, SUMOs, holocarboxylase synthetase and DNMTs are regarded as "writers"(Ganai 2019). Excluding DNMTs, all these enzymes write on DNA-underlying histones (Chen and Zhang 2020). Here I am using general statements as the theme is revolving around epigenetics. Indeed these "writers" have non-histone targets as well and thus may modulate the related pathways accordingly (Ganai 2018). The epigenetic signatures being dynamic are removed by specific enzymes meant for this purpose. Enzymes executing this task are known as "erasers" as predefined. So far 52 idiosyncratic "erasers" distributed among 6 different categories have been reported. Among the "erasers" HDACs (KDACs), HDMTs (KDMs), deubiquitinases and phosphatases are prominent (Biswas and Rao 2018). Erasers modify certain non-histone targets like α-tubulin, p53, heat shock proteins and cortactin (Ganai 2018).

As of now, I have discussed distinct epigenetic signatures occurring on histone proteins and the only modification (methylation) occurring on DNA. The specific

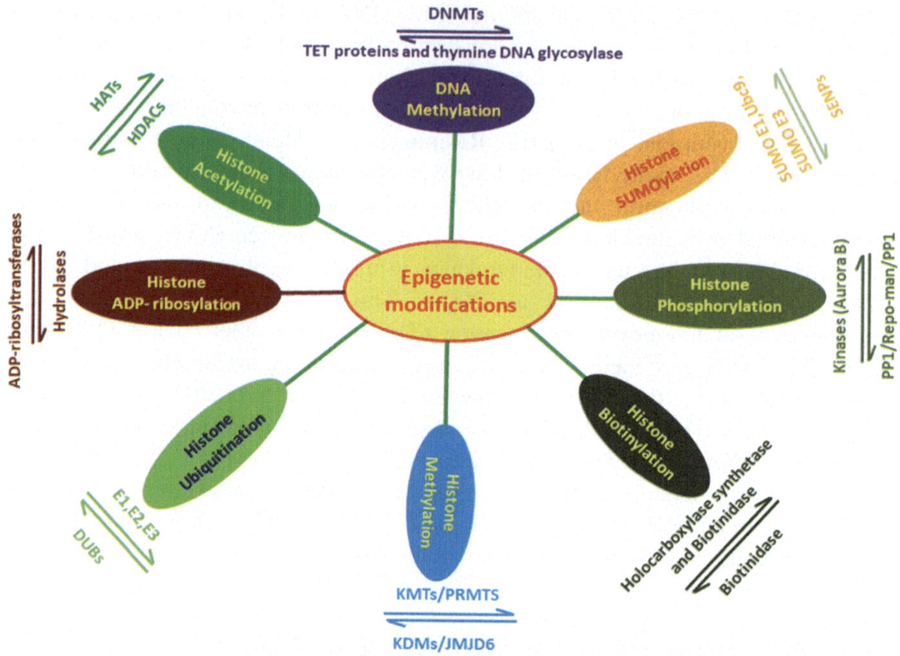

Fig. 1.1 Overview of different epigenetic modifications and their corresponding regulatory enzymes. Epigenetic modifications described here are DNA methylation and post-translational modifications of histone proteins. DNA methylation is performed by DNA methyltransferases (DNMTs) using *S*-adenosyl-L-methionine as cofactor. While DNMT1 is involved in maintenance methylation, DNMT3A and DNMT3B are responsible for de novo methylation. The process is reversed by ten-eleven translocation (TET) proteins and thymine-DNA glycosylase. Histone proteins undergo a variety of post-translational/epigenetic modifications such as acetylation, ADP-ribosylation, ubiquitination, methylation, biotinylation, phosphorylation and SUMOylation. Histone acetylation is controlled by functionally opposite enzyme families. While histone acetyl transferases (HATs) install this mark on ε-amino group of lysine residues and promote decondensation of chromatin, histone deacetylases erase this mark and augment chromatin condensation. ADP-ribosylation is regulated by writers ADP-ribosyltransferases and erased by hydrolases. Ubiquitination of histone proteins is also dynamic modification. Ubiquitination of histones occurs by interplay of ubiquitin-activating enzyme (E1), ubiquitin-conjugating enzyme (E2) and by ubiquitin ligase (E3). This process is reversed by enzymes deubiquitinases (DUBs). Histone methylation, another dynamic post-translational modification, occurs on lysine and arginine residues of histone substrates. Lysine methylation is regulated by lysine methyl transferases (KMTs) and lysine demethylases (KDMs). While arginine methylation is done by protein arginine methyltransferases (PRMTs), a single enzyme namely bifunctional arginine demethylase and lysyl-hydroxylase JMJD6 removing this methylation has been reported. Biotinylation, the addition of biotin to specific residue of histones, is done by holocarboxylase synthetase and biotinidase. Debiotinylation is also believed to be done by biotinidase. Histone phosphorylation occurring mainly in serine and threonine residues is regulated by specific enzymes. While histone phosphorylation is mediated by kinases such as Aurora B or death-associated protein (DAP)-like kinase (Dlk), dephosphorylation is performed by phosphatases, including protein phosphatase 1 (PP1) or Repo-man/PP1. SUMOylation like other histone post-translational modifications being dynamic is regulated by specific enzymes. SUMOylation occurs by cooperation of three enzymes namely SUMO E1 (activating enzyme), Ubc9 (conjugating enzyme) and SUMO E3 (SUMO ligase), while deSUMOylating enzymes namely sentrin-specific proteases (SENPs) reverse the process

sites of various histones undergoing such modifications have been extensively discussed based on solid experimental evidences. The dynamic regulation of these signatures by functionally antagonistic enzymes has also been described. Further, the impact of these modifications on the DNA-templated reactions has also been taken into consideration. Importantly, I have explained epigenetic modulating enzymes with strong emphasis on writers and erasers. Thus in the next chapter I will focus on the critical implications of epigenetic modifying enzymes in bellicose malignancies.

References

Abel KJ, Brody LC, Valdes JM, Erdos MR, McKinley DR, Castilla LH, Merajver SD, Couch FJ, Friedman LS, Ostermeyer EA, Lynch ED, King MC, Welcsh PL, Osborne-Lawrence S, Spillman M, Bowcock AM, Collins FS, Weber BL (1996) Characterization of EZH1, a human homolog of Drosophila enhancer of zeste near BRCA1. Genomics 37:161–171

Adams RR, Carmena M, Earnshaw WC (2001a) Chromosomal passengers and the (aurora) ABCs of mitosis. Trends Cell Biol 11:49–54

Adams RR, Maiato H, Earnshaw WC, Carmena M (2001b) Essential roles of Drosophila inner centromere protein (Incenp) and Aurora B in histone H3 phosphorylation, metaphase chromosome alignment, kinetochore disjunction, and chromosome segregation. J Cell Biol 153:865–880

Agger K, Cloos PA, Christensen J, Pasini D, Rose S, Rappsilber J, Issaeva I, Canaani E, Salcini AE, Helin K (2007) UTX and JMJD3 are histone H3K27 demethylases involved in HOX gene regulation and development. Nature 449:731–734

Aguilera O, Fernández AF, Muñoz A, Fraga MF (2010) Epigenetics and environment: a complex relationship. J Appl Physiol 109:243–251

Allis CD, Berger SL, Cote J, Dent S, Jenuwien T, Kouzarides T, Pillus L, Reinberg D, Shi Y, Shiekhattar R, Shilatifard A, Workman J, Zhang Y (2007) New nomenclature for chromatin-modifying enzymes. Cell 131:633–636

Alomer RM, da Silva EML, Chen J, Piekarz KM, McDonald K, Sansam CG, Sansam CL, Rankin S (2017) Esco1 and Esco2 regulate distinct cohesin functions during cell cycle progression. Proc Natl Acad Sci 114:9906–9911

Ansari KI, Mandal SS (2010) Mixed lineage leukemia: roles in gene expression, hormone signaling and mRNA processing. FEBS J 277:1790–1804

Avvakumov N, Côté J (2007) The MYST family of histone acetyltransferases and their intimate links to cancer. Oncogene 26:5395–5407

Ayaydin F, Dasso M (2004) Distinct in vivo dynamics of vertebrate SUMO paralogues. Mol Biol Cell 15:5208–5218

Baedke J (2013) The epigenetic landscape in the course of time: Conrad Hal Waddington's methodological impact on the life sciences. Stud Hist Philos Biol Biomed Sci 44:756–773

Ballard TD, Wolff J, Griffin JB, Stanley JS, van Calcar S, Zempleni J (2002) Biotinidase catalyzes debiotinylation of histones. Eur J Nutr 41:78–84

Bannister AJ, Kouzarides T (2011) Regulation of chromatin by histone modifications. Cell Res 21:381–395

Bannister AJ, Schneider R, Kouzarides T (2002) Histone methylation: dynamic or static? Cell 109:801–806

Bao B, Pestinger V, Hassan YI, Borgstahl GE, Kolar C, Zempleni J (2011) Holocarboxylase synthetase is a chromatin protein and interacts directly with histone H3 to mediate biotinylation of K9 and K18. J Nutr Biochem 22:470–475

Barau J, Teissandier A, Zamudio N, Roy S, Nalesso V, Herault Y, Guillou F, Bourc'his D (2016) The DNA methyltransferase DNMT3C protects male germ cells from transposon activity. Science (New York, N.Y.) 354:909–912

Barnes CE, English DM, Cowley SM (2019) Acetylation & Co: an expanding repertoire of histone acylations regulates chromatin and transcription. Essays Biochem 63:97–107

Baudat F, Buard J, Grey C, Fledel-Alon A, Ober C, Przeworski M, Coop G, de Massy B (2010) PRDM9 is a major determinant of meiotic recombination hotspots in humans and mice. Science (New York, N.Y.) 327:836–840

Bauer UM, Daujat S, Nielsen SJ, Nightingale K, Kouzarides T (2002) Methylation at arginine 17 of histone H3 is linked to gene activation. EMBO Rep 3:39–44

Bedford DC, Kasper LH, Fukuyama T, Brindle PK (2010) Target gene context influences the transcriptional requirement for the KAT3 family of CBP and p300 histone acetyltransferases. Epigenetics 5:9–15

Bernier-Villamor V, Sampson DA, Matunis MJ, Lima CD (2002) Structural basis for E2-mediated SUMO conjugation revealed by a complex between ubiquitin-conjugating enzyme Ubc9 and RanGAP1. Cell 108:345–356

Biswas S, Rao CM (2018) Epigenetic tools (the writers, the readers and the erasers) and their implications in cancer therapy. Eur J Pharmacol 837:8–24

Boulanger MC, Liang C, Russell RS, Lin R, Bedford MT, Wainberg MA, Richard S (2005) Methylation of Tat by PRMT6 regulates human immunodeficiency virus type 1 gene expression. J Virol 79:124–131

Boulikas T (1989) DNA strand breaks alter histone ADP-ribosylation. Proc Natl Acad Sci 86:3499–3503

Boulikas T (1990) Poly(ADP-ribosylated) histones in chromatin replication. J Biol Chem 265:14638–14647

Bourc'his D, Xu GL, Lin CS, Bollman B, Bestor TH (2001) Dnmt3L and the establishment of maternal genomic imprints. Science (New York, N.Y.) 294:2536–2539

Branscombe TL, Frankel A, Lee JH, Cook JR, Yang Z, Pestka S, Clarke S (2001) PRMT5 (Janus kinase-binding protein 1) catalyzes the formation of symmetric dimethylarginine residues in proteins. J Biol Chem 276:32971–32976

Brenner C (2002) Catalysis in the nitrilase superfamily. Curr Opin Struct Biol 12:775–782

Burbano HA (2006) Epigenetics and genetic determinism. História, Ciências, Saúde-Manguinhos 13:851–863

Burzio LO, Riquelme PT, Koide SS (1979) ADP ribosylation of rat liver nucleosomal core histones. J Biol Chem 254:3029–3037

Byvoet P, Shepherd GR, Hardin JM, Noland BJ (1972) The distribution and turnover of labeled methyl groups in histone fractions of cultured mammalian cells. Arch Biochem Biophys 148:558–567

Camporeale G, Shubert EE, Sarath G, Cerny R, Zempleni J (2004) K8 and K12 are biotinylated in human histone H4. Eur J Biochem 271:2257–2263

Cao J, Yan Q (2012) Histone ubiquitination and deubiquitination in transcription, DNA damage response, and cancer. Front Oncol 2:26

Cao R, Zhang Y (2004) The functions of E(Z)/EZH2-mediated methylation of lysine 27 in histone H3. Curr Opin Genet Dev 14:155–164

Caplan AI, Niedergang C, Okazaki H, Mandel P (1979) Poly ADP-ribose polymerase: self-ADP-ribosylation, the stimulation by DNA, and the effects on nucleosome formation and stability. Arch Biochem Biophys 198:60–69

Chan HM, La Thangue NB (2001) p300/CBP proteins: HATs for transcriptional bridges and scaffolds. J Cell Sci 114:2363–2373

Chang B, Chen Y, Zhao Y, Bruick RK (2007) JMJD6 is a histone arginine demethylase. Science (New York, N.Y.) 318:444–447

Chedin F (2011) The DNMT3 family of mammalian de novo DNA methyltransferases. Prog Mol Biol Transl Sci 101:255–285

Chen Z-X, Riggs AD (2011) DNA methylation and demethylation in mammals. J Biol Chem 286:18347–18353

Chen Z, Zhang Y (2020) Role of mammalian DNA methyltransferases in development. Annu Rev Biochem 89:135–158

Chen D, Dai C, Jiang Y (2015) Histone H2A and H2B deubiquitinase in developmental disease and cancer. Cancer Transl Med 1:170–175

Chew YC, Sarath G, Zempleni J (2007) An avidin-based assay for histone debiotinylase activity in human cell nuclei. J Nutr Biochem 18:475–481

Chou H-YE, Chou HT, Lee S-C (2006) CDK-dependent activation of poly(ADP-ribose) polymerase member 10 (PARP10). J Biol Chem 281:15201–15207

Christensen J, Agger K, Cloos PA, Pasini D, Rose S, Sennels L, Rappsilber J, Hansen KH, Salcini AE, Helin K (2007) RBP2 belongs to a family of demethylases, specific for tri- and dimethylated lysine 4 on histone 3. Cell 128:1063–1076

Cloos PA, Christensen J, Agger K, Maiolica A, Rappsilber J, Antal T, Hansen KH, Helin K (2006) The putative oncogene GASC1 demethylates tri- and dimethylated lysine 9 on histone H3. Nature 442:307–311

Cloos PA, Christensen J, Agger K, Helin K (2008) Erasing the methyl mark: histone demethylases at the center of cellular differentiation and disease. Genes Dev 22:1115–1140

Close P, Hawkes N, Cornez I, Creppe C, Lambert CA, Rogister B, Siebenlist U, Merville MP, Slaugenhaupt SA, Bours V, Svejstrup JQ, Chariot A (2006) Transcription impairment and cell migration defects in elongator-depleted cells: implication for familial dysautonomia. Mol Cell 22:521–531

Cribbs AP, Terlecki-Zaniewicz S, Philpott M, Baardman J, Ahern D, Lindow M, Obad S, Oerum H, Sampey B, Mander PK, Penn H, Wordsworth P, Bowness P, de Winther M, Prinjha RK, Feldmann M, Oppermann U (2020) Histone H3K27me3 demethylases regulate human Th17 cell development and effector functions by impacting on metabolism. Proc Natl Acad Sci 117:6056–6066

Crisp SE, Griffin JB, White BR, Toombs CF, Camporeale G, Said HM, Zempleni J (2004) Biotin supply affects rates of cell proliferation, biotinylation of carboxylases and histones, and expression of the gene encoding the sodium-dependent multivitamin transporter in JAr choriocarcinoma cells. Eur J Nutr 43:23–31

De Santa F, Totaro MG, Prosperini E, Notarbartolo S, Testa G, Natoli G (2007) The histone H3 lysine-27 demethylase Jmjd3 links inflammation to inhibition of polycomb-mediated gene silencing. Cell 130:1083–1094

DeVine T, Sears RC, Dai MS (2018) The ubiquitin-specific protease USP36 is a conserved histone H2B deubiquitinase. Biochem Biophys Res Commun 495:2363–2368

Di Bacco A, Ouyang J, Lee H-Y, Catic A, Ploegh H, Gill G (2006) The SUMO-specific protease SENP5 is required for cell division. Mol Cell Biol 26:4489–4498

Di Lorenzo A, Bedford MT (2011) Histone arginine methylation. FEBS Lett 585:2024–2031

Dillon SC, Zhang X, Trievel RC, Cheng X (2005) The SET-domain protein superfamily: protein lysine methyltransferases. Genome Biol 6:227

Dillon MBC, Rust HL, Thompson PR, Mowen KA (2013) Automethylation of protein arginine methyltransferase 8 (PRMT8) regulates activity by impeding S-adenosylmethionine sensitivity. J Biol Chem 288:27872–27880

Doi M, Hirayama J, Sassone-Corsi P (2006) Circadian regulator CLOCK is a histone acetyltransferase. Cell 125:497–508

D'Oto A, Tian Q-W, Davidoff AM, Yang J (2016) Histone demethylases and their roles in cancer epigenetics. J Med Oncol Ther 1:34–40

Dvoriantchikova G, Seemungal RJ, Ivanov D (2019) DNA methylation dynamics during the differentiation of retinal progenitor cells into retinal neurons reveal a role for the DNA Demethylation pathway. Front Mol Neurosci 12

Dyda F, Klein DC, Hickman AB (2000) GCN5-related N-acetyltransferases: a structural overview. Annu Rev Biophys Biomol Struct 29:81–103

El-Andaloussi N, Valovka T, Toueille M, Steinacher R, Focke F, Gehrig P, Covic M, Hassa PO, Schär P, Hübscher U, Hottiger MO (2006) Arginine methylation regulates DNA polymerase beta. Mol Cell 22:51–62

Eswaran J, Patnaik D, Filippakopoulos P, Wang F, Stein RL, Murray JW, Higgins JMG, Knapp S (2009) Structure and functional characterization of the atypical human kinase haspin. Proc Natl Acad Sci 106:20198–20203

Ewing RM, Chu P, Elisma F, Li H, Taylor P, Climie S, McBroom-Cerajewski L, Robinson MD, O'Connor L, Li M, Taylor R, Dharsee M, Ho Y, Heilbut A, Moore L, Zhang S, Ornatsky O, Bukhman YV, Ethier M, Sheng Y, Vasilescu J, Abu-Farha M, Lambert J-P, Duewel HS, Stewart II, Kuehl B, Hogue K, Colwill K, Gladwish K, Muskat B, Kinach R, Adams S-L, Moran MF, Morin GB, Topaloglou T, Figeys D (2007) Large-scale mapping of human protein–protein interactions by mass spectrometry. Mol Syst Biol 3:89

Fang R, Barbera AJ, Xu Y, Rutenberg M, Leonor T, Bi Q, Lan F, Mei P, Yuan GC, Lian C, Peng J, Cheng D, Sui G, Kaiser UB, Shi Y, Shi YG (2010) Human LSD2/KDM1b/AOF1 regulates gene transcription by modulating intragenic H3K4me2 methylation. Mol Cell 39:222–233

Feizbakhsh O, Place M, Fant X, Buron F, Routier S, Ruchaud S (2017) The mitotic protein kinase Haspin and its inhibitors. https://doi.org/10.5772/intechopen.70732

Felsenfeld G (2014) A brief history of epigenetics. Cold Spring Harb Perspect Biol 6

Feng Q, Wang H, Ng HH, Erdjument-Bromage H, Tempst P, Struhl K, Zhang Y (2002) Methylation of H3-lysine 79 is mediated by a new family of HMTases without a SET domain. Curr Biol 12:1052–1058

Fenley AT, Anandakrishnan R, Kidane YH, Onufriev AV (2018) Modulation of nucleosomal DNA accessibility via charge-altering post-translational modifications in histone core. Epigenetics Chromatin 11:11

Ferro AM, Olivera BM (1982) Poly(ADP-ribosylation) in vitro. Reaction parameters and enzyme mechanism. J Biol Chem 257:7808–7813

Fischle W, Emiliani S, Hendzel MJ, Nagase T, Nomura N, Voelter W, Verdin E (1999) A new family of human histone deacetylases related to Saccharomyces cerevisiae HDA1p. J Biol Chem 274:11713–11720

Fischle W, Franz H, Jacobs SA, Allis CD, Khorasanizadeh S (2008) Specificity of the chromodomain Y chromosome family of chromodomains for lysine-methylated ARK(S/T) motifs. J Biol Chem 283:19626–19635

Frankel A, Yadav N, Lee J, Branscombe TL, Clarke S, Bedford MT (2002) The novel human protein arginine N-methyltransferase PRMT6 is a nuclear enzyme displaying unique substrate specificity. J Biol Chem 277:3537–3543

Frescas D, Guardavaccaro D, Bassermann F, Koyama-Nasu R, Pagano M (2007) JHDM1B/FBXL10 is a nucleolar protein that represses transcription of ribosomal RNA genes. Nature 450:309–313

Frescas D, Guardavaccaro D, Kuchay SM, Kato H, Poleshko A, Basrur V, Elenitoba-Johnson KS, Katz RA, Pagano M (2008) KDM2A represses transcription of centromeric satellite repeats and maintains the heterochromatic state. Cell Cycle 7:3539–3547

Frontelo P, Leader JE, Yoo N, Potocki AC, Crawford M, Kulik M, Lechleider RJ (2004) Suv39h histone methyltransferases interact with Smads and cooperate in BMP-induced repression. Oncogene 23:5242–5251

Fullgrabe J, Hajji N, Joseph B (2010) Cracking the death code: apoptosis-related histone modifications. Cell Death Differ 17:1238–1243

Fulton MD, Brown T, Zheng YG (2019) The biological axis of protein arginine methylation and asymmetric dimethylarginine. Int J Mol Sci 20:3322

Gagné J-P, Hendzel MJ, Droit A, Poirier GG (2006) The expanding role of poly(ADP-ribose) metabolism: current challenges and new perspectives. Curr Opin Cell Biol 18:145–151

Ganai S (2015) In silico approaches towards safe targeting of class I histone deacetylases. https://doi.org/10.1007/978-1-4614-6436-5_459-1, pp 1–9

Ganai SA (2018) Histone deacetylase inhibitors modulating non-epigenetic players: the novel mechanism for small molecule based therapeutic intervention. Curr Drug Targets 19:593–601

Ganai SA (2019) Epigenetic enzymes and drawbacks of conventional therapeutic regimens. In: Ganai SA (ed) Histone deacetylase inhibitors — epidrugs for neurological disorders. Springer, Singapore, pp 11–19

Ganai SA, Shanmugam K, Mahadevan V (2015) Energy-optimised pharmacophore approach to identify potential hotspots during inhibition of class II HDAC isoforms. J Biomol Struct Dyn 33:374–387

Ganai SA, Banday S, Farooq Z, Altaf M (2016) Modulating epigenetic HAT activity for reinstating acetylation homeostasis: a promising therapeutic strategy for neurological disorders. Pharmacol Ther 166:106–122

Ganesh L, Yoshimoto T, Moorthy NC, Akahata W, Boehm M, Nabel EG, Nabel GJ (2006) Protein methyltransferase 2 inhibits NF-kappaB function and promotes apoptosis. Mol Cell Biol 26:3864–3874

García MA, Fueyo R, Martínez-Balbás MA (2016) Chapter 10 - Lysine demethylases: structure, function, and dysfunction. In: Binda O, Fernandez-Zapico ME (eds) Chromatin signaling and diseases. Academic Press, Boston, pp 179–194

Geiss-Friedlander R, Melchior F (2007) Concepts in sumoylation: a decade on. Nat Rev Mol Cell Biol 8:947–956

Giet R, Glover DM (2001) Drosophila Aurora B kinase is required for histone H3 phosphorylation and condensin recruitment during chromosome condensation and to organize the central spindle during cytokinesis. J Cell Biol 152:669–682

Gil RS, Vagnarelli P (2019) Protein phosphatases in chromatin structure and function. Biochim Biophys Acta Mol Cell Res 1866:90–101

Giri CP, West MHP, Smulson M (1978) Nuclear protein modification and chromatin substructure. 1. Differential poly(adenosine diphosphate) ribosylation of chromosomal proteins in nuclei versus isolated nucleosomes. Biochemistry 17:3495–3500

Glover DM, Leibowitz MH, McLean DA, Parry H (1995) Mutations in aurora prevent centrosome separation leading to the formation of monopolar spindles. Cell 81:95–105

Glowacki G, Braren R, Firner K, Nissen M, Kühl M, Reche P, Bazan F, Cetkovic-Cvrlje M, Leiter E, Haag F, Koch-Nolte F (2002) The family of toxin-related ecto-ADP-ribosyltransferases in humans and the mouse. Protein Sci 11:1657–1670

Goldberg AD, Allis CD, Bernstein E (2007) Epigenetics: a landscape takes shape. Cell 128:635–638

Goldknopf IL, Taylor CW, Baum RM, Yeoman LC, Olson MO, Prestayko AW, Busch H (1975) Isolation and characterization of protein A24, a "histone-like" non-histone chromosomal protein. J Biol Chem 250:7182–7187

Goll MG, Kirpekar F, Maggert KA, Yoder JA, Hsieh CL, Zhang X, Golic KG, Jacobsen SE, Bestor TH (2006) Methylation of tRNAAsp by the DNA methyltransferase homolog Dnmt2. Science (New York, N.Y.) 311:395–398

Gong L, Yeh ETH (2006) Characterization of a family of nucleolar SUMO-specific proteases with preference for SUMO-2 or SUMO-3. J Biol Chem 281:15869–15877

Gopalan G, Chan CSM, Donovan PJ (1997) A novel mammalian, mitotic spindle–associated kinase is related to yeast and fly chromosome segregation regulators. J Cell Biol 138:643–656

Goto H, Yasui Y, Nigg EA, Inagaki M (2002) Aurora-B phosphorylates histone H3 at serine28 with regard to the mitotic chromosome condensation. Genes Cells 7:11–17

Gowher H, Liebert K, Hermann A, Xu G, Jeltsch A (2005) Mechanism of stimulation of catalytic activity of Dnmt3A and Dnmt3B DNA-(cytosine-C5)-methyltransferases by Dnmt3L. J Biol Chem 280:13341–13348

Greenberg MVC, Bourc'his D (2019) The diverse roles of DNA methylation in mammalian development and disease. Nat Rev Mol Cell Biol 20:590–607

Gu B, Lee MG (2013) Histone H3 lysine 4 methyltransferases and demethylases in self-renewal and differentiation of stem cells. Cell Biosci 3:39–39

Gu Y, Jones AE, Yang W, Liu S, Dai Q, Liu Y, Swindle CS, Zhou D, Zhang Z, Ryan TM, Townes TM, Klug CA, Chen D, Wang H (2016) The histone H2A deubiquitinase Usp16 regulates hematopoiesis and hematopoietic stem cell function. Proc Natl Acad Sci 113:E51–E60
Guenther MG, Jenner RG, Chevalier B, Nakamura T, Croce CM, Canaani E, Young RA (2005) Global and Hox-specific roles for the MLL1 methyltransferase. Proc Natl Acad Sci U S A 102:8603–8608
Gujar H, Weisenberger DJ, Liang G (2019) The roles of human DNA methyltransferases and their isoforms in shaping the epigenome. Genes (Basel) 10:172
Guo X, Bai Y, Zhao M, Zhou M, Shen Q, Yun CH, Zhang H, Zhu WG, Wang J (2018) Acetylation of 53BP1 dictates the DNA double strand break repair pathway. Nucleic Acids Res 46:689–703
Hahn MA, Szabó PE, Pfeifer GP (2014) 5-Hydroxymethylcytosine: a stable or transient DNA modification? Genomics 104:314–323
Handy DE, Castro R, Loscalzo J (2011) Epigenetic modifications: basic mechanisms and role in cardiovascular disease. Circulation 123:2145–2156
Hashimoto H, Liu Y, Upadhyay AK, Chang Y, Howerton SB, Vertino PM, Zhang X, Cheng X (2012) Recognition and potential mechanisms for replication and erasure of cytosine hydroxymethylation. Nucleic Acids Res 40:4841–4849
Hassan YI, Zempleni J (2006) Epigenetic regulation of chromatin structure and gene function by biotin. J Nutr 136:1763–1765
Hay RT (2007) SUMO-specific proteases: a twist in the tail. Trends Cell Biol 17:370–376
He J, Kallin EM, Tsukada Y, Zhang Y (2008) The H3K36 demethylase Jhdm1b/Kdm2b regulates cell proliferation and senescence through p15(Ink4b). Nat Struct Mol Biol 15:1169–1175
He YF, Li BZ, Li Z, Liu P, Wang Y, Tang Q, Ding J, Jia Y, Chen Z, Li L, Sun Y, Li X, Dai Q, Song CX, Zhang K, He C, Xu GL (2011) Tet-mediated formation of 5-carboxylcytosine and its excision by TDG in mammalian DNA. Science (New York, N.Y.) 333:1303–1307
Heideker J, Wertz IE (2015) DUBs, the regulation of cell identity and disease. Biochem J 467:191
Heightman TD (2011) Chemical biology of lysine demethylases. Curr Chem Genomics 5:62–71
Hill PW, Amouroux R, Hajkova P (2014) DNA demethylation, Tet proteins and 5-hydroxymethylcytosine in epigenetic reprogramming: an emerging complex story. Genomics 104:324–333
Hochstrasser M (2001) SP-RING for SUMO: new functions bloom for a ubiquitin-like protein. Cell 107:5–8
Hochstrasser M (2009) Origin and function of ubiquitin-like proteins. Nature 458:422–429
Hojfeldt JW, Agger K, Helin K (2013) Histone lysine demethylases as targets for anticancer therapy. Nat Rev Drug Discov 12:917–930
Holliday R (2006) Epigenetics: a historical overview. Epigenetics 1:76–80
Holliday R, Pugh JE (1975) DNA modification mechanisms and gene activity during development. Science (New York, N.Y.) 187:226–232
Horton JR, Upadhyay AK, Qi HH, Zhang X, Shi Y, Cheng X (2010) Enzymatic and structural insights for substrate specificity of a family of jumonji histone lysine demethylases. Nat Struct Mol Biol 17:38–43
Hottiger MO (2011) ADP-ribosylation of histones by ARTD1: an additional module of the histone code? FEBS Lett 585:1595–1599
Hottiger MO, Hassa PO, Lüscher B, Schüler H, Koch-Nolte F (2010) Toward a unified nomenclature for mammalian ADP-ribosyltransferases. Trends Biochem Sci 35:208–219
Hsia DA, Tepper CG, Pochampalli MR, Hsia EY, Izumiya C, Huerta SB, Wright ME, Chen HW, Kung HJ, Izumiya Y (2010) KDM8, a H3K36me2 histone demethylase that acts in the cyclin A1 coding region to regulate cancer cell proliferation. Proc Natl Acad Sci U S A 107:9671–9676
Hsu J-Y, Sun Z-W, Li X, Reuben M, Tatchell K, Bishop DK, Grushcow JM, Brame CJ, Caldwell JA, Hunt DF, Lin R, Smith MM, Allis CD (2000) Mitotic phosphorylation of histone H3 is governed by Ipl1/aurora kinase and Glc7/PP1 phosphatase in budding yeast and nematodes. Cell 102:279–291

Hu X, Zhang L, Mao SQ, Li Z, Chen J, Zhang RR, Wu HP, Gao J, Guo F, Liu W, Xu GF, Dai HQ, Shi YG, Li X, Hu B, Tang F, Pei D, Xu GL (2014) Tet and TDG mediate DNA demethylation essential for mesenchymal-to-epithelial transition in somatic cell reprogramming. Cell Stem Cell 14:512–522

Huang OW, Cochran AG (2013) Regulation of deubiquitinase proteolytic activity. Curr Opin Struct Biol 23:806–811

Huletsky A, de Murcia G, Muller S, Hengartner M, Ménard L, Lamarre D, Poirier GG (1989) The effect of poly(ADP-ribosyl)ation on native and H1-depleted chromatin. A role of poly (ADP-ribosyl)ation on core nucleosome structure. J Biol Chem 264:8878–8886

Hymes J, Wolf B (1999) Human biotinidase isn't just for recycling biotin. J Nutr 129:485s–489s

Hymes J, Fleischhauer K, Wolf B (1995) Biotinylation of histones by human serum biotinidase: assessment of biotinyl-transferase activity in sera from normal individuals and children with biotinidase deficiency. Biochem Mol Med 56:76–83

Igarashi T, Izumi H, Uchiumi T, Nishio K, Arao T, Tanabe M, Uramoto H, Sugio K, Yasumoto K, Sasaguri Y, Wang KY, Otsuji Y, Kohno K (2007) Clock and ATF4 transcription system regulates drug resistance in human cancer cell lines. Oncogene 26:4749–4760

Iizuka M, Stillman B (1999) Histone acetyltransferase HBO1 interacts with the ORC1 subunit of the human initiator protein. J Biol Chem 274:23027–23034

Ito S, Shen L, Dai Q, Wu SC, Collins LB, Swenberg JA, He C, Zhang Y (2011) Tet proteins can convert 5-methylcytosine to 5-formylcytosine and 5-carboxylcytosine. Science (New York, N. Y.) 333:1300–1303

Iwase S, Lan F, Bayliss P, de la Torre-Ubieta L, Huarte M, Qi HH, Whetstine JR, Bonni A, Roberts TM, Shi Y (2007) The X-linked mental retardation gene SMCX/JARID1C defines a family of histone H3 lysine 4 demethylases. Cell 128:1077–1088

Jablonka E, Lamb MJ (2002) The changing concept of epigenetics. Ann N Y Acad Sci 981:82–96

Jain K, Clarke SG (2019) PRMT7 as a unique member of the protein arginine methyltransferase family: a review. Arch Biochem Biophys 665:36–45

Jenuwein T, Allis CD (2001) Translating the histone code. Science (New York, N.Y.) 293:1074–1080

Jeong H-J, Lee S-J, Lee H-J, Kim H-B, Anh Vuong T, Cho H, Bae G-U, Kang J-S (2020) Prmt7 promotes myoblast differentiation via methylation of p38MAPK on arginine residue 70. Cell Death Differ 27:573–586

Jin B, Robertson KD (2013) DNA methyltransferases, DNA damage repair, and cancer. Adv Exp Med Biol 754:3–29

Johnson ES (2004) Protein modification by SUMO. Annu Rev Biochem 73:355–382

Johnson ES, Schwienhorst I, Dohmen RJ, Blobel G (1997) The ubiquitin-like protein Smt3p is activated for conjugation to other proteins by an Aos1p/Uba2p heterodimer. EMBO J 16:5509–5519

Jones PL, Veenstra GJ, Wade PA, Vermaak D, Kass SU, Landsberger N, Strouboulis J, Wolffe AP (1998) Methylated DNA and MeCP2 recruit histone deacetylase to repress transcription. Nat Genet 19:187–191

Jump DB, Butt TR, Smulson M (1979) Nuclear protein modification and chromatin substructure. 3. Relationship between poly(adenosine diphosphate) ribosylation and different functional forms of chromatin. Biochemistry 18:983–990

Kareta MS, Botello ZM, Ennis JJ, Chou C, Chedin F (2006) Reconstitution and mechanism of the stimulation of de novo methylation by human DNMT3L. J Biol Chem 281:25893–25902

Karmodiya K, Anamika K, Muley V, Pradhan SJ, Bhide Y, Galande S (2014) Camello, a novel family of histone acetyltransferases that acetylate histone H4 and is essential for zebrafish development. Sci Rep 4:6076

Katsumoto T, Yoshida N, Kitabayashi I (2008) Roles of the histone acetyltransferase monocytic leukemia zinc finger protein in normal and malignant hematopoiesis. Cancer Sci 99:1523–1527

Kim JH, Lane WS, Reinberg D (2002) Human elongator facilitates RNA polymerase II transcription through chromatin. Proc Natl Acad Sci U S A 99:1241–1246

Kim W, Choi M, Kim J-E (2014) The histone methyltransferase Dot1/DOT1L as a critical regulator of the cell cycle. Cell Cycle 13:726–738

Kitabayashi I, Aikawa Y, Nguyen LA, Yokoyama A, Ohki M (2001) Activation of AML1-mediated transcription by MOZ and inhibition by the MOZ-CBP fusion protein. EMBO J 20:7184–7196

Kleine H, Poreba E, Lesniewicz K, Hassa PO, Hottiger MO, Litchfield DW, Shilton BH, Lüscher B (2008) Substrate-assisted catalysis by PARP10 limits its activity to mono-ADP-ribosylation. Mol Cell 32:57–69

Knight HC, Reynolds TR, Meyers GA, Pomponio RJ, Buck GA, Wolf B (1998) Structure of the human biotinidase gene. Mamm Genome 9:327–330

Koch-Nolte F, Kernstock S, Mueller-Dieckmann C, Weiss M, Haag F (2008) Mammalian ADP-ribosyltransferases and ADP-ribosylhydrolases. Front Biosci 13:6716–6729

Koh DW, Lawler AM, Poitras MF, Sasaki M, Wattler S, Nehls MC, Stöger T, Poirier GG, Dawson VL, Dawson TM (2004) Failure to degrade poly(ADP-ribose) causes increased sensitivity to cytotoxicity and early embryonic lethality. Proc Natl Acad Sci U S A 101:17699–17704

Kohli RM, Zhang Y (2013) TET enzymes, TDG and the dynamics of DNA demethylation. Nature 502:472–479

Kothapalli N, Camporeale G, Kueh A, Chew YC, Oommen AM, Griffin JB, Zempleni J (2005) Biological functions of biotinylated histones. J Nutr Biochem 16:446–448

Kouzarides T (2007) Chromatin modifications and their function. Cell 128:693–705

Krajewski WA, Reese JC (2010) SET domains of histone methyltransferases recognize ISWI-remodeled nucleosomal species. Mol Cell Biol 30:552–564

Kschonsak M, Haering CH (2015) Shaping mitotic chromosomes: from classical concepts to molecular mechanisms. BioEssays 37:755–766

Kurdistani SK, Grunstein M (2003) Histone acetylation and deacetylation in yeast. Nat Rev Mol Cell Biol 4:276–284

Labbé RM, Holowatyj A, Yang Z-Q (2013) Histone lysine demethylase (KDM) subfamily 4: structures, functions and therapeutic potential. Am J Transl Res 6:1–15

Lachner M, O'Carroll D, Rea S, Mechtler K, Jenuwein T (2001) Methylation of histone H3 lysine 9 creates a binding site for HP1 proteins. Nature 410:116–120

Lahn BT, Tang ZL, Zhou J, Barndt RJ, Parvinen M, Allis CD, Page DC (2002) Previously uncharacterized histone acetyltransferases implicated in mammalian spermatogenesis. Proc Natl Acad Sci U S A 99:8707–8712

Lakowski TM, Frankel A (2009) Kinetic analysis of human protein arginine N-methyltransferase 2: formation of monomethyl- and asymmetric dimethyl-arginine residues on histone H4. Biochem J 421:253–261

Lee J, Sayegh J, Daniel J, Clarke S, Bedford MT (2005a) PRMT8, a new membrane-bound tissue-specific member of the protein arginine methyltransferase family. J Biol Chem 280:32890–32896

Lee JH, Cook JR, Yang ZH, Mirochnitchenko O, Gunderson SI, Felix AM, Herth N, Hoffmann R, Pestka S (2005b) PRMT7, a new protein arginine methyltransferase that synthesizes symmetric dimethylarginine. J Biol Chem 280:3656–3664

Lee MG, Wynder C, Cooch N, Shiekhattar R (2005c) An essential role for CoREST in nucleosomal histone 3 lysine 4 demethylation. Nature 437:432–435

Lee DH, Kim GW, Jeon YH, Yoo J, Lee SW, Kwon SH (2020) Advances in histone demethylase KDM4 as cancer therapeutic targets. FASEB J 34:3461–3484

Liang G, Chan MF, Tomigahara Y, Tsai YC, Gonzales FA, Li E, Laird PW, Jones PA (2002) Cooperativity between DNA methyltransferases in the maintenance methylation of repetitive elements. Mol Cell Biol 22:480–491

Lin S-Y, Li TY, Liu Q, Zhang C, Li X, Chen Y, Zhang S-M, Lian G, Liu Q, Ruan K, Wang Z, Zhang C-S, Chien K-Y, Wu J, Li Q, Han J, Lin S-C (2012) GSK3-TIP60-ULK1 signaling pathway links growth factor deprivation to autophagy. Science (New York, N.Y.) 336:477–481

Lin Z, Jia L, Tomchick DR, Luo X, Yu H (2014) Substrate-specific activation of the mitotic kinase Bub1 through intramolecular autophosphorylation and kinetochore targeting. Structure 22:1616–1627

Liu H, Ling Y, Gong Y, Sun Y, Hou L, Zhang B (2007) DNA damage induces N-acetyltransferase NAT10 gene expression through transcriptional activation. Mol Cell Biochem 300:249–258

Liu G, Bollig-Fischer A, Kreike B, van de Vijver MJ, Abrams J, Ethier SP, Yang ZQ (2009) Genomic amplification and oncogenic properties of the GASC1 histone demethylase gene in breast cancer. Oncogene 28:4491–4500

Lowndes N, Toh G (2005) DNA repair: the importance of phosphorylating histone H2AX. Curr Biol 15:R99–R102

Luger K, Mader AW, Richmond RK, Sargent DF, Richmond TJ (1997) Crystal structure of the nucleosome core particle at 2.8 a resolution. Nature 389:251–260

Lüscher B, Bütepage M, Eckei L, Krieg S, Verheugd P, Shilton BH (2018) ADP-ribosylation, a multifaceted posttranslational modification involved in the control of cell physiology in health and disease. Chem Rev 118:1092–1136

Lv J, Liu H, Wang Q, Tang Z, Hou L, Zhang B (2003) Molecular cloning of a novel human gene encoding histone acetyltransferase-like protein involved in transcriptional activation of hTERT. Biochem Biophys Res Commun 311:506–513

Maeda K, Yoneda M, Nakagawa T, Ikeda K, Higashi M, Nakagawa K, Miyakoda M, Yui K, Oda H, Inoue S, Ito T (2018) Defects in centromeric/pericentromeric histone H2A T120 phosphorylation by hBUB1 cause chromosome missegregation producing multinucleated cells. Genes Cells 23:828–838

Manthey KC, Griffin JB, Zempleni J (2002) Biotin supply affects expression of biotin transporters, biotinylation of carboxylases and metabolism of interleukin-2 in Jurkat cells. J Nutr 132:887–892

Margueron R, Li G, Sarma K, Blais A, Zavadil J, Woodcock CL, Dynlacht BD, Reinberg D (2008) Ezh1 and Ezh2 maintain repressive chromatin through different mechanisms. Mol Cell 32:503–518

Martinez-Balbás MA, Bannister AJ, Martin K, Haus-Seuffert P, Meisterernst M, Kouzarides T (1998) The acetyltransferase activity of CBP stimulates transcription. EMBO J 17:2886–2893

Martinez-Zamudio R, Ha HC (2012) Histone ADP-ribosylation facilitates gene transcription by directly remodeling nucleosomes. Mol Cell Biol 32:2490–2502

Mashimo M, Kato J, Moss J (2014) Structure and function of the ARH family of ADP-ribosyl-acceptor hydrolases. DNA Repair 23:88–94

Matsui SI, Seon BK, Sandberg AA (1979) Disappearance of a structural chromatin protein A24 in mitosis: implications for molecular basis of chromatin condensation. Proc Natl Acad Sci U S A 76:6386–6390

Mayer W, Niveleau A, Walter J, Fundele R, Haaf T (2000) Demethylation of the zygotic paternal genome. Nature 403:501–502

McVean G, Myers S (2010) PRDM9 marks the spot. Nat Genet 42:821–822

Melchior F (2000) SUMO—nonclassical ubiquitin. Annu Rev Cell Dev Biol 16:591–626

Messner S, Hottiger MO (2011) Histone ADP-ribosylation in DNA repair, replication and transcription. Trends Cell Biol 21:534–542

Messner S, Altmeyer M, Zhao H, Pozivil A, Roschitzki B, Gehrig P, Rutishauser D, Huang D, Caflisch A, Hottiger MO (2010) PARP1 ADP-ribosylates lysine residues of the core histone tails. Nucleic Acids Res 38:6350–6362

Metzger E, Wissmann M, Yin N, Muller JM, Schneider R, Peters AH, Gunther T, Buettner R, Schule R (2005) LSD1 demethylates repressive histone marks to promote androgen-receptor-dependent transcription. Nature 437:436–439

Meulmeester E, Kunze M, Hsiao HH, Urlaub H, Melchior F (2008) Mechanism and consequences for paralog-specific sumoylation of ubiquitin-specific protease 25. Mol Cell 30:610–619

Miao F, Li S, Chavez V, Lanting L, Natarajan R (2006) Coactivator-associated arginine methyltransferase-1 enhances nuclear factor-kappaB-mediated gene transcription through methylation of histone H3 at arginine 17. Mol Endocrinol (Baltimore, MD) 20:1562–1573

Min J, Feng Q, Li Z, Zhang Y, Xu RM (2003) Structure of the catalytic domain of human DOT1L, a non-SET domain nucleosomal histone methyltransferase. Cell 112:711–723

Miotto B, Struhl K (2010) HBO1 histone acetylase activity is essential for DNA replication licensing and inhibited by Geminin. Mol Cell 37:57–66

Moore LD, Le T, Fan G (2013) DNA methylation and its basic function. Neuropsychopharmacology 38:23–38

Mosammaparast N, Shi Y (2010) Reversal of histone methylation: biochemical and molecular mechanisms of histone demethylases. Annu Rev Biochem 79:155–179

Moss J, Stanley SJ, Nightingale MS, Murtagh JJ Jr, Monaco L, Mishima K, Chen HC, Williamson KC, Tsai SC (1992) Molecular and immunological characterization of ADP-ribosylarginine hydrolases. J Biol Chem 267:10481–10488

Mukhopadhyay D, Dasso M (2007) Modification in reverse: the SUMO proteases. Trends Biochem Sci 32:286–295

Murnion ME, Adams RR, Callister DM, Allis CD, Earnshaw WC, Swedlow JR (2001) Chromatin-associated protein phosphatase 1 regulates Aurora-B and histone H3 phosphorylation. J Biol Chem 276:26656–26665

Nakagawa T, Kajitani T, Togo S, Masuko N, Ohdan H, Hishikawa Y, Koji T, Matsuyama T, Ikura T, Muramatsu M, Ito T (2008) Deubiquitylation of histone H2A activates transcriptional initiation via trans-histone cross-talk with H3K4 di- and trimethylation. Genes Dev 22:37–49

Nanney DL (1958) Epigenetic control systems. Proc Natl Acad Sci U S A 44:712–717

Narang MA, Dumas R, Ayer LM, Gravel RA (2004) Reduced histone biotinylation in multiple carboxylase deficiency patients: a nuclear role for holocarboxylase synthetase. Hum Mol Genet 13:15–23

Nathan D, Sterner DE, Berger SL (2003) Histone modifications: now summoning sumoylation. Proc Natl Acad Sci U S A 100:13118–13120

Nathan D, Ingvarsdottir K, Sterner DE, Bylebyl GR, Dokmanovic M, Dorsey JA, Whelan KA, Krsmanovic M, Lane WS, Meluh PB, Johnson ES, Berger SL (2006) Histone sumoylation is a negative regulator in Saccharomyces cerevisiae and shows dynamic interplay with positive-acting histone modifications. Genes Dev 20:966–976

Ng SS, Yue WW, Oppermann U, Klose RJ (2009) Dynamic protein methylation in chromatin biology. Cell Mol Life Sci 66:407–422

Niu G-J, Xu J-D, Yuan W-J, Sun J-J, Yang M-C, He Z-H, Zhao X-F, Wang J-X (2018) Protein inhibitor of activated STAT (PIAS) negatively regulates the JAK/STAT pathway by inhibiting STAT phosphorylation and translocation. Front Immunol 9:2392–2392

Nolan NL, Butt TR, Wong M, Lambrianidou A, Smulson ME (1980) Characterization of poly (ADP-ribose)--histone H1 complex formation in purified polynucleosomes and chromatin. Eur J Biochem 113:15–25

Ogata N, Ueda K, Hayaishi O (1980a) ADP-ribosylation of histone H2B. Identification of glutamic acid residue 2 as the modification site. J Biol Chem 255:7610–7615

Ogata N, Ueda K, Kagamiyama H, Hayaishi O (1980b) ADP-ribosylation of histone H1. Identification of glutamic acid residues 2, 14, and the COOH-terminal lysine residue as modification sites. J Biol Chem 255:7616–7620

Oka S, Kato J, Moss J (2006) Identification and characterization of a mammalian 39-kDa poly (ADP-ribose) glycohydrolase. J Biol Chem 281:705–713

Okada Y, Scott G, Ray MK, Mishina Y, Zhang Y (2007) Histone demethylase JHDM2A is critical for Tnp1 and Prm1 transcription and spermatogenesis. Nature 450:119–123

Okano M, Bell DW, Haber DA, Li E (1999) DNA methyltransferases Dnmt3a and Dnmt3b are essential for de novo methylation and mammalian development. Cell 99:247–257

Okazaki H, Niedergang C, Couppez M, Martinage A, Sautiere P, Mandel P (1980) In vitro ADP-ribosylation of histones by purified calf thymus polyadenosine diphosphate ribose polymerase. FEBS Lett 110:227–229
Oki M, Aihara H, Ito T (2007) Role of histone phosphorylation in chromatin dynamics and its implications in diseases. Subcell Biochem 41:319–336
Palvimo JJ (2007) PIAS proteins as regulators of small ubiquitin-related modifier (SUMO) modifications and transcription. Biochem Soc Trans 35:1405–1408
Peng J, Wysocka J (2008) It takes a PHD to SUMO. Trends Biochem Sci 33:191–194
Peters DM, Griffin JB, Stanley JS, Beck MM, Zempleni J (2002) Exposure to UV light causes increased biotinylation of histones in Jurkat cells. Am J Physiol Cell Physiol 283:C878–C884
Petersen J, Paris J, Willer M, Philippe M, Hagan IM (2001) The *S. pombe* aurora-related kinase Ark1 associates with mitotic structures in a stage dependent manner and is required for chromosome segregation. J Cell Sci 114:4371–4384
Pichler A, Gast A, Seeler JS, Dejean A, Melchior F (2002) The nucleoporin RanBP2 has SUMO1 E3 ligase activity. Cell 108:109–120
Poirier GG, Niedergang C, Champagne M, Mazen A, Mandel P (1982) Adenosine diphosphate ribosylation of chicken-erythrocyte histones H1, H5 and high-mobility-group proteins by purified calf-thymus poly(adenosinediphosphate-ribose) polymerase. Eur J Biochem 127:437–442
Pollack BP, Kotenko SV, He W, Izotova LS, Barnoski BL, Pestka S (1999) The human homologue of the yeast proteins Skb1 and Hsl7p interacts with Jak kinases and contains protein methyltransferase activity. J Biol Chem 274:31531–31542
Potts PR, Yu H (2005) Human MMS21/NSE2 is a SUMO ligase required for DNA repair. Mol Cell Biol 25:7021–7032
Preuss U, Landsberg G, Scheidtmann KH (2003) Novel mitosis-specific phosphorylation of histone H3 at Thr11 mediated by Dlk/ZIP kinase. Nucleic Acids Res 31:878–885
Prigent C, Dimitrov S (2003) Phosphorylation of serine 10 in histone H3, what for? J Cell Sci 116:3677–3685
Qian C, Zhou MM (2006) SET domain protein lysine methyltransferases: structure, specificity and catalysis. Cell Mol Life Sci 63:2755–2763
Qian J, Beullens M, Huang J, De Munter S, Lesage B, Bollen M (2015) Cdk1 orders mitotic events through coordination of a chromosome-associated phosphatase switch. Nat Commun 6:10215
Rasmussen KD, Helin K (2016) Role of TET enzymes in DNA methylation, development, and cancer. Genes Dev 30:733–750
Reich A, Yanai A, Mesilaty-Gross S, Chen-Moses A, Wides R, Motro B (1999) Cloning, mapping, and expression of ial, a novel Drosophila member of the Ipl1/aurora mitotic control kinase family. DNA Cell Biol 18:593–603
Reverter D, Lima CD (2005) Insights into E3 ligase activity revealed by a SUMO-RanGAP1-Ubc9-Nup358 complex. Nature 435:687–692
Robertson KD (2005) DNA methylation and human disease. Nat Rev Genet 6:597–610
Rossetto D, Avvakumov N, Cote J (2012) Histone phosphorylation: a chromatin modification involved in diverse nuclear events. Epigenetics 7:1098–1108
Roth SY, Allis CD (1992) Chromatin condensation: does histone H1 dephosphorylation play a role? Trends Biochem Sci 17:93–98
Rotili D, Mai A (2011) Targeting histone demethylases: a new avenue for the fight against cancer. Genes Cancer 2:663–679
Rulten SL, Fisher AEO, Robert I, Zuma MC, Rouleau M, Ju L, Poirier G, Reina-San-Martin B, Caldecott KW (2011) PARP-3 and APLF function together to accelerate nonhomologous end-joining. Mol Cell 41:33–45
Saitoh H, Hinchey J (2000) Functional heterogeneity of small ubiquitin-related protein modifiers SUMO-1 versus SUMO-2/3. J Biol Chem 275:6252–6258
Saitoh H, Sparrow DB, Shiomi T, Pu RT, Nishimoto T, Mohun TJ, Dasso M (1998) Ubc9p and the conjugation of SUMO-1 to RanGAP1 and RanBP2. Curr Biol 8:121–124

Saunders LR, Verdin E (2007) Sirtuins: critical regulators at the crossroads between cancer and aging. Oncogene. Oncogene 26:5489–5504
Sayegh J, Webb K, Cheng D, Bedford MT, Clarke SG (2007) Regulation of protein arginine methyltransferase 8 (PRMT8) activity by its N-terminal domain. J Biol Chem 282:36444–36453
Scheerger SB, Zempleni J (2003) Expression of oncogenes depends on biotin in human small cell lung cancer cells NCI-H69. Int J Vitam Nutr Res 73:461–467
Scheffner M, Nuber U, Huibregtse JM (1995) Protein ubiquitination involving an E1-E2-E3 enzyme ubiquitin thioester cascade. Nature 373:81–83
Schumacher JM, Ashcroft N, Donovan PJ, Golden A (1998) A highly conserved centrosomal kinase, AIR-1, is required for accurate cell cycle progression and segregation of developmental factors in Caenorhabditis elegans embryos. Development 125:4391–4402
Senthil Kumar G, Gokhan E, Munter S, Bollen M, Vagnarelli P, Peti W, Page R (2016) The Ki-67 and RepoMan mitotic phosphatases assemble via an identical, yet novel mechanism. elife 5
Shailesh H, Zakaria ZZ, Baiocchi R, Sif S (2018) Protein arginine methyltransferase 5 (PRMT5) dysregulation in cancer. Oncotarget 9:36705–36718
Shaytan AK, Armeev GA, Goncearenco A, Zhurkin VB, Landsman D, Panchenko AR (2016) Coupling between histone conformations and DNA geometry in nucleosomes on a microsecond timescale: atomistic insights into nucleosome functions. J Mol Biol 428:221–237
Sheikh BN (2014) Crafting the brain – role of histone acetyltransferases in neural development and disease. Cell Tissue Res 356:553–573
Shen L, Wu H, Diep D, Yamaguchi S, D'Alessio AC, Fung HL, Zhang K, Zhang Y (2013) Genome-wide analysis reveals TET- and TDG-dependent 5-methylcytosine oxidation dynamics. Cell 153:692–706
Shi YG, Tsukada Y (2013) The discovery of histone demethylases. Cold Spring Harb Perspect Biol 5:a017947
Shi Y, Lan F, Matson C, Mulligan P, Whetstine JR, Cole PA, Casero RA, Shi Y (2004) Histone demethylation mediated by the nuclear amine oxidase homolog LSD1. Cell 119:941–953
Shiio Y, Eisenman RN (2003) Histone sumoylation is associated with transcriptional repression. Proc Natl Acad Sci 100:13225–13230
Shuai K (2006) Regulation of cytokine signaling pathways by PIAS proteins. Cell Res 16:196–202
Shuai K, Liu B (2005) Regulation of gene-activation pathways by PIAS proteins in the immune system. Nat Rev Immunol 5:593–605
Simandi Z, Pajer K, Karolyi K, Sieler T, Jiang L-L, Kolostyak Z, Sari Z, Fekecs Z, Pap A, Patsalos A, Contreras GA, Reho B, Papp Z, Guo X, Horvath A, Kiss G, Keresztessy Z, Vámosi G, Hickman J, Xu H, Dormann D, Hortobagyi T, Antal M, Nógrádi A, Nagy L (2018) Arginine methyltransferase PRMT8 provides cellular stress tolerance in aging motoneurons. J Neurosci 38:7683–7700
Sims RJ 3rd, Nishioka K, Reinberg D (2003) Histone lysine methylation: a signature for chromatin function. Trends Genet 19:629–639
Sinha KM, Yasuda H, Coombes MM, Dent SY, de Crombrugghe B (2010) Regulation of the osteoblast-specific transcription factor Osterix by NO66, a Jumonji family histone demethylase. EMBO J 29:68–79
Stanley JS, Griffin JB, Zempleni J (2001) Biotinylation of histones in human cells. Effects of cell proliferation. Eur J Biochem 268:5424–5429
Stein C, Riedl S, Rüthnick D, Nötzold RR, Bauer U-M (2012) The arginine methyltransferase PRMT6 regulates cell proliferation and senescence through transcriptional repression of tumor suppressor genes. Nucleic Acids Res 40:9522–9533
Stephan AK, Kliszczak M, Morrison CG (2011) The Nse2/Mms21 SUMO ligase of the Smc5/6 complex in the maintenance of genome stability. FEBS Lett 585:2907–2913
Stone PR, Lorimer WS III, Kidwell WR (1977) Properties of the complex between histone H1 and poly(ADP-ribose) synthesised in HeLa cell nuclei. Eur J Biochem 81:9–18
Strahl BD, Allis CD (2000) The language of covalent histone modifications. Nature 403:41–45

Sun ZW, Allis CD (2002) Ubiquitination of histone H2B regulates H3 methylation and gene silencing in yeast. Nature 418:104–108

Sun Y, Jiang X, Chen S, Fernandes N, Price BD (2005) A role for the Tip60 histone acetyltransferase in the acetylation and activation of ATM. Proc Natl Acad Sci U S A 102:13182–13187

Sun Y, Jiang X, Xu Y, Ayrapetov MK, Moreau LA, Whetstine JR, Price BD (2009) Histone H3 methylation links DNA damage detection to activation of the tumour suppressor Tip60. Nat Cell Biol 11:1376–1382

Tahiliani M, Koh KP, Shen Y, Pastor WA, Bandukwala H, Brudno Y, Agarwal S, Iyer LM, Liu DR, Aravind L, Rao A (2009) Conversion of 5-methylcytosine to 5-hydroxymethylcytosine in mammalian DNA by MLL partner TET1. Science (New York, N.Y.) 324:930–935

Terada Y, Tatsuka M, Suzuki F, Yasuda Y, Fujita S, Otsu M (1998) AIM-1: a mammalian midbody-associated protein required for cytokinesis. EMBO J 17:667–676

Tewary SK, Zheng YG, Ho M-C (2019) Protein arginine methyltransferases: insights into the enzyme structure and mechanism at the atomic level. Cell Mol Life Sci 76:2917–2932

Toleman C, Paterson AJ, Whisenhunt TR, Kudlow JE (2004) Characterization of the histone acetyltransferase (HAT) domain of a bifunctional protein with activable O-GlcNAcase and HAT activities. J Biol Chem 279:53665–53673

Toleman CA, Paterson AJ, Kudlow JE (2006) The histone acetyltransferase NCOAT contains a zinc finger-like motif involved in substrate recognition. J Biol Chem 281:3918–3925

Torres IO, Fujimori DG (2015) Functional coupling between writers, erasers and readers of histone and DNA methylation. Curr Opin Struct Biol 35:68–75

Tronick E, Hunter RG (2016) Waddington, dynamic systems, and epigenetics. Front Behav Neurosci 10:107–107

Tsai W-C, Reineke LC, Jain A, Jung SY, Lloyd RE (2017) Histone arginine demethylase JMJD6 is linked to stress granule assembly through demethylation of the stress granule nucleating protein G3BP1. J Biol Chem

Tsukada Y, Fang J, Erdjument-Bromage H, Warren ME, Borchers CH, Tempst P, Zhang Y (2006) Histone demethylation by a family of JmjC domain-containing proteins. Nature 439:811–816

Turinetto V, Giachino C (2015) Multiple facets of histone variant H2AX: a DNA double-strand-break marker with several biological functions. Nucleic Acids Res 43:2489–2498

Upadhyay AK, Cheng X (2011) Dynamics of histone lysine methylation: structures of methyl writers and erasers. Prog Drug Res 67:107–124

Vagnarelli P, Ribeiro S, Sennels L, Sanchez-Pulido L, de Lima Alves F, Verheyen T, Kelly DA, Ponting CP, Rappsilber J, Earnshaw WC (2011) Repo-man coordinates chromosomal reorganization with nuclear envelope reassembly during mitotic exit. Dev Cell 21:328–342

Valinluck V, Sowers LC (2007) Endogenous cytosine damage products alter the site selectivity of human DNA maintenance methyltransferase DNMT1. Cancer Res 67:946–950

Van Speybroeck L (2002) From epigenesis to epigenetics: the case of C. H. Waddington. Ann N Y Acad Sci 981:61–81

Vento-Tormo R, Company C, Rodríguez-Ubreva J, de la Rica L, Urquiza JM, Javierre BM, Sabarinathan R, Luque A, Esteller M, Aran JM, Álvarez-Errico D, Ballestar E (2016) IL-4 orchestrates STAT6-mediated DNA demethylation leading to dendritic cell differentiation. Genome Biol 17:4

Verheugd P, Bütepage M, Eckei L, Lüscher B (2016) Players in ADP-ribosylation: readers and erasers. Curr Protein Pept Sci 17:654–667

Vertegaal ACO, Andersen JS, Ogg SC, Hay RT, Mann M, Lamond AI (2006) Distinct and overlapping sets of SUMO-1 and SUMO-2 target proteins revealed by quantitative proteomics. Mol Cell Proteomics 5:2298–2310

Villa F, Capasso P, Tortorici M, Forneris F, de Marco A, Mattevi A, Musacchio A (2009) Crystal structure of the catalytic domain of Haspin, an atypical kinase implicated in chromatin organization. Proc Natl Acad Sci 106:20204–20209

Waddington CH (1940) Organisers and genes. Cambridge biological studies. University press, Cambridge, x + 160 pp. p
Waddington CH (1942) The epigenotype. Endeavour 1:18–20
Waddington CH (1956) Embryology, epigenetics and biogenetics. Nature 177:1241–1241
Waddington CH (1968) Towards a theoretical biology. Nature 218:525–527
Waddington CH (2012) The epigenotype. 1942. Int J Epidemiol 41:10–13
Wagner GP (1993) Hall, B. K., 1992. Evolutionary developmental biology. Chapman and Hall, xii + 275pp. Price £29.95. ISBN: 0-412-27550-3. J Evol Biol 6:460–462
Wang Y, Dasso M (2009) SUMOylation and deSUMOylation at a glance. J Cell Sci 122:4249–4252
Wang H, Huang ZQ, Xia L, Feng Q, Erdjument-Bromage H, Strahl BD, Briggs SD, Allis CD, Wong J, Tempst P, Zhang Y (2001) Methylation of histone H4 at arginine 3 facilitating transcriptional activation by nuclear hormone receptor. Science (New York, N.Y.) 293:853–857
Wang XX, Fu L, Li X, Wu X, Zhu Z, Fu L, Dong JT (2011) Somatic mutations of the mixed-lineage leukemia 3 (MLL3) gene in primary breast cancers. Pathol Oncol Res 17:429–433
Watt F, Molloy PL (1988) Cytosine methylation prevents binding to DNA of a HeLa cell transcription factor required for optimal expression of the adenovirus major late promoter. Genes Dev 2:1136–1143
Weger S, Hammer E, Heilbronn R (2005) Topors acts as a SUMO-1 E3 ligase for p53 in vitro and in vivo. FEBS Lett 579:5007–5012
Whetstine JR, Nottke A, Lan F, Huarte M, Smolikov S, Chen Z, Spooner E, Li E, Zhang G, Colaiacovo M, Shi Y (2006) Reversal of histone lysine trimethylation by the JMJD2 family of histone demethylases. Cell 125:467–481
Wolf B (2005) Biotinidase: its role in biotinidase deficiency and biotin metabolism. J Nutr Biochem 16:441–445
Wong M, Polly P, Liu T (2015) The histone methyltransferase DOT1L: regulatory functions and a cancer therapy target. Am J Cancer Res 5:2823–2837
Wood K, Tellier M, Murphy S (2018) DOT1L and H3K79 methylation in transcription and genomic stability. Biomol Ther 8:11
Wotton D, Merrill JC (2007) Pc2 and SUMOylation. Biochem Soc Trans 35:1401–1404
Wu C, Morris JR (2001) Genes, genetics, and epigenetics: a correspondence. Science (New York, N.Y.) 293:1103–1105
Wu H, Zhang Y (2014) Reversing DNA methylation: mechanisms, genomics, and biological functions. Cell 156:45–68
Wu C-S, Zou L (2016) The SUMO (small ubiquitin-like modifier) ligase PIAS3 primes ATR for checkpoint activation. J Biol Chem 291:279–290
Xiang Y, Zhu Z, Han G, Lin H, Xu L, Chen CD (2007a) JMJD3 is a histone H3K27 demethylase. Cell Res 17:850–857
Xiang Y, Zhu Z, Han G, Ye X, Xu B, Peng Z, Ma Y, Yu Y, Lin H, Chen AP, Chen CD (2007b) JARID1B is a histone H3 lysine 4 demethylase up-regulated in prostate cancer. Proc Natl Acad Sci U S A 104:19226–19231
Yamane K, Toumazou C, Tsukada Y, Erdjument-Bromage H, Tempst P, Wong J, Zhang Y (2006) JHDM2A, a JmjC-containing H3K9 demethylase, facilitates transcription activation by androgen receptor. Cell 125:483–495
Yamane K, Tateishi K, Klose RJ, Fang J, Fabrizio LA, Erdjument-Bromage H, Taylor-Papadimitriou J, Tempst P, Zhang Y (2007) PLU-1 is an H3K4 demethylase involved in transcriptional repression and breast cancer cell proliferation. Mol Cell 25:801–812
Yang M, Culhane JC, Szewczuk LM, Jalili P, Ball HL, Machius M, Cole PA, Yu H (2007) Structural basis for the inhibition of the LSD1 histone demethylase by the antidepressant trans-2-phenylcyclopropylamine. Biochemistry 46:8058–8065
Yang Y, He Y, Wang X, Liang Z, He G, Zhang P, Zhu H, Xu N, Liang S (2017) Protein SUMOylation modification and its associations with disease. Open Biol 7:170167

Yunus AA, Lima CD (2006) Lysine activation and functional analysis of E2-mediated conjugation in the SUMO pathway. Nat Struct Mol Biol 13:491–499

Zempleni J, Li Y, Xue J, Cordonier EL (2011) The role of holocarboxylase synthetase in genome stability is mediated partly by epigenomic synergies between methylation and biotinylation events. Epigenetics 6:892–894

Zeng L, Zhou M-M (2002) Bromodomain: an acetyl-lysine binding domain. FEBS Lett 513:124–128

Zeng L, Yap KL, Ivanov AV, Wang X, Mujtaba S, Plotnikova O, Rauscher FJ 3rd, Zhou MM (2008) Structural insights into human KAP1 PHD finger-bromodomain and its role in gene silencing. Nat Struct Mol Biol 15:626–633

Zhang X, Bruice TC (2008) Enzymatic mechanism and product specificity of SET-domain protein lysine methyltransferases. Proc Natl Acad Sci 105:5728–5732

Zhao X, Sternsdorf T, Bolger TA, Evans RM, Yao T-P (2005) Regulation of MEF2 by histone deacetylase 4- and SIRT1 deacetylase-mediated lysine modifications. Mol Cell Biol 25:8456–8464

Zhao Q, Rank G, Tan YT, Li H, Moritz RL, Simpson RJ, Cerruti L, Curtis DJ, Patel DJ, Allis CD, Cunningham JM, Jane SM (2009) PRMT5-mediated methylation of histone H4R3 recruits DNMT3A, coupling histone and DNA methylation in gene silencing. Nat Struct Mol Biol 16:304–311

Zhu Z, Wang Y, Li X, Wang Y, Xu L, Wang X, Sun T, Dong X, Chen L, Mao H, Yu Y, Li J, Chen PA, Chen CD (2010) PHF8 is a histone H3K9me2 demethylase regulating rRNA synthesis. Cell Res 20:794–801

Epigenetic Regulator Enzymes and Their Implications in Distinct Malignancies

2

Histone protein modifications and DNA methylation come within the confines of epigenetic modifications (Jarmasz et al. 2019; Tolsma and Hansen 2019). These modifications being dynamic are perfectly tuned by the functionally antagonistic enzymes (Allis et al. 2007; Kang et al. 2017; Tian et al. 2013; Yang and Seto 2007). Histone proteins unlike DNA undergo a variety of post-translational modifications. While histone proteins undergo acetylation (Eberharter and Becker 2002), phosphorylation (Rossetto et al. 2012), methylation (Luo 2018), SUMOylation (Flotho and Melchior 2013), ubiquitination (Cao and Yan 2012), biotinylation and ADP-ribosylation (Palazzo et al. 2019), DNA undergoes only one epigenetic modification, that is methylation (Jarmasz et al. 2019).

2.1 Misregulation of Epigenetic Modifying Enzymes Triggers Cancer Onset

Both histone modifications and DNA methylation have the potential to modulate chromatin topology through different mechanisms (Cheng 2014). These modifications and modifiers under balanced condition play a key role in executing the gene expression programs precisely, besides regulating the biological outcome (Zhao and Shilatifard 2019). Misregulation of these modifications due to aberrant activity of epigenetic modifying enzymes, results in transcriptional dysregulation which in turn primes the cells for diseases onset and advancement (Peschle et al. 1967; Piunti and Shilatifard 2016; Zhao and Shilatifard 2019). Multiple chromatin-modulating enzymes contribute significantly to different malignancies, and it will be more understandable by taking the enzymes of a particular modification into consideration at a time (Cheng et al. 2019; Ellis et al. 2009; Wang et al. 2016b).

S. A. Ganai, *Histone Deacetylase Inhibitors in Combinatorial Anticancer Therapy*,
https://doi.org/10.1007/978-981-15-8179-3_2

2.1.1 Implications of Histone AcetylTransferases (HATs) in Cancer

Histone acetyltransferases (HATs) or lysine acetyltransferases (KATs) as discussed in the previous chapter hyperacetylate histone substrates and thus promote transcription by powering electrostatic repulsion between polyanioinic DNA and polycationic histones (Barnes et al. 2019; Roth et al. 2001; Strahl and Allis 2000). If this hyperacetylation occurs in proto-oncogenes, this might provide impetus to cancer progression while hypoacetylation may offer protection by silencing such genes. This statement receives its support from the findings where histone hyperacetylation has been reported in hepatocellular carcinoma (Bai et al. 2008) and acetylation of lysine 18 of H3 has been linked to recurrence of prostate cancer (Bianco-Miotto et al. 2010).

Overexpression of HAT p300 in colorectal cancer patients has been related to poor prognosis (Ishihama et al. 2007). Mammalian GCN5 or TIP60 may promote cancer as they stabilize oncoprotein c-MYC (Patel et al. 2004; Wapenaar and Dekker 2016). Evidence-based studies suggest that p300 is potentially involved in breast cancer recurrence and has been correlated to poor prognosis (Xiao et al. 2011). Further adverse survival in patients with resectable non-small cell lung cancers has been ascribed to escalated expression of p300 (Hou et al. 2012b). This HAT has also been related to increased proliferation of prostate cancer cells and altered nuclear morphology. Under androgen-deprived conditions, p300 has been found to be essential for IL-6-provoked activation of androgen receptor (Heemers et al. 2008). Another study has also shown the involvement of p300 in prostate cancer cell proliferation and advancement. Study on prostate cancer samples showed that high expression of this HAT is positively related to high proliferation and large tumour volume apart from involvement of seminal vesicles. Small interfering RNA-based disruption of p300 transcripts restrained prostate cancer cell proliferation even on stimulation by interleukin 6 (Debes et al. 2003). Moreover, in patients of nasopharyngeal carcinoma (Liao et al. 2012) and aggressive hepatocellular carcinomas elevated expression of p300 HAT has been reported (Li et al. 2011a).

Overexpression of ornithine decarboxylase and enhanced polyamine synthesis are the authentication marks for epithelial tumorigenesis. It has been reported that polyamine-mediated tumour progression may be supported by Tip 60 (MYST family HAT) upregulation (Hobbs et al. 2006). Elevated expression of HATs has also been proved in malignant pleural mesothelioma, a belligerent but rare cancer of pleura. RT-PCR results have shown that all the variants of KAT5 (another name of Tip60) compared to benign pleura are substantially elevated in malignant tumours (Cregan et al. 2016). Histone acetyltransferase (MST4 or MORF or KAT6B) overexpression has been identified in ovarian cancer through SAGE (serial analysis of gene expression). This HAT was markedly overexpressed in HGSCs (ovarian high grade serous carcinomas) compared to endometroid and ovarian clear cell carcinomas. This high expression of this HAT in advanced HGSCs has also been linked to the alleviated patient survival (Liu et al. 2019a). Another histone acetyltransferase namely MYST2 or Hbo1 serves as positive regulator during DNA duplication. Strong expression of this HAT has been revealed in various carcinomas including ovary, testis, stomach, breast and bladder (Iizuka et al. 2009).

As of now I have discussed the overexpression of some HATs in different cancers. Plenty of evidences are also available where alleviated expression of HATs has been linked to different malignancies (Demetriadou and Kirmizis 2017). Human MOF (hMOF/MYST1) a member of MYST family has been found to be downregulated in patients of primary breast carcinoma and medulloblastoma. Above twofold downregulation was noted in 41% and 79% patients of respective cancers. Patients with insufficient expression of this acetyltransferase show markedly adverse overall survival (Pfister et al. 2008). Similar trend of hMOF expression has been reported in ovarian cancer (Cai et al. 2015), hepatocellular carcinoma (Zhang et al. 2014), renal cell carcinoma (Wang et al. 2013), colorectal (Cao et al. 2014) and gastric cancer (Cao et al. 2014; Zhu et al. 2015). Among the patients of gastric and hepatocellular carcinoma, patients with low MOF/MYST1 level showed shorter survival time compared to those of high MOF (Su et al. 2016; Zhang et al. 2014). Abnormal expression of MYST1, a HAT responsible for acetylating H4 at lysine 16 (H4K16) has implications in some primary cancers. Studies performed on renal cell carcinoma samples and cell lines revealed the downregulation of this HAT in majority (above 90%) of samples. Alleviated expression of this HAT both in cell models and patient samples has strongly correlated with H4K16 acetylation (Wang et al. 2013).

Only recently, it has been seen that p300 deletion accentuated leukaemogenesis in transgenic mice model of myelodysplastic syndrome. Deletion of this HAT reinstated the self-renewal ability of haematopoietic stem and progenitor cells. Thus p300 plays a crucial role in conversion of myelodysplastic syndrome to acute myeloid leukaemia. However, no such effect was seen with another HAT namely CBP (Cheng et al. 2017). Although p300 is expressed in both nucleus and cytoplasm, the former is the premier location. During the transformation of dysplastic nevi to metastatic melanoma via the primary melanoma, nuclear p300 expression diminishes while its cytoplasmic expression escalates. Further, the loss of nuclear p300 expression promoted metastasis and corresponded with poor survival of melanoma patients (Demetriadou and Kirmizis 2017; Rotte et al. 2013).

2.1.2 Role of Histone Deacetylases (HDACs) in Tumorigenesis

As discussed in previous chapter that HDACs deinstall acetylation marks deposited by HATs on specific residues of histone substrates. HDACs are surely having a strong role in cancer fuelling. However, the detailed involvement will be provided in Chap. 4 that is solely dedicated to implications of HDACs in cancer.

2.1.3 Histone Methyltransferases (HMTs) and Their Involvement in Cancer

This modification occurring on the lysine and arginine residues of histone proteins has strong involvement in regulating gene expression programs. Enzymes methylating lysine residues are lysine methyltransferases (KMTs), while those methylating arginine residues are protein arginine methyltransferases (PRMTs). Thus KATs and PRMTs come under the confines of histone methyltransferases. Lysine methylation is critical as it has strong impact on protein stability and function. Thus it will be more interesting to discuss the involvement of these enzymes in tumorigenesis under separate headings.

2.1.3.1 Lysine Methyltransferases (KATs) in Cancer Signalling

Several writers of histone lysine methylation on aberrant expression promote cancer onset and advancement. Enhancer of zeste homolog 2 (EZH2), the polycomb group protein, is overexpressed in prostate cancer. Elevated expression of this KMT has been both at message and protein level in metastatic prostate cancer (Varambally et al. 2002). This methyltransferase has shown high expression at both levels in invasive breast carcinoma as compared to normal breast epithelia. Aggressiveness of breast cancer has positive correlation with enhanced EZH2 protein levels as identified through tissue microarray analysis (Kleer et al. 2003). EZH2 being the crucial part of polycomb-repressive complex 2 underpins gene silencing by installing methyl groups on lysine 27 of histone H3 (H3K27me3). While bellicose solid tumours notably breast, prostate and bladder are associated with heightened levels of EZH2, the benign epithelial cells show reverse trend. Enhanced levels of EZH2 downregulate the expression of E-cadherin (tumour suppressor) by promoting H3K27me3 (Cao et al. 2008). Mutation (activating one), in the SET domain of this enzyme, has been reported in 7% and 22% patients of follicular lymphomas and diffuse large B-cell lymphoma, respectively (Bennett and Licht 2018; Morin et al. 2010).

SUV39H1 and G9a are two distinct KMTs acting as writers for histone H3 lysine methylation (SUV39H1 for H3K9me3, G9a for H3K9me1 and H3K9me2). These two enzymes sustain the malignant phenotype as knockdown of both of them in human prostate cancer cell line hampered proliferation (Kondo et al. 2008). Above 40% of non-small cell lung cancer tissues (NSCLC) G9a elevation has been reported. This overexpression triggers Wnt signalling pathway and subsequent proliferation of NSCLC. Selective inhibitor-based intervention of this KMT or siRNA-mediated knockdown markedly restrained tumour growth and subdued Wnt signalling. G9a alleviates the expression of APC regulator of WNT signalling pathway 2 (APC2), the effect being mediated by promoter demethylation and HP1α (Zhang et al. 2018).

It has been seen in patients (advanced gastric carcinoma) that tumour metastasis is the prime death cause. The higher expression of this methyltransferase corresponded with advanced stage of disease in addition to reduced overall survival. G9a (enhanced expression) triggers gastric carcinoma metastasis by facilitating the integrin beta-3 expression and for this function, methyltransferase activity of G9a has been found to be immaterial (Hu et al. 2018). This hypoxia-regulated methyltransferase accentuates tumour growth and its loss of function has been reported to attenuate neoplastic growth. This methyltransferase contributes significantly towards mediating the hypoxic response by lowering the expression of certain specific genes including GATA2, ARNTL, HHEX, OGN and KLRG1 (Casciello et al. 2017). G9a along with HDAC1 has been reported to suppress the expression of RUNX3, a tumour suppressor when ovexpressed by hypoxia. Following hypoxia these enzymes enhance the H3K9me2 and hypoacetylate the promoter of defined tumour suppressor, respectively. Thus, in the course of gastric cancer progression, hypoxia suppresses RUNX3 by altering methylation and acetylation levels in its promoter region (Lee et al. 2009). Further elevated expression of G9a has also been demonstrated in lung cancer cells (Watanabe et al. 2008).

The MLL family of methyltransferases includes various histone methyltransferase enzymes (MLL1–MLL5) meant for methylation of histone H3 at lysine 4. Studies have shown the downregulation of genes encoding these enzymes in breast cancer cell lines. Although all the five genes were also suppressed in breast tumour samples in comparison to normal breast tissue, substantial downregulation was seen in MML2 and to certain extent in MLL5. These findings suggest the possible contribution of MLL2 and MLL5 towards breast cancer progression (Rabello et al. 2013). In a broad range of solid and haematological malignancies MLL3 and MLL4 have been identified as cancer drivers through cancer genome sequencing and animal-based studies (Meeks and Shilatifard 2017). MLL1/MLL/KTM2A shows alterations in 0.17% of all cancers with highest frequency in breast carcinomas (invasive ductal and invasive breast) (2017). KMT2A accentuates human melanoma cell growth and its knockdown substantially attenuated viability and migration finally leading to apoptosis. Further the overexpression of this methyltransferase rescued these effects and thereby facilitated proliferation in melanoma cell lines. This effect of K2MT2A is mediated by the human telomerase reverse transcriptase (hTERT) as overexpression latter rescued the knockdown effects of former (Zhang et al. 2017a). This suggests KMT2A as a candidate therapeutic target for ameliorating melanoma proliferation and metastasis. Moreover, KMT2A upregulation has also been identified to power the colorectal cancer as evidenced by the studies on colorectal cancer samples. Colorectal cancer cells (KMT2A knockdown ones) expectedly showed hampered cell invasion and migration. This enzyme along with p53 mediates its effect through transcriptional activation of cathepsin Z/CTSZ (Hidaka et al. 2000), a lysosomal cysteine proteinase contributing to colorectal cancer progression (Fang et al. 2019).

Histone-lysine N-methyl transferase SETDB1, a specific methyltransferase for trimethylation of H3K9, is quite often heightened in lung cancers and melanoma and has been linked to proliferation facilitation (Ceol et al. 2011; Rodriguez-Paredes et al. 2014). Studies with cancer cells of various lineages have shown escalation of H3K9me3 installed by SETDB1 and SETDB2 post exposure to hypoxia, chemotherapy and targeted therapies. These findings suggest the possible implications of these KMTs in post-treatment resistance mechanisms (Torrano et al. 2019). Further, overexpression of SETDB1 occurs in a panel of solid tumours and is inversely proportional to survival rate of colorectal cancer patients. Apart from triggering the proliferation and migration of cancer cells, its elevated expression makes these cells resistant to 5-fluorouracil-induced cytotoxic effect (Chen et al. 2017b). Cellular tumour antigen p53 encoded by TP53 gene is a well-known tumour suppressor (Harris 1996), and SETDB1 restrains its expression through H3K9me3 culminating in apoptosis inhibition (Chen et al. 2017b). These evidences suggest the cardinal role of defined enzyme in colorectal carcinogenesis and suggest SETDB1 as a possible target for therapeutic intervention.

Histone methylation at lysine 36 (H3K36) being multifunctional regulates transcription, DNA repair and alternative splicing (Li 2013; Li et al. 2019). Genes encoding the methyltransferases specific for these positions either are mutated, are overexpressed or have role in chromosomal translocation, the crucial events in cancer. These methyltransferases control oncogenic transcriptional events as revealed by molecular-level studies (Rogawski et al. 2016). H3K36 methylation being involved in a panel of processes suggests that the respective transferases may function as either oncogenes or tumour suppressors in cancer events. Minimum eight H3K36 KMTs have been reported to be encoded by human genome that can methylate histone H3 at mentioned site. However, these methyltransferases differ not only in number of methyl groups transferred but also in additional substrates they can methylate. Certain methyltransferases like ASH1L, NSD1–3 and SETD2 exhibit H3K36 methylation specificity as their SET domains are agnate (Rogawski et al. 2016). Among H3K36-related KMTs the well characterized oncoproteins are NSD proteins as they have implications in various cancer types. They occur as fusion proteins in multiple myeloma and acute myeloid leukaemia because of chromosomal translocations (Kuo et al. 2011; Wang et al. 2007). Overexpression of SMYD2 contributes substantially to proliferation and invasion of gastric cancer cells (Komatsu et al. 2015). Heightened protein levels of this methyltransferase were detected in 76.5% primary tumour samples of oesophageal squamous cell carcinoma (Komatsu et al. 2009). Patients victimized with SMYD2-overexpressing tumours showed poorer survival rate compared to non-expressing ones (Komatsu et al. 2009). Besides higher expression of SMYD2 has been reported in acute lymphoblastic leukaemia of children (Sakamoto et al. 2014).

NSD2 (Nuclear Receptor Binding SET Domain Protein 2), another methyltransferase of H3K36, also contributes to various malignancies either through mutations or through overexpression. In childhood acute lymphocytic leukaemia, an activating mutation (E1099K) of this enzyme has been seen. This mutation accentuates the H3K36me2 concurrently alleviating H3K27me3 especially on nucleosomes containing variant of histone H1 (H1.3). Cells harbouring this mutated enzyme show hampered apoptosis but elevated proliferation and migration (Swaroop et al. 2019). Its overexpression has also been seen in multiple myeloma and has been linked to myelomagenesis. While its methyltransferase activity is critical for adherence, proliferation and other purposes, its PHD domain is crucial for its biological function and cellular activity. Through this domain its gets recruited to respective oncogenic target genes and prompts their transcription (Huang et al.

2013). The prime chromatin regulatory activity of this enzyme being H3K36me2 is enough for accelerating transcriptional events. Distribution of this methylation (both gene specific and genome wide) is disorganized in myeloma cells (t (4; 14)-positive) favouring the chromatin topology conducive for activation of genes driving myeloma (Kuo et al. 2011). Last but not least, this methyltransferase is also overrepresented in tumours of prostate cancer where it plays a key role in tumour growth (Yang et al. 2012).

Another methyltransferase namely disruptor of telomeric silencing 1-like (DOT1L) has potent implications in cancer. This atypical methyltransferase meant for the mono-, di- and trimethylation of histone H3K79 functions primarily in telomeric gene silencing and in addition interacts with silent information regulator (SIR) proteins (Feng et al. 2002; Lacoste et al. 2002; Ng et al. 2002). DOT1L possesses *S*-adenosyl-L-methionine (AdoMet)-binding motif resembling arginine and DNA methyltransferases, as an alternative to SET domain (Sawada et al. 2004). Only recently, in yeast model DOT1 through its histone chaperone activity has been reported to regulate nucleosome dynamics (Lee et al. 2018). While mono- and dimethylation of H3K79 prompts gene activation, trimethylation at this position promotes gene silencing (Feng et al. 2002; Wong et al. 2015; Zhang et al. 2004). Central involvement of this enzyme has been identified in diverse cellular processes ranging from development, somatic reprogramming to repair of DNA damage (Sarno et al. 2019; Wong et al. 2015).

DOT1L has been reported to initiate and sustain oncofusion protein (MLL-AF9)-driven leukaemogenesis (Nguyen et al. 2011). Chromosomal translocations of MLL (mixed lineage leukaemia) gene form a usual aetiology of acute leukaemias. Studies have shown that DOT1L interacts directly with AF9 (C-terminal domain), and mistargeting of this methyltransferase to *Hoxa* and *Meis1* genes enhances their transcription through H3K79 methylation thereby contributing to MLL-AF9-induced leukaemia (Nguyen et al. 2011; Zhang et al. 2006). DOT1L also has implications in breast cancer and it regulates the transcriptional activity of oestrogen receptor alpha in such cell lines and obstructing this enzyme attenuated proliferation of hormone responsive breast cancer cells under both in vitro and in vivo set-up (Nassa et al. 2019; Salvati et al. 2019). Higher expression of this methyltransferase has been observed in colorectal cancer (Yang et al. 2019a). It is known that c-Myc is the principal regulator of cell cycle associated factors (Miller et al. 2012). DOT1L promotes the proliferation of colorectal cancer cells by activating c-Myc transcription through epigenetic mechanism (H3K9me2) (Yang et al. 2019a).

Thus these evidences are sufficient for justifying the role of lysine methyltransferases in the aetiology of multiple cancers. Now I will discuss the involvement of histone arginine methyltransferases in various cancers.

2.1.3.2 Arginine Methyltransferases in Driving Cancer

Methylation of specific histone arginine residues has role in distinct cellular processes such as transcription, splicing of mRNA and DNA repair apart from signal transduction (Guccione and Richard 2019). As of now nine methyltransferases belonging to protein arginine methyltransferase (PRMT) family have been identified but only seven of them have tendency to methylate (Di Lorenzo and Bedford 2011). Although the mutations of these enzymes have not been reported in neoplasms, surely their overexpression imparts cancer signalling (Poulard et al. 2016). This shows that these enzymes may be exploited as therapeutic targets for effective epigenetic-based therapeutic intervention.

Arginine methyltransferases impact cellular activity through epigenetic mechanism. Certain arginine methyltransferases have been implicated in cancer. For instance PRMT1 in breast cancer acts as a master regulator of epithelial to mesenchymal transition. Migratory and invasive behaviour of breast cancer cells is facilitated by PRMT1 (Gao et al. 2016). Under in vivo conditions abrogation of its expression attenuated metastasis. This methyl transferase mediated epithelial to mesenchymal transition through activation of (zinc-finger E-box-binding homeobox 1) (ZEB1) transcription by epigenetic mechanism (asymmetrical dimethylation of histone H4 at arginine 3/H4R3me2a) (Gao et al. 2016). PRMT1 in cooperation with lysine demethylase (KDM4C) makes the ground fertile for leukaemic transformation (Cheung et al. 2016).

Breast cancer stem cells are the originating points for treatment resistance, progression and relapse. Proliferation and self-renewal of these cancer stem cells is crucially regulated by arginine methyltransferase PRMT5 (Chiang et al. 2017). On recruitment to promoter of forkhead Box P1 (FOXP1), PRMT5 activates the transcription of former by enhancing symmetrical dimethylation of histone H3 at arginine 2 (H3R2me2). This arginine dimethylation in turn recruits SET1, thereby installing histone H3 lysine 4 trimethylation and subsequent transcription of FOXP1 (Chiang et al. 2017). This all suggests that breast cancer stem cell maintenance is due to PRMT5 and as such this arginine methyltransferase may prove as promising target for intervention against therapeutically challenging breast cancer.

Mounting evidences suggest that PRMT5 may function as an oncogene and its overexpression has been demonstrated to play a principal role in different malignancies such as lung cancer (Shilo et al. 2013), hepatocellular cancer (Jeon et al. 2018), breast cancer (Wu et al. 2017) and oropharyngeal squamous cell carcinoma (Kumar et al. 2017). Through symmetrical dimethylation of H4R3, PRMT5 silenced expression of miR-99 family genes, which in turn activated expression of fibroblast growth factor receptor 3 (FGFR3) triggering downstream signalling culminating in lung cancer proliferation and metastasis (Jing et al. 2018). Higher expression of PRMT5 in oral squamous cell carcinoma has been seen during the course of oncogenesis and progression certifying its role in cell invasion (Amano et al. 2018). This arginine methyltransferase also controls pathological processes associated with gastrointestinal cancer by facilitating proliferation and metastatic events (Liu et al. 2018; Xiao et al. 2019). Compared to nearby normal tissue, expression levels of PRMT5 were found to be markedly higher in gastric cancer tissues (Kanda et al. 2016).

Elevated expression of PRMT5 has been linked to multiple myeloma proliferation. Post-transcriptional gene silencing of PRMT5 through siRNAs or its selective inhibition by EPZ015666 alleviates proliferation of multiple myeloma cells (Gullà et al. 2018). In acute myeloid leukaemia cells increased expression of this enzyme markedly increased proliferation and colony formation. Special inhibitor against PRMT5 mitigated the growth and proliferation of leukaemic cells (Serio et al. 2018). Moreover similar trend of PRMT5 has been seen in glioblastoma patient-derived tumours and cell lines. Although high expression of this methyltransferase has been detected in glioblastoma cells, its levels were undetectable or low depending on the grade of astrocytoma. Moreover, cell proliferation and highly aggressive phenotype has been positively correlated with the expression of this type II arginine methyltransferase (Han et al. 2014).

PRMT2, a type I arginine methyltransferase, acts as ERα coactivator. Increased expression of this arginine methyltransferase has been reported in breast cancer tissues (Zhong et al. 2011). Overexpression of this enzyme has also been reported in glioblastoma and inactivation or silencing of PRM2 impeded glioblastoma cell growth. This enzyme being involved in asymmetric dimethylation of histone H3 at arginine 8 (H3R8me2a) plays a crucial role in sustaining expression of target genes. Thus PRMT2 has implications in pathogenesis of glioblastoma as it facilitates oncogenic gene expression by serving the function of transcriptional coactivator (Dong et al. 2018). Its nuclear loss in tumour samples of breast was associated with enhanced expression of cyclin D suggesting its crucial role in proliferation of breast tumour cells (Zhong et al. 2014).

Biological role of another arginine methyltransferase PRMT3 is not fully known as meagre physiological targets modulated by this enzyme have been recognized up to this time. Studies have shown that DAL-1/4.1B, the tumour suppressor protein, binds to this enzyme and regulates its methyltransferase activity negatively (Heller et al. 2007; Singh et al. 2004; Takahashi et al. 2012). The expression of this tumour suppressor being lost in breast tumour samples quite often suggests the higher PRMT3 activity in such samples (Morettin and Baldwin 2015). Only recently upregulation of PRMT3 has been related to pancreatic cancer. PRMT3 binds to GAPDH and methylates it at arginine residue 248, thereby enhancing the activity of latter. Thus overexpression of PRMT3 triggers cellular proliferation in addition to metabolic reprogramming by methylating GAPDH at the defined spot (Hsu et al. 2019). Higher expression of PRMT3 has also been connected to gemcitabine resistance in pancreatic cancer cells. Reduced expression of this enzyme reinstates gemcitabine sensitivity in resistant models. Once overexpressed, PRMT3 upregulates ATP-binding cassette subfamily G member 2 (key player in drug resistance), by stabilizing its mRNA (Hsu et al. 2018).

Coactivator-associated arginine methyltransferase 1 (CARM1) also known as PRMT4 acts as coregulator of ERα. This methyltransferase in many cancers performs as oncogene. PRMT4/CARM1 elevated expression has been found in osteosarcoma, and post-transcriptional gene silencing of CARM1 hampered the proliferation of osteosarcoma cells. CARM1-induced proliferation of these cells has been ascribed to pGSK3β/β-catenin/cyclinD1 signalling (Li et al. 2017). Studies have shown that in epithelial ovarian cancer CARM1 relies on the activity of lysine methyltransferase EZH2. This conclusion has been drawn as CARM1-expressing epithelial cancer responded to small molecule-based therapeutic intervention against EZH2 while CARM1-deficient did not. This also suggests the possible cross-talk between arginine and lysine methyltransferases in facilitating certain malignancies (Karakashev et al. 2018). Further, in breast tumours enhanced expression of CARM1 has been reported. Higher protein levels of this methyltransferase were seen in ERA (−) human breast cancer cell lines relative to ERA (+). Enhanced proliferation was seen in ERA (+) MCF-7 cells on ectopic expression of CARM1. Besides CARM1-induced breast cancer proliferation has been attributed to Her-2/Neu/ErbB2 signalling (Zhang et al. 2005).

In invasive breast cancer elevated status of CARM1 has been reported (Cheng et al. 2013). This enzyme in colorectal cancer is post-transcriptionally regulated by microRNA-195-5p. In human colorectal cancer tissues, it has been verified through expression analysis that microRNA-195-5p undergoes marked downregulation. This reduction elevates the CARM1 levels. Substantial reduction of cell proliferation has been noted in colorectal cancer cells expressing increased levels of this microRNA. However, restoring the CARM1 levels in microRNA-195-5p-transfected cells partly rescued the effects of latter on proliferation and colony formation (Zhang et al. 2017b). From these observations it is quite evident that microRNA-195-5p has antitumour function that is mediated through negative regulation of CARM1. Being the androgen receptors transcriptional coactivator CARM1 modulates the biological functions of former. Higher expression of CARM1 has been implicated in the prostate carcinoma and in androgen-independent prostate carcinoma as well (Hong et al. 2004). This enzyme being functionally distinct from other transcriptional coactivators of above-mentioned receptor may prove as a promising target for ameliorating hormone-independent prostate carcinoma.

Another arginine methyltransferase PRMT6, asymmetrically dimethylating histone H3 at arginine 2 position (H3R2me2a) has potent involvement in various cancers. DNA methylation has great impact on transcriptional events, and it has been reported that aberrant DNA methylation has strong cross-talk with cancer and oncogenic events (Veland et al. 2017). Experimental findings suggest that PRMT6 alleviates DNA methylation and escalated expression of this arginine methyltransferase augments global DNA hypomethylation, thereby triggering cancer. PRMT6 overexpression mediates passive DNA demethylation by hindering the association of DNMT1 accessory factor (UHRF1) to chromatin (Veland et al. 2017). Speaking briefly, enhanced dimethylation of H3R2 impedes the UHRF1–histone H3 interaction, which in turn promotes global DNA hypomethylation resulting in cancer. Worse prognosis of lung cancer has strong cross-talk with PRMT6 expression. Lung-targeted enhanced expression of this arginine methyltransferase facilitates cell proliferation and powered urethane (chemical carcinogen)-induced tumour growth.

Moreover, PRMT6 depletion resulted in attenuation of cell proliferation and migration of NSCLC cells. In originally distinct tumours interleukin-enhancer binding protein 2 (ILF2) has been linked to poor gross survival (Bi et al. 2017; Cheng et al. 2016; Wan et al. 2015). Evidences suggest that PRMT6 in lungs serves as regulator of ILF2 (protumorigenic protein). Drastic increase in the protein levels of ILF2 occurs on increased expression of defined methyltransferase in bronchial epithelial cells (Avasarala et al. 2020). Apart from this PRMT6 has been proved to have oncogenic role in prostate cancer, and higher expression of this arginine methyltransferase both at message and protein level has been reported in this malignancy. Attenuation of malignant phenotype, migration and invasion of prostate cancer cells has been observed on stable knockdown of PRMT6. Enhanced MLL and MYD3 expression has been noted in knockdown cells (Almeida-Rios et al. 2016).

PRMT7, another member of arginine methyltransferases family, has strong link with onset and progression of various cancers. This PRMT acts as metastasis inducer in case of breast cancer. E-cadherin restrains metastasis and its loss is considered as an indication of epithelial to mesenchymal transition (Onder et al. 2008). PRMT6 being highly expressed in breast carcinoma cells restrains the expression of this cadherin through epigenetic mechanism (Yao et al. 2014). Multiple gene expression studies have proved the involvement of PRMT7 in metastasis and poor survival in patients of breast cancer. In addition to primary breast tumour tissue, substantial overexpression of this arginine methyltransferase has been observed in breast cancer lymph node metastasis. PRMT7 attenuation through post-transcriptional gene silencing markedly reported invasion and metastasis under in vitro and in vivo conditions, respectively. Enhanced invasiveness was seen on PRMT7 overexpression in non-aggressive breast cancer cells. This effect of PRMT7 has been attributed to matrix metalloproteinase 9 (MMP9) (Baldwin et al. 2015). This metalloproteinase is very closely involved in breast cancer metastasis (Mehner et al. 2014). Overexpression of PRMT7 promotes MMP9 expression, thereby accentuating breast cancer metastasis.

In non-small cell lung cancer cells (A549 and SPC-A1) the hyperexpression of PRMT7 not only facilitated invasion but also colony formation. Accentuation of metastasis in these cells on increased PRMT5 expression was found to be due to its interaction with HSPAS (78-kDa glucose-regulated protein) and elongation factor 2 (EEF2) (Cheng et al. 2018). In various cancer types invasion of cancer cells is associated with attenuated proliferation. PRMT7 has been identified as the arginine methyltransferase that has the central role in facilitating invasion but hampering proliferation of breast cancer cell models. This effect of PRMT7 is mediated through dimethylation of SH3 and multiple ankyrin repeat domains protein 2 (Shank2). From this finding, it is obvious that that reciprocal switching in breast cancer cells is mediated by PRMT7-dependent Shank2 methylation (Liu et al. 2019b). Certain members of PRMTs are known to undergo self-methylation (Dillon et al. 2013). It has been validated that PRMT7 undergoes auto-methylation at arginine 531 (R531). This self-methylation plays a critical role in epithelial to mesenchymal transition besides powering the invasive behaviour of breast cancer cells. While canonical PRMT7 facilitated metastasis, contrasting effect was seen with PRMT7 mutant (R531K) in nude mouse model. Speaking mechanistically, auto-methylation of this arginine methyltransferase impedes its recruitment to promoter region of E-cadherin. Subsequently decrease in methylation levels occurs, which triggers the expression of E-cadherin, thereby facilitating breast cancer metastasis (Geng et al. 2017).

PRMT8, one of the members of type I PRMTs, has been less studied compared to other members. While its high expression has been related to increased patient survival in breast and ovarian cancer patients, contrasting situation is seen in gastric cancer patients (Hernandez et al. 2017). Overexpression of PRMT8 in colon cancer cells induced aggressive traits including enhanced proliferation and invasion. Moreover, this overexpression was associated with chemoresistance (drug resistance). PRMT8 may facilitate the formation of colon cancer stem cells by regulating pluripotent transcription factors (Lin et al. 2018).

2.1.4 Histone Demethylases in Cancer Pathogenesis

These enzymes are functionally antagonistic to histone methyltransferase and thus erase the methyl tags deposited on specific lysine or arginine residues by these enzymes. Histone demethylases include both lysine and arginine demethylases (Cloos et al. 2008). Tremendous efforts have been taken by various research groups for the delineating the implications of these demethylases in cancer signalling.

2.1.4.1 Lysine Demethylases and Their Crosstalk with Cancer

Among lysine demethylases, LSD1/KDM1A and many members of Jumonji family including KDM5A/JARID1A, KDM2B/JHDM1B, KDM3A/JHDM2A/JMJD1A, KDM4A/JMJD2A, KDM4C/JMJD2C, KDM6B/JMJD3 and KDM6A/UTX have cancer implications (Sainathan et al. 2015). These erasers have importance in embryonic development. From pathophysiological perspective, potential links exist between histone demethylase expression and onset and sustenance of metastatic tumours. Multiple evidences regarding the overexpression of histone demethylases in tumours are currently available (Hoffmann et al. 2012). Indeed the oncogenic power of some demethylases was known prior to their demethylase activity. For instance, JMJD2C was earlier known as GASC1 (gene amplified in squamous cell carcinoma 1) as in oesophageal cancer cells its amplification was seen (Yang et al. 2000).

Various knockdown assays have validated the implications of histone demethylases in aggressive cancer phenotype. Post-transcriptional gene silencing of LSD1 product attenuated proliferation of bladder carcinoma, authenticated to overexpress this demethylase. These effects were rescued on exogenous overexpression of LSD1 (Hayami et al. 2011). It is interesting that prostate cancer cells require androgen receptor activation for proliferation only initially and thus respond to anti-androgen therapy. At later stages these cells do not respond to anti-androgens as they no longer remain androgen dependent (Willmann et al. 2012). LSD1 modulates the expression of multiple genes crucial for cancer proliferation and progression. Obstructing the expression or activity of this demethylase hampers the oestrogen mediated signalling in breast cancer suggesting its role in cancer pathogenesis. Moreover, LSD1 intervention-triggered cytostatic effect has been found to be independent of the ER status (Pollock et al. 2012). From these findings, it is clear that targeting LSD1 may prove as an effective strategy against ER negative breast cancers as well. Demethylases like LSD1 and JMJD2C have the role of androgen receptor coactivation. These enzymes jointly demethylate the trimethylated lysine 9 of histone H3. First JMJD2C removes the methyl group from H3K9me3 following which the remaining groups are erased by LSD1. Thus through epigenetic mechanism these enzymes accentuate the derepression of androgen-dependent genes (Wissmann et al. 2007). In association with androgen receptor, change in substrate specificity of LSD1 occurs and instead of H3K4me2 it accepts H3K9me2 as substrate (Metzger et al. 2005).

Among the known isoforms of JARID1, overrepresented in cancers, the link of JARID1B/KDM5B with tumorigenesis is highly studied. While JARID1B is chiefly considered as oncogene in breast cancer (Yamane et al. 2007), there is evidence that its overexpression attenuates invasion of breast cancer cells (triple-negative) (Li et al. 2011b). Besides its upregulation has been reported in several cancers including ovarian (Wang et al. 2015), hepatocellular (Wang et al. 2016a) and prostate (Xiang et al. 2007). On overexpression this demethylase impedes proliferation and DNA duplication of melanoma cells. Further the expression of melanoma progression-associated genes gets attenuated on higher expression of JARID1B (Roesch et al. 2006). JARID1A/KDM5A also shows escalated expression in various tumours including breast cancer. Knockdown of KDM5A through shRNA hampered proliferation of breast cancer cells (KDM5A-amplified). Overexpression of this demethylase also induced drug resistance in breast cancer cells (Hou et al. 2012a). JARID1C has dual role and thus functions oncogene in certain malignancies such as clear renal cell carcinoma (Ricketts and Linehan 2015) and tumour suppressor in human papilloma virus-related malignancies (Harmeyer et al. 2017; Smith et al. 2014).

JMJD2B/KDM2B is highly expressed in gastric cancer cells where it facilitates proliferation in addition to survival and tumorigenesis. Knocking down its expression in gastric cancer cells and restrained proliferation and induced apoptosis in certain cases as well. Attenuation of xenograft tumour growth following the JMJD2B knockdown suggests the crucial role of this demethylase in succouring proliferation (Li et al. 2011c). Elevated expression of this demethylase showed antiproliferative effect in triple negative breast cancer cells. This effect of KDM2B involved epigenetic mechanism (H3K4me3 and H3K36me2 demethylation) resulting in the downregulation of transcription of cell cycle inhibitors including p57KIP2, p15INK4B and p16INK4A (Zheng et al. 2018). This demethylase is significantly enhanced in pancreatic cancer and its enrichment occurs in invasive cancer cells. Obstructing KDM2B has antiproliferative effect and prevents xenograft tumour formation. This demethylase interacts with Polycomb group proteins to restrain cellular differentiation epigenetically while on the other hand promotes the transcription of MYC and KDM5A for enhancing pancreatic ductal adenocarcinoma progression (Tzatsos et al. 2013). Besides KDM2B has role in gastric cancer (Zhao et al. 2017) and it has been revealed by ChIP study that this demethylase acts directly on promoter of MYC following which the attenuation of glycolysis occurs (Hong et al. 2016). Moreover, this demethylase has implications in several other cancers including nasopharyngeal carcinoma (Ren et al. 2015), ovarian cancer (Kuang et al.

2017), lung, bladder cancer (Kottakis et al. 2011) and leukaemias (Andersson et al. 2007; He et al. 2011; Yan et al. 2018).

Another demethylase KDM3A dimethylating H3K9me1/2 has involvement in colorectal cancer. Delaying of cancer cell growth and migration has been noted on its deficiency. For hippo target genes KDM3A serves as positive regulator. It facilitates gene expression on one hand by increasing Yes Associated Protein 1 (YAP1) expression and on the other hand by promoting H3K27 acetylation (Wang et al. 2019). Upregulation of this demethylase has also been seen in pancreatic tumours. Human pancreatic ductal adenocarcinoma samples showed enhanced expression of both KDM3A and another protein positively regulated by this demethylase doublecortin-like kinase 1 (DCLK1) (Dandawate et al. 2019). Apart from this KDM3A has involvement in ovarian cancer as it regulates ovarian cancer stemness and cisplatin resistance. This effect of KDM3A has been ascribed to its ability of provoking Sox2 (SRY-Box Transcription Factor 2)/Nanog (Nanog Homeobox) and Bcl-2 (B-cell lymphoma 2), respectively (Ramadoss et al. 2017).

KDM4C demethylates H3K9me3 (Yuan et al. 2016) and has been reported to facilitate prostate cancer cell proliferation. Knockdown of this enzyme impeded proliferation, colony formation and androgen receptor transcriptional activity besides hampering growth of pancreatic xenotransplants (Lin et al. 2019). This demethylase has contribution in colorectal cancer as it substantially promotes colonosphere formation. Sphere from this cancer show enhanced expression of KDM4C and its knockdown blocked colonosphere formation. β-catenin has a critical role in oncogenesis of colorectal cancer. The defined protein enhances the sphere formation by transcriptional activation of KDM4C (Yamamoto et al. 2013).

KDM6B, a histone H3K27 demethylase, epigenetically activates differentiation of neuroblastoma cells. KDM6B depletion in neuroblastoma cell lines facilitates proliferation while its upregulated expression apart from inducing neuronal differentiation restrains cell proliferation. KDM6B derepresses the expression of epigenetically silenced neuronal genes by erasing H3K27 trimethylation (Yang et al. 2019b). Elevated expression of this H3K27 trimethylation demethylase hampered cell growth by inducing mitochondria-dependent apoptosis and by impairing invasion-metastasis signalling in non-small cell lung cancer cells. These effects of KDM6B were found to be mediated by FOXO1 as knockdown of latter attenuated the KDM6B-induced apoptosis and metastasis (Ma et al. 2015). High levels of this demethylase have also been linked to the promotion of Hodgkin's lymphoma (Anderton et al. 2011), AML (Li et al. 2018), medulloblastoma (Chen et al. 2017a), melanoma (Park et al. 2016), liver cancer (Tang et al. 2016), ovarian cancer (Mo et al. 2017), cervical cancer (McLaughlin-Drubin et al. 2013), breast cancer (Xun et al. 2017; Yan et al. 2017) and renal cell carcinoma (Chen et al. 2019; Shen et al. 2012).

KDM6A along with LSD1 seems to promote proliferation of breast cancer cells. This crux has been taken from the study where the dual inhibitor (MC3324) of these demethylases was found to induce marked growth arrest and programmed cell death (apoptosis) in breast cancer model (hormone-responsive) (Benedetti et al. 2019). Studies on human pancreatic ductal adenocarcinoma cells have shown that this demethylase acts as tumour suppressor and its loss makes these cells more vulnerable to HDAC inhibitor-induced effects (Watanabe et al. 2019). Both KDM6A and KDM6B were found to be transcriptionally upregulated in malignant pleural mesothelioma cancer cell lines and patient samples (Cregan et al. 2017).

2.1.4.2 Histone Arginine Demethylases and Their Link with Cancer

Till date a single arginine demethylase namely JMJD6 (bifunctional arginine demethylase and lysyl-hydroxylase) has been reported. This enzyme demethylates H3R2 and H4R3, and these findings are well proved by biochemical and cell-based assays (Chang et al. 2007). This arginine demethylase has potential implications in various cancers. Only recently this demethylase has been linked to oral squamous cell carcinoma. Compared to normal oral epithelia, higher expression of JMJD6 was not in oral squamous cell carcinoma tissues suggesting its potential role in driving oral carcinogenesis. On knockdown of endogenous JMJD6, impaired self-renewal capacity has been observed in oral squamous cell carcinoma cells. This anticancer effect of JMJD6 has been related to induction of interleukin 4 as this interleukin rescued the self-renewal capacity even in JMJD6 knockdown cells (Lee et al. 2015).

Evidences suggest that JMJD6 promotes proliferation of colon cancer cells in a p53-dependent manner. JMJD6 catalyses the hydroxylation of p53, facilitating its association with MDMX (negative regulator of p53), which in turn leads to attenuation of p53 transcriptional activity. Depleting this enzyme facilitates cell apoptosis and makes them sensitive to DNA damaging agents (Wang et al. 2014). JMJD6 via facilitating cancer cell proliferation and motility may accentuate the cancer virulence under in vivo conditions. In advanced and more aggressive breast tumours this arginine demethylase has been found to be highly expressed (Lee et al. 2012). JMJD6 plays a crucial role in hepatocellular carcinoma carcinogenesis. The higher expression of JMJD6 serves as an indicator for poor prognosis and aggressive phenotype in hepatocellular carcinoma. Migration and proliferation of hepatoma cells was impaired in JMJD6 knockdown cells. Mechanistically, JMJD6 promotes proliferation and carcinogenesis by enhancing the expression of CDK4 through epigenetic route (Wan et al. 2019).

2.1.5 DNA Methyltransferases (DNMTs) and Their Involvement in Cancer

These enzymes methylate DNA by transferring the methyl group from *S*-adenosyl-L-methionine (universal methyl donor) to 5-position of cytosine residues present in DNA (Jin and Robertson 2013; Robertson 2005). Almost six DNMTs have been reported from mammalian systems including DNMT1, DNMT2, DNMT3A, DNMT3B, DNMT3C and DNMT3L (Gowher and Jeltsch 2018). While DNMT1 is maintenance DNMT, DNMT3A and DNMT3B are de novo methyltransferases. DNMT3L is devoid of inherent methyltransferase activity but significantly escalates the activities of de novo DNA methyltransferases (Chen and Zhang 2020). DNMT2 methylates tRNA instead of DNA, whereas DNMT3C (found in rodent genomes and earlier thought to be pseudogene) methylates young retrotransposons (Barau et al. 2016). Thus DNMT1, DNMT3A and DNMT3B being real DNMTs will be focussed while discussing the implications of DNMTs in cancer.

Enhanced expression of DNMT1 is associated with mammary tumours and gland-specific deletion of this enzyme rescued mice from tumorigenesis by alleviating the cancer stem cell pool. This effect of DNMT1 was mediated by ISL1 as either DNMT inhibition or ISL1 expression reduces cancer stem cell concentration (Pathania et al. 2015). DNMT1 in association with EZH2 silences miR-484, which in turn facilitates cervical cancer tumorigenesis (Hu et al. 2019). DNMT1 plays a critical role in cancer stem cell maintenance, and knockdown of this enzyme resulted in attenuation of malignant phenotypes in oesophageal squamous cell carcinoma cells (Teng et al. 2018). Higher protein levels of this DNMT were reported from breast cancer cells and breast cancer tissues (Agoston et al. 2005).

In 80% of the sporadic breast tumours substantial expression of DNMT3B has been reported (Butcher and Rodenhiser 2007). Studies have shown that the expression of DNMT1 and DNMT3B is regulated by MYC in Burkitt's lymphoma in addition to T-cell acute lymphoblastic leukaemia (T-ALL). Overexpression of these DNMTs by MYC oncogene plays a crucial role in maintenance of tumours. These genes underwent downregulation on MYC inactivation suggesting that DNMTs mediate the effects of this oncogene (Poole et al. 2017). Enhanced expression of DNMT3B and low expression of miR-29b have been seen in pancreatic cancer tissues compared to pancreatic tissues. Overexpression of miR-29b reduced cell viability and facilitated apoptosis by restraining DNMT3B (Wang et al. 2018). Breast cancer cell showed hypermethylator phenotype, which has been attributed to enhanced expression of DNMT3B (Roll et al. 2008). The expression of DNMT1, along with DNMT3A and DNMT3B, has been found to be elevated in retinoblastomas as compared to normal retinas. While frequency of DNMT1 was found to be 100%, frequency of DNMT3A and DNMT3B was found to be 98% and 92%, respectively (Qu et al. 2010). In AML and patients of other haematological malignancies mutations in the gene encoding DNMT3A have been reported suggesting its crucial tumour suppressor role (Yang et al. 2015).

2.1.6 DNA Demethylases and Cancer

As mentioned in previous chapter, DNA methylation was thought to be an irreversible post-translational modification for long. However, with the discovery of ten-eleven translocation (TET) enzymes this notion lost its ground forever. The TET family is of three members ranging from TET1 to TET3 (Rasmussen and Helin 2016; Tahiliani et al. 2009). These enzymes have strong implications in haematological malignancies. In MLL-rearranged AML this protein TET1 is fused to mixed lineage leukaemia (MLL) protein (Lorsbach et al. 2003). In 15% of the patients having various myeloid malignancies somatic mutations in the enzyme TET2 have been seen (Delhommeau et al. 2009). Further the mutations in TET2 have also been reported in patients of CMML, AML, MDS, PTCL and angioimmunoblastic T-cell lymphoma (AITL) (Rasmussen and Helin 2016; Scourzic et al. 2015).

2.1.7 Kinases and Phosphatases in Cancer

Histone phosphorylation and dephosphorylation are regulated by kinases and phosphatases, respectively. Aberrant activities of these enzymes result in oncogenic transformation and thus are emerging as novel targets for therapeutic intervention. Aurora family of serine/threonine kinases in mammals are key controllers of mitotic progression and in human cancers are quite often overexpressed. Studies have proved that mammalian Aurora A-protein is composed of 403 amino acid residues and is encoded by BTAK/STK15 (breast tumour amplified kinase) gene. In carcinoma and cervical intraepithelial neoplasm 3, the expression of Aurora A and B was substantially elevated when compared to normal cervix (Twu et al. 2009). A variety of solid tumours show overexpression of Aurora-A suggesting its role in driving tumorigenesis (Fukuda et al. 2005; Zhou et al. 1998). Further, in several gastrointestinal malignancies amplification and escalated expression of Aurora kinase A have been reported. From these studies it became evident that this kinases regulates not only cell cycle and mitosis but also crucial signalling pathways related to oncogenesis (Katsha et al. 2015). BTAK gene encoding this kinase has been found to be amplified and transcriptionally overactive in breast tumour cells suggesting its involvement in oncogenic transformation (Sen et al. 1997).

Overexpression of Aurora B has been seen in glioblastoma. Its overexpression was found to be directly proportional to aggressive behaviour and shortened survival (Zeng et al. 2007). Moreover, enhanced expression of this kinase has been reported in a number of cancers including prostate (Chieffi 2018; Chieffi et al. 2006), colon (Tatsuka et al. 1998), thyroid (Sorrentino et al. 2005), hepatocellular carcinoma (Yasen et al. 2009), oral cancer (Qi et al. 2007), mesothelioma (López-Ríos et al. 2006) and non-small cell lung carcinoma (Chieffi 2018; Vischioni et al. 2006). Aurora B overexpression contributes to tumorigenesis on one hand by causing chromosomal instability and on the other hand by restraining the $p21^{Cip1}$ (cell cycle inhibitor) (González-Loyola et al. 2015).

Haspin, a protein kinases installing phosphorylating H3R3, is pivotal for mitotic progression (Dai et al. 2005). Its overexpression in pancreatic ductal adenocarcinoma cells triggers mitosis and subsequent tumour cell proliferation. Haspin intervention disrupts histone H3-survivin protein complex culminating in growth blockade and apoptosis (Bastea et al. 2019). Haspin drives proliferation and drug resistance of melanoma cells. Therapeutic intervention with on-target Haspin inhibitor (CX-6258) attenuated proliferation and cytotoxicity in sensitive and resistant melanoma cells (Melms et al. 2020).

Serine/threonine protein phosphatase-1 (PP1), a vital regulator of cell cycle, may have implications in cancer. It interacts with a large number of proteins and forms heterodimeric or heterotrimeric protein complexes. These interacting proteins not only determine its specificity but also control its activity (Ludlow et al. 1993). In

HeLa cells, hyperexpression of NIPP1 (nuclear inhibitor of PP1) resulted in prometaphase arrest. Thus it has been proposed that cancer cell death can be induced by selective PP1 inhibition by way of mitotic catastrophe (Winkler et al. 2015). PP1 isoform namely protein phosphatase 1α regulates critical events including cell cycle advancement and programmed cell death (apoptosis). In 21% of the oral squamous cell carcinoma cell lines, the gene encoding PP1 isoform (PPP1CA) was present in high copy number. Post-transcriptional gene silencing of PP1α attenuated squamous cell carcinoma cell growth partly by modifying phosphorylation of retinoblastoma protein (Hsu et al. 2006).

Thus I have discussed the implications of main epigenetic modification enzymes in various cancers. During this discussion light was shed on HATs, lysine methyltransferases and lysine demethylases, arginine methyltransferases and arginine demethylases, DNMTs and DNA demethylases. Further the role of kinases and phosphatases in distinct cancers was thoroughly discussed. However, HDACs were only mentioned but not discussed as the upcoming chapter is solely dedicated to these erasers of acetylation marks.

References

Agoston AT, Argani P, Yegnasubramanian S, De Marzo AM, Ansari-Lari MA, Hicks JL, Davidson NE, Nelson WG (2005) Increased protein stability causes DNA methyltransferase 1 dysregulation in breast cancer. J Biol Chem 280:18302–18310

Allis CD, Berger SL, Cote J, Dent S, Jenuwien T, Kouzarides T, Pillus L, Reinberg D, Shi Y, Shiekhattar R, Shilatifard A, Workman J, Zhang Y (2007) New nomenclature for chromatin-modifying enzymes. Cell 131:633–636

Almeida-Rios D, Graça I, Vieira FQ, Ramalho-Carvalho J, Pereira-Silva E, Martins AT, Oliveira J, Gonçalves CS, Costa BM, Henrique R, Jerónimo C (2016) Histone methyltransferase PRMT6 plays an oncogenic role of in prostate cancer. Oncotarget 7:53018–53028

Amano Y, Matsubara D, Yoshimoto T, Tamura T, Nishino H, Mori Y, Niki T (2018) Expression of protein arginine methyltransferase-5 in oral squamous cell carcinoma and its significance in epithelial-to-mesenchymal transition. Pathol Int 68:359–366

Andersson A, Ritz C, Lindgren D, Edén P, Lassen C, Heldrup J, Olofsson T, Råde J, Fontes M, Porwit-Macdonald A, Behrendtz M, Höglund M, Johansson B, Fioretos T (2007) Microarray-based classification of a consecutive series of 121 childhood acute leukemias: prediction of leukemic and genetic subtype as well as of minimal residual disease status. Leukemia 21:1198–1203

Anderton JA, Bose S, Vockerodt M, Vrzalikova K, Wei W, Kuo M, Helin K, Christensen J, Rowe M, Murray PG, Woodman CB (2011) The H3K27me3 demethylase, KDM6B, is induced by Epstein–Barr virus and over-expressed in Hodgkin's Lymphoma. Oncogene 30:2037–2043

Avasarala S, Wu P-Y, Khan SQ, Yanlin S, Van Scoyk M, Bao J, Di Lorenzo A, David O, Bedford MT, Gupta V, Winn RA, Bikkavilli RK (2020) PRMT6 promotes lung tumor progression via the alternate activation of tumor-associated macrophages. Mol Cancer Res 18:166–178

Bai X, Wu L, Liang T, Liu Z, Li J, Li D, Xie H, Yin S, Yu J, Lin Q, Zheng S (2008) Overexpression of myocyte enhancer factor 2 and histone hyperacetylation in hepatocellular carcinoma. J Cancer Res Clin Oncol 134:83–91

Baldwin RM, Haghandish N, Daneshmand M, Amin S, Paris G, Falls TJ, Bell JC, Islam S, Côté J (2015) Protein arginine methyltransferase 7 promotes breast cancer cell invasion through the induction of MMP9 expression. Oncotarget 6

Barau J, Teissandier A, Zamudio N, Roy S, Nalesso V, Herault Y, Guillou F, Bourc'his D (2016) The DNA methyltransferase DNMT3C protects male germ cells from transposon activity. Science (New York, N.Y.) 354:909–912

Barnes CE, English DM, Cowley SM (2019) Acetylation & Co: an expanding repertoire of histone acylations regulates chromatin and transcription. Essays Biochem 63:97–107

Bastea LI, Hollant LMA, Döppler HR, Reid EM, Storz P (2019) Sangivamycin and its derivatives inhibit Haspin-histone H3-survivin signaling and induce pancreatic cancer cell death. Sci Rep 9:16588

Benedetti R, Dell'Aversana C, De Marchi T, Rotili D, Liu NQ, Novakovic B, Boccella S, Di Maro S, Cosconati S, Baldi A, Niméus E, Schultz J, Höglund U, Maione S, Papulino C, Chianese U, Iovino F, Federico A, Mai A, Stunnenberg HG, Nebbioso A, Altucci L (2019) Inhibition of histone demethylases LSD1 and UTX regulates ERα signaling in breast cancer. Cancers 11

Bennett RL, Licht JD (2018) Targeting epigenetics in cancer. Annu Rev Pharmacol Toxicol 58:187–207

Bi Y, Shen W, Min M, Liu Y (2017) MicroRNA-7 functions as a tumor-suppressor gene by regulating ILF2 in pancreatic carcinoma. Int J Mol Med 39:900–906

Bianco-Miotto T, Chiam K, Buchanan G, Jindal S, Day TK, Thomas M, Pickering MA, O'Loughlin MA, Ryan NK, Raymond WA, Horvath LG, Kench JG, Stricker PD, Marshall VR, Sutherland RL, Henshall SM, Gerald WL, Scher HI, Risbridger GP, Clements JA, Butler LM, Tilley WD, Horsfall DJ, Ricciardelli C (2010) Global levels of specific histone modifications and an epigenetic gene signature predict prostate cancer progression and development. Cancer Epidemiol Biomarkers Prev 19:2611–2622

Butcher DT, Rodenhiser DI (2007) Epigenetic inactivation of BRCA1 is associated with aberrant expression of CTCF and DNA methyltransferase (DNMT3B) in some sporadic breast tumours. Eur J Cancer (Oxford, England: 1990) 43:210–219

Cai M, Hu Z, Liu J, Gao J, Tan M, Zhang D, Zhu L, Liu S, Hou R, Lin B (2015) Expression of hMOF in different ovarian tissues and its effects on ovarian cancer prognosis. Oncol Rep 33:685–692

Cao J, Yan Q (2012) Histone ubiquitination and deubiquitination in transcription, DNA damage response, and cancer. Front Oncol 2:26

Cao Q, Yu J, Dhanasekaran SM, Kim JH, Mani RS, Tomlins SA, Mehra R, Laxman B, Cao X, Yu J, Kleer CG, Varambally S, Chinnaiyan AM (2008) Repression of E-cadherin by the polycomb group protein EZH2 in cancer. Oncogene 27:7274–7284

Cao L, Zhu L, Yang J, Su J, Ni J, Du Y, Liu D, Wang Y, Wang F, Jin J, Cai Y (2014) Correlation of low expression of hMOF with clinicopathological features of colorectal carcinoma, gastric cancer and renal cell carcinoma. Int J Oncol 44:1207–1214

Casciello F, Al-Ejeh F, Kelly G, Brennan DJ, Ngiow SF, Young A, Stoll T, Windloch K, Hill MM, Smyth MJ, Gannon F, Lee JS (2017) G9a drives hypoxia-mediated gene repression for breast cancer cell survival and tumorigenesis. Proc Natl Acad Sci U S A 114:7077–7082

Ceol CJ, Houvras Y, Jane-Valbuena J, Bilodeau S, Orlando DA, Battisti V, Fritsch L, Lin WM, Hollmann TJ, Ferré F, Bourque C, Burke CJ, Turner L, Uong A, Johnson LA, Beroukhim R, Mermel CH, Loda M, Ait-Si-Ali S, Garraway LA, Young RA, Zon LI (2011) The histone methyltransferase SETDB1 is recurrently amplified in melanoma and accelerates its onset. Nature 471:513–517

Chang B, Chen Y, Zhao Y, Bruick RK (2007) JMJD6 is a histone arginine demethylase. Science (New York, N.Y.) 318:444–447

Chen Z, Zhang Y (2020) Role of mammalian DNA methyltransferases in development. Annu Rev Biochem 89:135–158

Chen J, Zhao J, Zhou X, Liu S, Yan Y, Wang Y, Cao C, Han S, Zhou N, Xu Y, Zhao J, Yan Y, Cui H (2017a) Immunohistochemical investigation of topoIIβ, H3K27me3 and JMJD3 expressions in medulloblastoma. Pathol Res Pract 213:975–981

Chen K, Zhang F, Ding J, Liang Y, Zhan Z, Zhan Y, Chen L-H, Ding Y (2017b) Histone methyltransferase SETDB1 promotes the progression of colorectal cancer by inhibiting the expression of TP53. J Cancer 8:3318–3330
Chen X, Xiao X, Guo F (2019) Roles of H3K27me3 demethylase JMJD3 in inflammation and cancers. J Rare Dis Res Treat 4:71–76
Cheng X (2014) Structural and functional coordination of DNA and histone methylation. Cold Spring Harb Perspect Biol 6
Cheng H, Qin Y, Fan H, Su P, Zhang X, Zhang H, Zhou G (2013) Overexpression of CARM1 in breast cancer is correlated with poorly characterized clinicopathologic parameters and molecular subtypes. Diagn Pathol 8:129–129
Cheng S, Jiang X, Ding C, Du C, Owusu-Ansah KG, Weng X, Hu W, Peng C, Lv Z, Tong R, Xiao H, Xie H, Zhou L, Wu J, Zheng S (2016) Expression and critical role of interleukin enhancer binding factor 2 in hepatocellular carcinoma. Int J Mol Sci 17
Cheng G, Liu F, Asai T, Lai F, Man N, Xu H, Chen S, Greenblatt S, Hamard PJ, Ando K, Chen X, Wang L, Martinez C, Tadi M, Wang L, Xu M, Yang FC, Shiekhattar R, Nimer SD (2017) Loss of p300 accelerates MDS-associated leukemogenesis. Leukemia 31:1382–1390
Cheng D, He Z, Zheng L, Xie D, Dong S, Zhang P (2018) PRMT7 contributes to the metastasis phenotype in human non-small-cell lung cancer cells possibly through the interaction with HSPA5 and EEF2. Onco Targets Ther 11:4869–4876
Cheng Y, He C, Wang M, Ma X, Mo F, Yang S, Han J, Wei X (2019) Targeting epigenetic regulators for cancer therapy: mechanisms and advances in clinical trials. Signal Transduct Target Ther 4:62
Cheung N, Fung TK, Zeisig BB, Holmes K, Rane JK, Mowen KA, Finn MG, Lenhard B, Chan LC, So CWE (2016) Targeting aberrant epigenetic networks mediated by PRMT1 and KDM4C in acute myeloid leukemia. Cancer Cell 29:32–48
Chiang K, Zielinska AE, Shaaban AM, Sanchez-Bailon MP, Jarrold J, Clarke TL, Zhang J, Francis A, Jones LJ, Smith S, Barbash O, Guccione E, Farnie G, Smalley MJ, Davies CC (2017) PRMT5 is a critical regulator of breast cancer stem cell function via histone methylation and FOXP1 expression. Cell Rep 21:3498–3513
Chieffi P (2018) Aurora B: a new promising therapeutic target in cancer. Intractable Rare Dis Res 7:141–144
Chieffi P, Cozzolino L, Kisslinger A, Libertini S, Staibano S, Mansueto G, De Rosa G, Villacci A, Vitale M, Linardopoulos S, Portella G, Tramontano D (2006) Aurora B expression directly correlates with prostate cancer malignancy and influence prostate cell proliferation. Prostate 66:326–333
Cloos PA, Christensen J, Agger K, Helin K (2008) Erasing the methyl mark: histone demethylases at the center of cellular differentiation and disease. Genes Dev 22:1115–1140
Cregan S, McDonagh L, Gao Y, Barr M, Byrne K, Finn S, Cuffe S, Gray S (2016) KAT5 (Tip60) is a potential therapeutic target in malignant pleural mesothelioma. Int J Oncol 48
Cregan S, Breslin M, Roche G, Wennstedt S, MacDonagh L, Albadri C, Gao Y, O'Byrne KJ, Cuffe S, Finn SP, Gray SG (2017) Kdm6a and Kdm6b: altered expression in malignant pleural mesothelioma. Int J Oncol 50:1044–1052
Dai J, Sultan S, Taylor SS, Higgins JM (2005) The kinase haspin is required for mitotic histone H3 Thr 3 phosphorylation and normal metaphase chromosome alignment. Genes Dev 19:472–488
Dandawate P, Ghosh C, Palaniyandi K, Paul S, Rawal S, Pradhan R, Sayed AAA, Choudhury S, Standing D, Subramaniam D, Padhye SB, Gunewardena S, Thomas SM, Neil MO, Tawfik O, Welch DR, Jensen RA, Maliski S, Weir S, Iwakuma T, Anant S, Dhar A (2019) The histone demethylase KDM3A, increased in human pancreatic tumors, regulates expression of DCLK1 and promotes tumorigenesis in mice. Gastroenterology 157:1646–1659.e1611
Debes JD, Sebo TJ, Lohse CM, Murphy LM, Haugen DA, Tindall DJ (2003) p300 in prostate cancer proliferation and progression. Cancer Res 63:7638–7640
Delhommeau F, Dupont S, Della Valle V, James C, Trannoy S, Massé A, Kosmider O, Le Couedic JP, Robert F, Alberdi A, Lécluse Y, Plo I, Dreyfus FJ, Marzac C, Casadevall N, Lacombe C,

Romana SP, Dessen P, Soulier J, Viguié F, Fontenay M, Vainchenker W, Bernard OA (2009) Mutation in TET2 in myeloid cancers. N Engl J Med 360:2289–2301

Demetriadou C, Kirmizis A (2017) Histone acetyltransferases in cancer: guardians or hazards? Crit Rev Oncog 22

Di Lorenzo A, Bedford MT (2011) Histone arginine methylation. FEBS Lett 585:2024–2031

Dillon MBC, Rust HL, Thompson PR, Mowen KA (2013) Automethylation of protein arginine methyltransferase 8 (PRMT8) regulates activity by impeding S-adenosylmethionine sensitivity. J Biol Chem 288:27872–27880

Dong F, Li Q, Yang C, Huo D, Wang X, Ai C, Kong Y, Sun X, Wang W, Zhou Y, Liu X, Li W, Gao W, Liu W, Kang C, Wu X (2018) PRMT2 links histone H3R8 asymmetric dimethylation to oncogenic activation and tumorigenesis of glioblastoma. Nat Commun 9:4552

Eberharter A, Becker PB (2002) Histone acetylation: a switch between repressive and permissive chromatin. Second in review series on chromatin dynamics. EMBO Rep 3:224–229

Ellis L, Atadja PW, Johnstone RW (2009) Epigenetics in cancer: targeting chromatin modifications. Mol Cancer Ther 8:1409–1420

Fang Y, Zhang D, Hu T, Zhao H, Zhao X, Lou Z, He Y, Qin W, Xia J, Zhang X, Ye LC (2019) KMT2A histone methyltransferase contributes to colorectal cancer development by promoting cathepsin Z transcriptional activation. Cancer Med 8:3544–3552

Feng Q, Wang H, Ng HH, Erdjument-Bromage H, Tempst P, Struhl K, Zhang Y (2002) Methylation of H3-lysine 79 is mediated by a new family of HMTases without a SET domain. Curr Biol 12:1052–1058

Flotho A, Melchior F (2013) Sumoylation: a regulatory protein modification in health and disease. Annu Rev Biochem 82:357–385

Fukuda T, Mishina Y, Walker MP, DiAugustine RP (2005) Conditional transgenic system for mouse aurora a kinase: degradation by the ubiquitin proteasome pathway controls the level of the transgenic protein. Mol Cell Biol 25:5270–5281

Gao Y, Zhao Y, Zhang J, Lu Y, Liu X, Geng P, Huang B, Zhang Y, Lu J (2016) The dual function of PRMT1 in modulating epithelial-mesenchymal transition and cellular senescence in breast cancer cells through regulation of ZEB1. Sci Rep 6:19874

Geng P, Zhang Y, Liu X, Zhang N, Liu Y, Liu X, Lin C, Yan X, Li Z, Wang G, Li Y, Tan J, Liu D-X, Huang B, Lu J (2017) Automethylation of protein arginine methyltransferase 7 and its impact on breast cancer progression. FASEB J 31:2287–2300

González-Loyola A, Fernández-Miranda G, Trakala M, Partida D, Samejima K, Ogawa H, Cañamero M, de Martino A, Martínez-Ramírez Á, de Cárcer G, Pérez de Castro I, Earnshaw WC, Malumbres M (2015) Aurora B overexpression causes aneuploidy and p21Cip1 repression during tumor development. Mol Cell Biol 35:3566–3578

Gowher H, Jeltsch A (2018) Mammalian DNA methyltransferases: new discoveries and open questions. Biochem Soc Trans 46:1191–1202

Guccione E, Richard S (2019) The regulation, functions and clinical relevance of arginine methylation. Nat Rev Mol Cell Biol 20:642–657

Gullà A, Hideshima T, Bianchi G, Fulciniti M, Kemal Samur M, Qi J, Tai YT, Harada T, Morelli E, Amodio N, Carrasco R, Tagliaferri P, Munshi NC, Tassone P, Anderson KC (2018) Protein arginine methyltransferase 5 has prognostic relevance and is a druggable target in multiple myeloma. Leukemia 32:996–1002

Han X, Li R, Zhang W, Yang X, Wheeler CG, Friedman GK, Province P, Ding Q, You Z, Fathallah-Shaykh HM, Gillespie GY, Zhao X, King PH, Nabors LB (2014) Expression of PRMT5 correlates with malignant grade in gliomas and plays a pivotal role in tumor growth in vitro. J Neuro-Oncol 118:61–72

Harmeyer KM, Facompre ND, Herlyn M, Basu D (2017) JARID1 histone demethylases: emerging targets in cancer. Trends Cancer 3:713–725

Harris CC (1996) Structure and function of the p53 tumor suppressor gene: clues for rational cancer therapeutic strategies. J Natl Cancer Inst 88:1442–1455

Hayami S, Kelly JD, Cho H-S, Yoshimatsu M, Unoki M, Tsunoda T, Field HI, Neal DE, Yamaue H, Ponder BAJ, Nakamura Y, Hamamoto R (2011) Overexpression of LSD1 contributes to human carcinogenesis through chromatin regulation in various cancers. Int J Cancer 128:574–586

He J, Nguyen AT, Zhang Y (2011) KDM2b/JHDM1b, an H3K36me2-specific demethylase, is required for initiation and maintenance of acute myeloid leukemia. Blood 117:3869–3880

Heemers HV, Debes JD, Tindall DJ (2008) The role of the transcriptional coactivator p300 in prostate cancer progression. Adv Exp Med Biol 617:535–540

Heller G, Geradts J, Ziegler B, Newsham I, Filipits M, Markis-Ritzinger EM, Kandioler D, Berger W, Stiglbauer W, Depisch D, Pirker R, Zielinski CC, Zochbauer-Muller S (2007) Downregulation of TSLC1 and DAL-1 expression occurs frequently in breast cancer. Breast Cancer Res Treat 103:283–291

Hernandez SJ, Dolivo DM, Dominko T (2017) PRMT8 demonstrates variant-specific expression in cancer cells and correlates with patient survival in breast, ovarian and gastric cancer. Oncol Lett 13:1983–1989

Hidaka S, Yasutake T, Takeshita H, Kondo M, Tsuji T, Nanashima A, Sawai T, Yamaguchi H, Nakagoe T, Ayabe H, Tagawa Y (2000) Differences in 20q13.2 copy number between colorectal cancers with and without liver metastasis. Clin Cancer Res 6:2712–2717

Hobbs CA, Wei G, DeFeo K, Paul B, Hayes CS, Gilmour SK (2006) Tip60 protein isoforms and altered function in skin and tumors that overexpress ornithine decarboxylase. Cancer Res 66:8116–8122

Hoffmann I, Roatsch M, Schmitt ML, Carlino L, Pippel M, Sippl W, Jung M (2012) The role of histone demethylases in cancer therapy. Mol Oncol 6:683–703

Hong H, Kao C, Jeng MH, Eble JN, Koch MO, Gardner TA, Zhang S, Li L, Pan CX, Hu Z, MacLennan GT, Cheng L (2004) Aberrant expression of CARM1, a transcriptional coactivator of androgen receptor, in the development of prostate carcinoma and androgen-independent status. Cancer 101:83–89

Hong X, Xu Y, Qiu X, Zhu Y, Feng X, Ding Z, Zhang S, Zhong L, Zhuang Y, Su C, Hong X, Cai J (2016) MiR-448 promotes glycolytic metabolism of gastric cancer by downregulating KDM2B. Oncotarget 7:22092–22102

Hou J, Wu J, Dombkowski A, Zhang K, Holowatyj A, Boerner JL, Yang ZQ (2012a) Genomic amplification and a role in drug-resistance for the KDM5A histone demethylase in breast cancer. Am J Transl Res 4:247–256

Hou X, Li Y, Luo RZ, Fu JH, He JH, Zhang LJ, Yang HX (2012b) High expression of the transcriptional co-activator p300 predicts poor survival in resectable non-small cell lung cancers. Eur J Surg Oncol 38:523–530

Hsu LC, Huang X, Seasholtz S, Potter DM, Gollin SM (2006) Gene amplification and overexpression of protein phosphatase 1α in oral squamous cell carcinoma cell lines. Oncogene 25:5517–5526

Hsu MC, Pan MR, Chu PY, Tsai YL, Tsai CH, Shan YS, Chen LT, Hung WC (2018) Protein arginine methyltransferase 3 enhances chemoresistance in pancreatic cancer by Methylating hnRNPA1 to increase ABCG2 expression. Cancers 11

Hsu MC, Tsai YL, Lin CH, Pan MR, Shan YS, Cheng TY, Cheng SH, Chen LT, Hung WC (2019) Protein arginine methyltransferase 3-induced metabolic reprogramming is a vulnerable target of pancreatic cancer. J Hematol Oncol 12:79

Hu L, Zang M-D, Wang H-X, Zhang B-G, Wang Z-Q, Fan Z-Y, Wu H, Li J-F, Su L-P, Yan M, Zhu Z-Q, Yang Q-M, Huang Q, Liu B-Y, Zhu Z-G (2018) G9A promotes gastric cancer metastasis by upregulating ITGB3 in a SET domain-independent manner. Cell Death Dis 9:278

Hu Y, Wu F, Liu Y, Zhao Q, Tang H (2019) DNMT1 recruited by EZH2-mediated silencing of miR-484 contributes to the malignancy of cervical cancer cells through MMP14 and HNF1A. Clin Epigenetics 11:186

Huang Z, Wu H, Chuai S, Xu F, Yan F, Englund N, Wang Z, Zhang H, Fang M, Wang Y, Gu J, Zhang M, Yang T, Zhao K, Yu Y, Dai J, Yi W, Zhou S, Li Q, Wu J, Liu J, Wu X, Chan H, Lu C,

Atadja P, Li E, Wang Y, Hu M (2013) NSD2 is recruited through its PHD domain to oncogenic gene loci to drive multiple myeloma. Cancer Res 73:6277–6288

Iizuka M, Takahashi Y, Mizzen CA, Cook RG, Fujita M, Allis CD, Frierson HF Jr, Fukusato T, Smith MM (2009) Histone acetyltransferase Hbo1: catalytic activity, cellular abundance, and links to primary cancers. Gene 436:108–114

Ishihama K, Yamakawa M, Semba S, Takeda H, Kawata S, Kimura S, Kimura W (2007) Expression of HDAC1 and CBP/p300 in human colorectal carcinomas. J Clin Pathol 60:1205–1210

Jarmasz JS, Stirton H, Davie JR, Del Bigio MR (2019) DNA methylation and histone post-translational modification stability in post-mortem brain tissue. Clin Epigenetics 11:5–5

Jeon JY, Lee JS, Park ER, Shen YN, Kim MY, Shin HJ, Joo HY, Cho EH, Moon SM, Shin US, Park SH, Han CJ, Choi DW, Gu MB, Kim SB, Lee KH (2018) Protein arginine methyltransferase 5 is implicated in the aggressiveness of human hepatocellular carcinoma and controls the invasive activity of cancer cells. Oncol Rep 40:536–544

Jin B, Robertson KD (2013) DNA methyltransferases, DNA damage repair, and cancer. Adv Exp Med Biol 754:3–29

Jing P, Zhao N, Ye M, Zhang Y, Zhang Z, Sun J, Wang Z, Zhang J, Gu Z (2018) Protein arginine methyltransferase 5 promotes lung cancer metastasis via the epigenetic regulation of miR-99 family/FGFR3 signaling. Cancer Lett 427:38–48

Kanda M, Shimizu D, Fujii T, Tanaka H, Shibata M, Iwata N, Hayashi M, Kobayashi D, Tanaka C, Yamada S, Nakayama G, Sugimoto H, Koike M, Fujiwara M, Kodera Y (2016) Protein arginine methyltransferase 5 is associated with malignant phenotype and peritoneal metastasis in gastric cancer. Int J Oncol 49:1195–1202

Kang MK, Mehrazarin S, Park NH, Wang CY (2017) Epigenetic gene regulation by histone demethylases: emerging role in oncogenesis and inflammation. Oral Dis 23:709–720

Karakashev S, Zhu H, Wu S, Yokoyama Y, Bitler BG, Park PH, Lee JH, Kossenkov AV, Gaonkar KS, Yan H, Drapkin R, Conejo-Garcia JR, Speicher DW, Ordog T, Zhang R (2018) CARM1-expressing ovarian cancer depends on the histone methyltransferase EZH2 activity. Nat Commun 9:631

Katsha A, Belkhiri A, Goff L, El-Rifai W (2015) Aurora kinase A in gastrointestinal cancers: time to target. Mol Cancer 14:106

Kleer CG, Cao Q, Varambally S, Shen R, Ota I, Tomlins SA, Ghosh D, Sewalt RG, Otte AP, Hayes DF, Sabel MS, Livant D, Weiss SJ, Rubin MA, Chinnaiyan AM (2003) EZH2 is a marker of aggressive breast cancer and promotes neoplastic transformation of breast epithelial cells. Proc Natl Acad Sci U S A 100:11606–11611

Komatsu S, Imoto I, Tsuda H, Kozaki KI, Muramatsu T, Shimada Y, Aiko S, Yoshizumi Y, Ichikawa D, Otsuji E, Inazawa J (2009) Overexpression of SMYD2 relates to tumor cell proliferation and malignant outcome of esophageal squamous cell carcinoma. Carcinogenesis 30:1139–1146

Komatsu S, Ichikawa D, Hirajima S, Nagata H, Nishimura Y, Kawaguchi T, Miyamae M, Okajima W, Ohashi T, Konishi H, Shiozaki A, Fujiwara H, Okamoto K, Tsuda H, Imoto I, Inazawa J, Otsuji E (2015) Overexpression of SMYD2 contributes to malignant outcome in gastric cancer. Br J Cancer 112:357–364

Kondo Y, Shen L, Ahmed S, Boumber Y, Sekido Y, Haddad BR, Issa J-PJ (2008) Downregulation of histone H3 lysine 9 methyltransferase G9a induces centrosome disruption and chromosome instability in cancer cells. PLoS One 3:e2037–e2037

Kottakis F, Polytarchou C, Foltopoulou P, Sanidas I, Kampranis SC, Tsichlis PN (2011) FGF-2 regulates cell proliferation, migration, and angiogenesis through an NDY1/KDM2B-miR-101-EZH2 pathway. Mol Cell 43:285–298

Kuang Y, Lu F, Guo J, Xu H, Wang Q, Xu C, Zeng L, Yi S (2017) Histone demethylase KDM2B upregulates histone methyltransferase EZH2 expression and contributes to the progression of ovarian cancer in vitro and in vivo. Onco Targets Ther 10:3131–3144

Kumar B, Yadav A, Brown NV, Zhao S, Cipolla MJ, Wakely PE, Schmitt AC, Baiocchi RA, Teknos TN, Old M, Kumar P (2017) Nuclear PRMT5, cyclin D1 and IL-6 are associated with poor outcome in oropharyngeal squamous cell carcinoma patients and is inversely associated with p16-status. Oncotarget 8:14847–14859

Kuo AJ, Cheung P, Chen K, Zee BM, Kioi M, Lauring J, Xi Y, Park BH, Shi X, Garcia BA, Li W, Gozani O (2011) NSD2 links dimethylation of histone H3 at lysine 36 to oncogenic programming. Mol Cell 44:609–620

Lacoste N, Utley RT, Hunter JM, Poirier GG, Côte J (2002) Disruptor of telomeric silencing-1 is a chromatin-specific histone H3 methyltransferase. J Biol Chem 277:30421–30424

Lee SH, Kim J, Kim WH, Lee YM (2009) Hypoxic silencing of tumor suppressor RUNX3 by histone modification in gastric cancer cells. Oncogene 28:184–194

Lee YF, Miller LD, Chan XB, Black MA, Pang B, Ong CW, Salto-Tellez M, Liu ET, Desai KV (2012) JMJD6 is a driver of cellular proliferation and motility and a marker of poor prognosis in breast cancer. Breast Cancer Res 14:3001

Lee C-R, Lee SH, Rigas NK, Kim RH, Kang MK, Park N-H, Shin K-H (2015) Elevated expression of JMJD6 is associated with oral carcinogenesis and maintains cancer stemness properties. Carcinogenesis 37:119–128

Lee S, Oh S, Jeong K, Jo H, Choi Y, Seo HD, Kim M, Choe J, Kwon CS, Lee D (2018) Dot1 regulates nucleosome dynamics by its inherent histone chaperone activity in yeast. Nat Commun 9:240

Li GM (2013) Decoding the histone code: role of H3K36me3 in mismatch repair and implications for cancer susceptibility and therapy. Cancer Res 73:6379–6383

Li M, Luo RZ, Chen JW, Cao Y, Lu JB, He JH, Wu QL, Cai MY (2011a) High expression of transcriptional coactivator p300 correlates with aggressive features and poor prognosis of hepatocellular carcinoma. J Transl Med 9:5

Li Q, Shi L, Gui B, Yu W, Wang J, Zhang D, Han X, Yao Z, Shang Y (2011b) Binding of the JmjC demethylase JARID1B to LSD1/NuRD suppresses angiogenesis and metastasis in breast cancer cells by repressing chemokine CCL14. Cancer Res 71:6899–6908

Li W, Zhao L, Zang W, Liu Z, Chen L, Liu T, Xu D, Jia J (2011c) Histone demethylase JMJD2B is required for tumor cell proliferation and survival and is overexpressed in gastric cancer. Biochem Biophys Res Commun 416:372–378

Li S, Cheng D, Zhu B, Yang Q (2017) The overexpression of CARM1 promotes human osteosarcoma cell proliferation through the pGSK3β/β-catenin/cyclinD1 signaling pathway. Int J Biol Sci 13:976–984

Li Y, Zhang M, Sheng M, Zhang P, Chen Z, Xing W, Bai J, Cheng T, Yang F-C, Zhou Y (2018) Therapeutic potential of GSK-J4, a histone demethylase KDM6B/JMJD3 inhibitor, for acute myeloid leukemia. J Cancer Res Clin Oncol 144:1065–1077

Li J, Ahn JH, Wang GG (2019) Understanding histone H3 lysine 36 methylation and its deregulation in disease. Cell Mol Life Sci 76:2899–2916

Liao ZW, Zhou TC, Tan XJ, Song XL, Liu Y, Shi XY, Huang WJ, Du LL, Tu BJ, Lin XD (2012) High expression of p300 is linked to aggressive features and poor prognosis of nasopharyngeal carcinoma. J Transl Med 10:110

Lin H, Wang B, Yu J, Wang J, Li Q, Cao B (2018) Protein arginine methyltransferase 8 gene enhances the colon cancer stem cell (CSC) function by upregulating the pluripotency transcription factor. J Cancer 9:1394–1402

Lin C-Y, Wang B-J, Chen B-C, Tseng J-C, Jiang SS, Tsai KK, Shen Y-Y, Yuh CH, Sie Z-L, Wang W-C, Kung H-J, Chuu C-P (2019) Histone demethylase KDM4C stimulates the proliferation of prostate cancer cells via activation of AKT and c-Myc. Cancers 11:1785

Liu X, Zhang J, Liu L, Jiang Y, Ji J, Yan R, Zhu Z, Yu Y (2018) Protein arginine methyltransferase 5-mediated epigenetic silencing of IRX1 contributes to tumorigenicity and metastasis of gastric cancer. Biochim Biophys Acta (BBA) - Mol Basis Dis 1864:2835–2844

Liu CL, Sheu JJ, Lin HP, Jeng YM, Chang CY, Chen CM, Cheng J, Mao TL (2019a) The overexpression of MYST4 in human solid tumors is associated with increased aggressiveness and decreased overall survival. Int J Clin Exp Pathol 12:431–442

Liu Y, Liu X, Lingxia L, Wang Y, Peng L, Liu J, Li L, Zhang L, Wong G, Li H, Huang B, Lu J, Zhang Y (2019b) The PRMT7-dependent methylation of shank2 modulates invasion-proliferation switching during breast cancer metastasis

López-Ríos F, Chuai S, Flores R, Shimizu S, Ohno T, Wakahara K, Illei PB, Hussain S, Krug L, Zakowski MF, Rusch V, Olshen AB, Ladanyi M (2006) Global gene expression profiling of pleural mesotheliomas: overexpression of aurora kinases and P16/CDKN2A deletion as prognostic factors and critical evaluation of microarray-based prognostic prediction. Cancer Res 66:2970–2979

Lorsbach RB, Moore J, Mathew S, Raimondi SC, Mukatira ST, Downing JR (2003) TET1, a member of a novel protein family, is fused to MLL in acute myeloid leukemia containing the t(10;11)(q22;q23). Leukemia 17:637–641

Ludlow JW, Glendening CL, Livingston DM, DeCarprio JA (1993) Specific enzymatic dephosphorylation of the retinoblastoma protein. Mol Cell Biol 13:367–372

Luo M (2018) Chemical and biochemical perspectives of protein lysine methylation. Chem Rev 118:6656–6705

Ma J, Wang N, Zhang Y, Wang C, Ge T, Jin H, Deng X, Huo X, Gu D, Ge Z, Chu W, Jiang L, Qin W (2015) KDM6B elicits cell apoptosis by promoting nuclear translocation of FOXO1 in non-small cell lung cancer. Cell Physiol Biochem 37:201–213

McLaughlin-Drubin ME, Park D, Munger K (2013) Tumor suppressor p16INK4A is necessary for survival of cervical carcinoma cell lines. Proc Natl Acad Sci U S A 110:16175–16180

Meeks JJ, Shilatifard A (2017) Multiple roles for the MLL/COMPASS family in the epigenetic regulation of gene expression and in cancer. Annu Rev Cancer Biol 1:425–446

Mehner C, Hockla A, Miller E, Ran S, Radisky DC, Radisky ES (2014) Tumor cell-produced matrix metalloproteinase 9 (MMP-9) drives malignant progression and metastasis of basal-like triple negative breast cancer. Oncotarget 5:2736–2749

Melms JC, Vallabhaneni S, Mills CE, Yapp C, Chen JY, Morelli E, Waszyk P, Kumar S, Deming D, Moret N, Rodriguez S, Subramanian K, Rogava M, Cartwright ANR, Luoma A, Mei S, Brinker TJ, Miller DM, Spektor A, Schadendorf D, Riggi N, Wucherpfennig KW, Sorger PK, Izar B (2020) Inhibition of Haspin kinase promotes cell-intrinsic and extrinsic antitumor activity. Cancer Res 80:798–810

Metzger E, Wissmann M, Yin N, Muller JM, Schneider R, Peters AH, Gunther T, Buettner R, Schule R (2005) LSD1 demethylates repressive histone marks to promote androgen-receptor-dependent transcription. Nature 437:436–439

Miller DM, Thomas SD, Islam A, Muench D, Sedoris K (2012) c-Myc and cancer metabolism. Clin Cancer Res 18:5546–5553

Mo J, Wang L, Huang X, Lu B, Zou C, Wei L, Chu J, Eggers PK, Chen S, Raston CL, Wu J, Lim LY, Zhao W (2017) Multifunctional nanoparticles for co-delivery of paclitaxel and carboplatin against ovarian cancer by inactivating the JMJD3-HER2 axis. Nanoscale 9:13142–13152

Morettin A, Baldwin RM, Côté J (2015) Arginine methyltransferases as novel therapeutic targets for breast cancer. Mutagenesis 30:177–189

Morin RD, Johnson NA, Severson TM, Mungall AJ, An J, Goya R, Paul JE, Boyle M, Woolcock BW, Kuchenbauer F, Yap D, Humphries RK, Griffith OL, Shah S, Zhu H, Kimbara M, Shashkin P, Charlot JF, Tcherpakov M, Corbett R, Tam A, Varhol R, Smailus D, Moksa M, Zhao Y, Delaney A, Qian H, Birol I, Schein J, Moore R, Holt R, Horsman DE, Connors JM, Jones S, Aparicio S, Hirst M, Gascoyne RD, Marra MA (2010) Somatic mutations altering EZH2 (Tyr641) in follicular and diffuse large B-cell lymphomas of germinal-center origin. Nat Genet 42:181–185

Nassa G, Salvati A, Tarallo R, Gigantino V, Alexandrova E, Memoli D, Sellitto A, Rizzo F, Malanga D, Mirante T, Morelli E, Nees M, Akerfelt M, Kangaspeska S, Nyman TA, Milanesi L, Giurato G, Weisz A (2019) Inhibition of histone methyltransferase DOT1L silences ERalpha gene and blocks proliferation of antiestrogen-resistant breast cancer cells. Science Adv 5: eaav5590

Ng HH, Feng Q, Wang H, Erdjument-Bromage H, Tempst P, Zhang Y, Struhl K (2002) Lysine methylation within the globular domain of histone H3 by Dot1 is important for telomeric silencing and Sir protein association. Genes Dev 16:1518–1527

Nguyen AT, Taranova O, He J, Zhang Y (2011) DOT1L, the H3K79 methyltransferase, is required for MLL-AF9-mediated leukemogenesis. Blood 117:6912–6922

Onder TT, Gupta PB, Mani SA, Yang J, Lander ES, Weinberg RA (2008) Loss of E-cadherin promotes metastasis via multiple downstream transcriptional pathways. Cancer Res 68:3645–3654

Palazzo L, Mikolčević P, Mikoč A, Ahel I (2019) ADP-ribosylation signalling and human disease. Open Biol 9:190041

Park WY, Hong BJ, Lee J, Choi C, Kim MY (2016) H3K27 demethylase JMJD3 employs the NF-κB and BMP signaling pathways to modulate the tumor microenvironment and promote melanoma progression and metastasis. Cancer Res 76:161–170

Patel JH, Du Y, Ard PG, Phillips C, Carella B, Chen CJ, Rakowski C, Chatterjee C, Lieberman PM, Lane WS, Blobel GA, McMahon SB (2004) The c-MYC oncoprotein is a substrate of the acetyltransferases hGCN5/PCAF and TIP60. Mol Cell Biol 24:10826–10834

Pathania R, Ramachandran S, Elangovan S, Padia R, Yang P, Cinghu S, Veeranan-Karmegam R, Arjunan P, Gnana-Prakasam JP, Sadanand F, Pei L, Chang C-S, Choi J-H, Shi H, Manicassamy S, Prasad PD, Sharma S, Ganapathy V, Jothi R, Thangaraju M (2015) DNMT1 is essential for mammary and cancer stem cell maintenance and tumorigenesis. Nat Commun 6:6910

Peschle C, Tallarida G, Semprini A, Ravetta M, Condorelli M (1967) Research on the renal erythropoietic factor. I. Behavior of erythropoietic activity in rats treated with purified extracts of swine kidney with renin activity. Bollettino della Societa italiana di biologia sperimentale 43:755–759

Pfister S, Rea S, Taipale M, Mendrzyk F, Straub B, Ittrich C, Thuerigen O, Sinn HP, Akhtar A, Lichter P (2008) The histone acetyltransferase hMOF is frequently downregulated in primary breast carcinoma and medulloblastoma and constitutes a biomarker for clinical outcome in medulloblastoma. Int J Cancer 122:1207–1213

Piunti A, Shilatifard A (2016) Epigenetic balance of gene expression by Polycomb and COMPASS families. Science (New York, N.Y.) 352:aad9780

Pollock JA, Larrea MD, Jasper JS, McDonnell DP, McCafferty DG (2012) Lysine-specific histone demethylase 1 inhibitors control breast cancer proliferation in ERα-dependent and -independent manners. ACS Chem Biol 7:1221–1231

Poole CJ, Zheng W, Lodh A, Yevtodiyenko A, Liefwalker D, Li H, Felsher DW, van Riggelen J (2017) DNMT3B overexpression contributes to aberrant DNA methylation and MYC-driven tumor maintenance in T-ALL and Burkitt's lymphoma. Oncotarget 8

Poulard C, Corbo L, Le Romancer M (2016) Protein arginine methylation/demethylation and cancer. Oncotarget 7:67532–67550

Qi G, Ogawa I, Kudo Y, Miyauchi M, Siriwardena BS, Shimamoto F, Tatsuka M, Takata T (2007) Aurora-B expression and its correlation with cell proliferation and metastasis in oral cancer. Virchows Arch 450:297–302

Qu Y, Mu G, Wu Y, Dai X, Zhou F, Xu X, Wang Y, Wei F (2010) Overexpression of DNA methyltransferases 1, 3a, and 3b significantly correlates with retinoblastoma tumorigenesis. Am J Clin Pathol 134:826–834

Rabello D, Moura C, Andrade R, Motoyama A, Pittella Silva F (2013) Altered expression of MLL methyltransferase family genes in breast cancer. Int J Oncol 43

Ramadoss S, Sen S, Ramachandran I, Roy S, Chaudhuri G, Farias-Eisner R (2017) Lysine-specific demethylase KDM3A regulates ovarian cancer stemness and chemoresistance. Oncogene 36:1537–1545

Rasmussen KD, Helin K (2016) Role of TET enzymes in DNA methylation, development, and cancer. Genes Dev 30:733–750
Ren Y, Wu L, Li X, Li W, Yang Y, Zhang M (2015) FBXL10 contributes to the progression of nasopharyngeal carcinoma via involving in PI3K/mTOR pathway. Neoplasma 62:925–931
Ricketts CJ, Linehan WM (2015) Gender specific mutation incidence and survival associations in clear cell renal cell carcinoma (CCRCC). PLoS One 10:e0140257
Robertson KD (2005) DNA methylation and human disease. Nat Rev Genet 6:597–610
Rodriguez-Paredes M, Martinez de Paz A, Simó-Riudalbas L, Sayols S, Moutinho C, Moran S, Villanueva A, Vázquez-Cedeira M, Lazo PA, Carneiro F, Moura CS, Vieira J, Teixeira MR, Esteller M (2014) Gene amplification of the histone methyltransferase SETDB1 contributes to human lung tumorigenesis. Oncogene 33:2807–2813
Roesch A, Becker B, Schneider-Brachert W, Hagen I, Landthaler M, Vogt T (2006) Re-expression of the retinoblastoma-binding protein 2-homolog 1 reveals tumor-suppressive functions in highly metastatic melanoma cells. J Invest Dermatol 126:1850–1859
Rogawski DS, Grembecka J, Cierpicki T (2016) H3K36 methyltransferases as cancer drug targets: rationale and perspectives for inhibitor development. Future Med Chem 8:1589–1607
Roll JD, Rivenbark AG, Jones WD, Coleman WB (2008) DNMT3b overexpression contributes to a hypermethylator phenotype in human breast cancer cell lines. Mol Cancer 7:15
Rossetto D, Avvakumov N, Cote J (2012) Histone phosphorylation: a chromatin modification involved in diverse nuclear events. Epigenetics 7:1098–1108
Roth SY, Denu JM, Allis CD (2001) Histone acetyltransferases. Annu Rev Biochem 70:81–120
Rotte A, Bhandaru M, Cheng Y, Sjoestroem C, Martinka M, Li G (2013) Decreased expression of nuclear p300 is associated with disease progression and worse prognosis of melanoma patients. PLoS One 8:e75405
Sainathan S, Paul S, Ramalingam S, Baranda J, Anant S, Dhar A (2015) Histone demethylases in cancer. Curr pharmacol rep 1:234–244
Sakamoto LH, Andrade RV, Felipe MS, Motoyama AB, Pittella Silva F (2014) SMYD2 is highly expressed in pediatric acute lymphoblastic leukemia and constitutes a bad prognostic factor. Leuk Res 38:496–502
Salvati A, Gigantino V, Nassa G, Giurato G, Alexandrova E, Rizzo F, Tarallo R, Weisz A (2019) The histone methyltransferase DOT1L is a functional component of estrogen receptor alpha signaling in ovarian Cancer cells. Cancers 11
Sarno F, Nebbioso A, Altucci L (2019) DOT1L: a key target in normal chromatin remodelling and in mixed-lineage leukaemia treatment. Epigenetics:1–15
Sawada K, Yang Z, Horton JR, Collins RE, Zhang X, Cheng X (2004) Structure of the conserved core of the yeast Dot1p, a nucleosomal histone H3 lysine 79 methyltransferase. J Biol Chem 279:43296–43306
Scourzic L, Mouly E, Bernard OA (2015) TET proteins and the control of cytosine demethylation in cancer. Genome Med 7:9
Sen S, Zhou H, White RA (1997) A putative serine/threonine kinase encoding gene BTAK on chromosome 20q13 is amplified and overexpressed in human breast cancer cell lines. Oncogene 14:2195–2200
Serio J, Ropa J, Chen W, Mysliwski M, Saha N, Chen L, Wang J, Miao H, Cierpicki T, Grembecka J, Muntean AG (2018) The PAF complex regulation of Prmt5 facilitates the progression and maintenance of MLL fusion leukemia. Oncogene 37:450–460
Shen Y, Guo X-Q, Wang Y, Qiu W, Chang Y-Z, Zhang A, Duan X (2012) Expression and significance of histone H3K27 demethylases in renal cell carcinoma. BMC Cancer 12:470
Shilo K, Wu X, Sharma S, Welliver M, Duan W, Villalona-Calero M, Fukuoka J, Sif S, Baiocchi R, Hitchcock CL, Zhao W, Otterson GA (2013) Cellular localization of protein arginine methyltransferase-5 correlates with grade of lung tumors. Diagn Pathol 8:201
Singh V, Miranda TB, Jiang W, Frankel A, Roemer ME, Robb VA, Gutmann DH, Herschman HR, Clarke S, Newsham IF (2004) DAL-1/4.1B tumor suppressor interacts with protein arginine

N-methyltransferase 3 (PRMT3) and inhibits its ability to methylate substrates in vitro and in vivo. Oncogene 23:7761–7771

Smith JA, Haberstroh FS, White EA, Livingston DM, DeCaprio JA, Howley PM (2014) SMCX and components of the TIP60 complex contribute to E2 regulation of the HPV E6/E7 promoter. Virology 468-470:311–321

Sorrentino R, Libertini S, Pallante PL, Troncone G, Palombini L, Bavetsias V, Spalletti-Cernia D, Laccetti P, Linardopoulos S, Chieffi P, Fusco A, Portella G (2005) Aurora B overexpression associates with the thyroid carcinoma undifferentiated phenotype and is required for thyroid carcinoma cell proliferation. J Clin Endocrinol Metab 90:928–935

Strahl BD, Allis CD (2000) The language of covalent histone modifications. Nature 403:41–45

Su J, Wang F, Cai Y, Jin J (2016) The functional analysis of histone acetyltransferase MOF in tumorigenesis. Int J Mol Sci 17:99

Swaroop A, Oyer JA, Will CM, Huang X, Yu W, Troche C, Bulic M, Durham BH, Wen QJ, Crispino JD, MacKerell AD, Bennett RL, Kelleher NL, Licht JD (2019) An activating mutation of the NSD2 histone methyltransferase drives oncogenic reprogramming in acute lymphocytic leukemia. Oncogene 38:671–686

Tahiliani M, Koh KP, Shen Y, Pastor WA, Bandukwala H, Brudno Y, Agarwal S, Iyer LM, Liu DR, Aravind L, Rao A (2009) Conversion of 5-methylcytosine to 5-hydroxymethylcytosine in mammalian DNA by MLL partner TET1. Science (New York, N.Y.) 324:930–935

Takahashi Y, Iwai M, Kawai T, Arakawa A, Ito T, Sakurai-Yageta M, Ito A, Goto A, Saito M, Kasumi F, Murakami Y (2012) Aberrant expression of tumor suppressors CADM1 and 4.1B in invasive lesions of primary breast cancer. Breast Cancer (Tokyo, Japan) 19:242–252

Tang B, Qi G, Tang F, Yuan S, Wang Z, Liang X, Li B, Yu S, Liu J, Huang Q, Wei Y, Zhai R, Lei B, Yu H, Tomlinson S, He S (2016) Aberrant JMJD3 expression upregulates slug to promote migration, invasion, and stem cell-like behaviors in hepatocellular carcinoma. Cancer Res 76:6520–6532

Tatsuka M, Katayama H, Ota T, Tanaka T, Odashima S, Suzuki F, Terada Y (1998) Multinuclearity and increased ploidy caused by overexpression of the aurora- and Ipl1-like midbody-associated protein mitotic kinase in human cancer cells. Cancer Res 58:4811–4816

Teng Y, Yu X, Yuan H, Guo L, Jiang W, Lu S-H (2018) DNMT1 ablation suppresses tumorigenesis by inhibiting the self-renewal of esophageal cancer stem cells. Oncotarget 9:18896–18907

Tian X, Zhang S, Liu HM, Zhang YB, Blair CA, Mercola D, Sassone-Corsi P, Zi X (2013) Histone lysine-specific methyltransferases and demethylases in carcinogenesis: new targets for cancer therapy and prevention. Curr Cancer Drug Targets 13:558–579

Tolsma TO, Hansen JC (2019) Post-translational modifications and chromatin dynamics. Essays Biochem 63:89–96

Torrano J, Al Emran A, Hammerlindl H, Schaider H (2019) Emerging roles of H3K9me3, SETDB1 and SETDB2 in therapy-induced cellular reprogramming. Clin Epigenetics 11:43

Twu NF, Yuan CC, Yen MS, Lai CR, Chao KC, Wang PH, Wu HH, Chen YJ (2009) Expression of Aurora kinase A and B in normal and malignant cervical tissue: high Aurora A kinase expression in squamous cervical cancer. Eur J Obstet Gynecol Reprod Biol 142:57–63

Tzatsos A, Paskaleva P, Ferrari F, Deshpande V, Stoykova S, Contino G, Wong KK, Lan F, Trojer P, Park PJ, Bardeesy N (2013) KDM2B promotes pancreatic cancer via Polycomb-dependent and -independent transcriptional programs. J Clin Invest 123:727–739

Varambally S, Dhanasekaran SM, Zhou M, Barrette TR, Kumar-Sinha C, Sanda MG, Ghosh D, Pienta KJ, Sewalt RG, Otte AP, Rubin MA, Chinnaiyan AM (2002) The polycomb group protein EZH2 is involved in progression of prostate cancer. Nature 419:624–629

Veland N, Hardikar S, Zhong Y, Gayatri S, Dan J, Strahl BD, Rothbart SB, Bedford MT, Chen T (2017) The arginine methyltransferase PRMT6 regulates DNA methylation and contributes to global DNA hypomethylation in cancer. Cell Rep 21:3390–3397

Vischioni B, Oudejans JJ, Vos W, Rodriguez JA, Giaccone G (2006) Frequent overexpression of aurora B kinase, a novel drug target, in non-small cell lung carcinoma patients. Mol Cancer Ther 5:2905–2913

Wan C, Gong C, Ji L, Liu X, Wang Y, Wang L, Shao M, Yang L, Fan S, Xiao Y, Wang X, Li M, Zhou G, Zhang Y (2015) NF45 overexpression is associated with poor prognosis and enhanced cell proliferation of pancreatic ductal adenocarcinoma. Mol Cell Biochem 410:25–35

Wan J, Liu H, Yang L, Ma L, Liu J, Ming L (2019) JMJD6 promotes hepatocellular carcinoma carcinogenesis by targeting CDK4. Int J Cancer 144:2489–2500

Wang GG, Cai L, Pasillas MP, Kamps MP (2007) NUP98-NSD1 links H3K36 methylation to Hox-A gene activation and leukaemogenesis. Nat Cell Biol 9:804–812

Wang Y, Zhang R, Wu D, Lu Z, Sun W, Cai Y, Wang C, Jin J (2013) Epigenetic change in kidney tumor: downregulation of histone acetyltransferase MYST1 in human renal cell carcinoma. J Exp Clin Cancer Res 32:8

Wang F, He L, Huangyang P, Liang J, Si W, Yan R, Han X, Liu S, Gui B, Li W, Miao D, Jing C, Liu Z, Pei F, Sun L, Shang Y (2014) JMJD6 promotes colon carcinogenesis through negative regulation of p53 by hydroxylation. PLoS Biol 12:e1001819–e1001819

Wang L, Mao Y, Du G, He C, Han S (2015) Overexpression of JARID1B is associated with poor prognosis and chemotherapy resistance in epithelial ovarian cancer. Tumour Biol 36:2465–2472

Wang D, Han S, Peng R, Jiao C, Wang X, Yang X, Yang R, Li X (2016a) Depletion of histone demethylase KDM5B inhibits cell proliferation of hepatocellular carcinoma by regulation of cell cycle checkpoint proteins p15 and p27. J Exp Clin Cancer Res 35:37

Wang R, Xin M, Li Y, Zhang P, Zhang M (2016b) The functions of histone modification enzymes in cancer. Curr Protein Pept Sci 17:438–445

Wang LH, Huang J, Wu CR, Huang LY, Cui J, Xing ZZ, Zhao CY (2018) Downregulation of miR-29b targets DNMT3b to suppress cellular apoptosis and enhance proliferation in pancreatic cancer. Mol Med Rep 17:2113–2120

Wang H-Y, Long Q-Y, Tang S-B, Xiao Q, Gao C, Zhao Q-Y, Li Q-L, Ye M, Zhang L, Li L-Y, Wu M (2019) Histone demethylase KDM3A is required for enhancer activation of hippo target genes in colorectal cancer. Nucleic Acids Res 47:2349–2364

Wapenaar H, Dekker FJ (2016) Histone acetyltransferases: challenges in targeting bi-substrate enzymes. Clin Epigenetics 8:59

Watanabe H, Soejima K, Yasuda H, Kawada I, Nakachi I, Yoda S, Naoki K, Ishizaka A (2008) Deregulation of histone lysine methyltransferases contributes to oncogenic transformation of human bronchoepithelial cells. Cancer Cell Int 8:15

Watanabe S, Shimada S, Akiyama Y, Ishikawa Y, Ogura T, Ogawa K, Ono H, Mitsunori Y, Ban D, Kudo A, Yamaoka S, Tanabe M, Tanaka S (2019) Loss of KDM6A characterizes a poor prognostic subtype of human pancreatic cancer and potentiates HDAC inhibitor lethality. Int J Cancer 145:192–205

Willmann D, Lim S, Wetzel S, Metzger E, Jandausch A, Wilk W, Jung M, Forne I, Imhof A, Janzer A, Kirfel J, Waldmann H, Schüle R, Buettner R (2012) Impairment of prostate cancer cell growth by a selective and reversible lysine-specific demethylase 1 inhibitor. Int J Cancer 131:2704–2709

Winkler C, De Munter S, Van Dessel N, Lesage B, Heroes E, Boens S, Beullens M, Van Eynde A, Bollen M (2015) The selective inhibition of protein phosphatase-1 results in mitotic catastrophe and impaired tumor growth. J Cell Sci 128:4526–4537

Wissmann M, Yin N, Müller JM, Greschik H, Fodor BD, Jenuwein T, Vogler C, Schneider R, Günther T, Buettner R, Metzger E, Schüle R (2007) Cooperative demethylation by JMJD2C and LSD1 promotes androgen receptor-dependent gene expression. Nat Cell Biol 9:347–353

Wong M, Polly P, Liu T (2015) The histone methyltransferase DOT1L: regulatory functions and a cancer therapy target. Am J Cancer Res 5:2823–2837

Wu Y, Wang Z, Zhang J, Ling R (2017) Elevated expression of protein arginine methyltransferase 5 predicts the poor prognosis of breast cancer. Tumour Biol 39:1010428317695917

Xiang Y, Zhu Z, Han G, Ye X, Xu B, Peng Z, Ma Y, Yu Y, Lin H, Chen AP, Chen CD (2007) JARID1B is a histone H3 lysine 4 demethylase up-regulated in prostate cancer. Proc Natl Acad Sci U S A 104:19226–19231

Xiao XS, Cai MY, Chen JW, Guan XY, Kung HF, Zeng YX, Xie D (2011) High expression of p300 in human breast cancer correlates with tumor recurrence and predicts adverse prognosis. Chinese J Cancer Res 23:201–207

Xiao W, Chen X, Liu L, Shu Y, Zhang M, Zhong Y (2019) Role of protein arginine methyltransferase 5 in human cancers. Biomed Pharmacother 114:108790

Xun J, Wang D, Shen L, Gong J, Gao R, Du L, Chang A, Song X, Xiang R, Tan X (2017) JMJD3 suppresses stem cell-like characteristics in breast cancer cells by downregulation of Oct4 independently of its demethylase activity. Oncotarget 8:21918–21929

Yamamoto S, Tateishi K, Kudo Y, Yamamoto K, Isagawa T, Nagae G, Nakatsuka T, Asaoka Y, Ijichi H, Hirata Y, Otsuka M, Ikenoue T, Aburatani H, Omata M, Koike K (2013) Histone demethylase KDM4C regulates sphere formation by mediating the cross talk between Wnt and Notch pathways in colonic cancer cells. Carcinogenesis 34:2380–2388

Yamane K, Tateishi K, Klose RJ, Fang J, Fabrizio LA, Erdjument-Bromage H, Taylor-Papadimitriou J, Tempst P, Zhang Y (2007) PLU-1 is an H3K4 demethylase involved in transcriptional repression and breast cancer cell proliferation. Mol Cell 25:801–812

Yan N, Xu L, Wu X, Zhang L, Fei X, Cao Y, Zhang F (2017) GSKJ4, an H3K27me3 demethylase inhibitor, effectively suppresses the breast cancer stem cells. Exp Cell Res 359:405–414

Yan M, Yang X, Wang H, Shao Q (2018) The critical role of histone lysine demethylase KDM2B in cancer. Am J Transl Res 10:2222–2233

Yang XJ, Seto E (2007) HATs and HDACs: from structure, function and regulation to novel strategies for therapy and prevention. Oncogene 26:5310–5318

Yang ZQ, Imoto I, Fukuda Y, Pimkhaokham A, Shimada Y, Imamura M, Sugano S, Nakamura Y, Inazawa J (2000) Identification of a novel gene, GASC1, within an amplicon at 9p23-24 frequently detected in esophageal cancer cell lines. Cancer Res 60:4735–4739

Yang P, Guo L, Duan ZJ, Tepper CG, Xue L, Chen X, Kung H-J, Gao AC, Zou JX, Chen H-W (2012) Histone methyltransferase NSD2/MMSET mediates constitutive NF-κB signaling for cancer cell proliferation, survival, and tumor growth via a feed-forward loop. Mol Cell Biol 32:3121–3131

Yang L, Rau R, Goodell MA (2015) DNMT3A in haematological malignancies. Nat Rev Cancer 15:152–165

Yang L, Lei Q, Li L, Yang J, Dong Z, Cui H (2019a) Silencing or inhibition of H3K79 methyltransferase DOT1L induces cell cycle arrest by epigenetically modulating c-Myc expression in colorectal cancer. Clin Epigenetics 11:199

Yang L, Zha Y, Ding J, Ye B, Liu M, Yan C, Dong Z, Cui H, Ding H-F (2019b) Histone demethylase KDM6B has an anti-tumorigenic function in neuroblastoma by promoting differentiation. Oncogenesis 8:3–3

Yao R, Jiang H, Ma Y, Wang L, Wang L, Du J, Hou P, Gao Y, Zhao L, Wang G, Zhang Y, Liu D-X, Huang B, Lu J (2014) PRMT7 induces epithelial-to-mesenchymal transition and promotes metastasis in breast cancer. Cancer Res 74:5656–5667

Yasen M, Mizushima H, Mogushi K, Obulhasim G, Miyaguchi K, Inoue K, Nakahara I, Ohta T, Aihara A, Tanaka S, Arii S, Tanaka H (2009) Expression of Aurora B and alternative variant forms in hepatocellular carcinoma and adjacent tissue. Cancer Sci 100:472–480

Yuan X, Kong J, Ma Z, Li N, Jia R, Liu Y, Zhou F, Zhan Q, Liu G, Gao S (2016) KDM4C, a H3K9me3 histone demethylase, is involved in the maintenance of human ESCC-initiating cells by epigenetically enhancing SOX2 expression. Neoplasia 18:594–609

Zeng WF, Navaratne K, Prayson RA, Weil RJ (2007) Aurora B expression correlates with aggressive behaviour in glioblastoma multiforme. J Clin Pathol 60:218–221

Zhang W, Hayashizaki Y, Kone BC (2004) Structure and regulation of the mDot1 gene, a mouse histone H3 methyltransferase. Biochem J 377:641–651

Zhang Q, Kao C, Zhang J, Vieth E, Gao H, Cai A, Kim B-O, Cheng L, Juliar BE, Li L, Goulet RJ, Miller KD, Sledge GW, Stallcup MR, Jeng M-H (2005) Overexpression of CARM1 in human breast carcinoma stimulated breast cancer cell proliferation. Cancer Res 65:298–298

Zhang W, Xia X, Reisenauer MR, Hemenway CS, Kone BC (2006) Dot1a-AF9 complex mediates histone H3 Lys-79 hypermethylation and repression of ENaCalpha in an aldosterone-sensitive manner. J Biol Chem 281:18059–18068

Zhang J, Liu H, Pan H, Yang Y, Huang G, Yang Y, Zhou WP, Pan ZY (2014) The histone acetyltransferase hMOF suppresses hepatocellular carcinoma growth. Biochem Biophys Res Commun 452:575–580

Zhang C, Song C, Liu T, Tang R, Chen M, Gao F, Xiao B, Qin G, Shi F, Li W, Li Y, Fu X, Shi D, Xiao X, Kang L, Huang W, Wu X, Tang B, Deng W (2017a) KMT2A promotes melanoma cell growth by targeting hTERT signaling pathway. Cell Death Dis 8:e2940–e2940

Zhang M, Wu W, Gao M, Zhang J, Ding X, Zhu R, Chen H, Fei Z (2017b) Coactivator-associated arginine methyltransferase 1 promotes cell growth and is targeted by microRNA-195-5p in human colorectal cancer. Tumour Biol 39:1010428317694305

Zhang K, Wang J, Yang L, Yuan Y-C, Tong TR, Wu J, Yun X, Bonner M, Pangeni R, Liu Z, Yuchi T, Kim JY, Raz DJ (2018) Targeting histone methyltransferase G9a inhibits growth and Wnt signaling pathway by epigenetically regulating HP1α and APC2 gene expression in non-small cell lung cancer. Mol Cancer 17:153–153

Zhao Z, Shilatifard A (2019) Epigenetic modifications of histones in cancer. Genome Biol 20:245

Zhao E, Tang C, Jiang X, Weng X, Zhong X, Zhang D, Hou J, Wang F, Huang M, Cui H (2017) Inhibition of cell proliferation and induction of autophagy by KDM2B/FBXL10 knockdown in gastric cancer cells. Cell Signal 36:222–229

Zheng Q, Fan H, Meng Z, Yuan L, Liu C, Peng Y, Zhao W, Wang L, Li J, Feng J (2018) Histone demethylase KDM2B promotes triple negative breast cancer proliferation by suppressing p15INK4B, p16INK4A, and p57KIP2 transcription. Acta Biochim Biophys Sin 50:897–904

Zhong J, Cao R-X, Hong T, Yang J, Zu X-Y, Xiao X-H, Liu J-H, Wen G-B (2011) Identification and expression analysis of a novel transcript of the human PRMT2 gene resulted from alternative polyadenylation in breast cancer. Gene 487:1–9

Zhong J, Cao RX, Liu JH, Liu YB, Wang J, Liu LP, Chen YJ, Yang J, Zhang QH, Wu Y, Ding WJ, Hong T, Xiao XH, Zu XY, Wen GB (2014) Nuclear loss of protein arginine N-methyltransferase 2 in breast carcinoma is associated with tumor grade and overexpression of cyclin D1 protein. Oncogene 33:5546–5558

Zhou H, Kuang J, Zhong L, Kuo W-L, Gray J, Sahin A, Brinkley B, Sen S (1998) Tumour amplified kinase STK15/BTAK induces centrosome amplification, aneuploidy and transformation. Nat Genet 20:189–193

Zhu L, Yang J, Zhao L, Yu X, Wang L, Wang F, Cai Y, Jin J (2015) Expression of hMOF, but not HDAC4, is responsible for the global histone H4K16 acetylation in gastric carcinoma. Int J Oncol 46:2535–2545

Summa of Erasers of Histone Acetylation with Special Emphasis on Classical Histone Deacetylases (HDACs)

3

The human genome in order to get accommodated within the constrained nuclear space requires remarkably high level of condensation (Tseng and Yang 2013). This condensation results in the formation of chromatin, a supremely organized nucleoprotein structure (Olins and Olins 1974). Nucleosome forms the fundamental unit of this structural polymer (chromatin). Each nucleosome has a nucleosome core formed from an octameric complex made of polycationic core histones around which 145–147 bp of DNA are wrapped (Davey et al. 2002; Korolev et al. 2018; Luger et al. 1997; McGinty and Tan 2015). Adjacent nucleosome cores of two nucleosomes are connected by linker DNA which is frequently in contact with linker histone H1 (or H5 in birds) (Andreeva et al. 1978; Simpson 1978). Apart from the significant role in genomic compaction, nucleosomes serve as signalling focal points for chromatin-templated processes (McGinty and Tan 2015).

Polycationic histone proteins play a significant role not only in chromatin compaction but also regulate gene expression by undergoing different post-translational modifications particularly on their amino-terminal unstructured tails (Bannister and Kouzarides 2011; Tolsma and Hansen 2019). Among these modifications, histone acetylation occurring on the lysine residues of histone proteins favours DNA-templated reactions by way of passive chromatin remodelling (Allfrey et al. 1964; Ganai 2016). Histone acetylation being extremely dynamic is rightly controlled by functionally opposing enzyme families. While enzymes installing this covalent modification are histone acetyltransferases (HATs), the enzymes erasing this covalent tag are termed as histone deacetylases (HDACs) (Barnes et al. 2019; Yang and Seto 2007). Thus it is obvious that unlike HATs, HDACs by favouring closed chromatin state make the conditions non-conducive for transcriptional events.

S. A. Ganai, *Histone Deacetylase Inhibitors in Combinatorial Anticancer Therapy*,
https://doi.org/10.1007/978-981-15-8179-3_3

3.1 Different Classes of Histone Deacetylase Enzymes

As of now 18 human enzymes have been identified to have deacetylase activity. These HDACs being conjugated proteins require cofactor for their catalytic activity. Based on cofactor required these 18 HDACs have been grouped into two broad categories. Majority of HDACs (11) being metalloenzymes require zinc for their function and are known as zinc-dependent HDACs. While the remaining 7 HDACs requiring NAD+ for their activity are described as NAD+-dependent HDACs (Lawlor and Yang 2019). Another classification based on resemblance to yeast HDACs classifies HDACs into four main classes ranging from Class I to Class IV. Class I has four members, namely HDAC1, 2, 3 and 8 within its confinement (Bradner et al. 2010; Ganai 2015). These enzymes resemble yeast transcriptional repressor RPD3 and exhibit ubiquitous tissue expression. While HDAC1 and HDAC2 are confined to nucleus, HDAC3 (Yang et al. 2002) and HDAC8 show shuttling ability (Khan et al. 2007; Mottamal et al. 2015). All the members of Class I have a single catalytic domain. Class II HDACs resemble yeast HDA1and these enzymes have been subdivided into Class IIa and Class IIb HDACs. While Class IIa holds HDAC4, 5, 7 and 9, Class IIb includes two members: only one tubulin deacetylase HDAC6 and another polyamine deacetylase HDAC10 (Hubbert et al. 2002; Shinsky and Christianson 2018; Yang and Grégoire 2005). Class II HDACs unlike Class I HDACs exhibit tissue-restricted distribution (Verdin et al. 2003). Further majority of these HDACs have shuttling ability. In mammalian cells, Class II HDACs have been found to recruit Class I HDACs as they possess strong deacetylase activity. Low deacetylase activity of Class IIa HDACs has been ascribed to the presence of a special histidine residue in their catalytic domain unlike Class I HDACs which possess tyrosine residue (Lahm et al. 2007).

Class III HDACs also known as Sirtuins that are silent information regulator 2 (Sir2) like and act primarily as NAD^+-dependent deacetylases. Plethora of studies have proved that Sir2 (first discovered sirtuin in yeast) is crucial for gene silencing in budding yeast. These enzymes are mechanistically different from remaining HDACs and couple lysine deacetylation to NAD+ hydrolysis (Denu 2005). Certain sirtuins remove succinyl or malonyl from target proteins by a mechanism closely resembling deacetylation. Other sirtuins do not show deacetylase activity but instead function as ADP-ribosyltransferases (Houtkooper et al. 2012). Sirtuins are expressed ubiquitously and till date seven sirtuins (SIRT1–SIRT7) have been discovered (Fig. 3.1). These enzymes in addition to subcellular localization differ in targets and enzymatic activity. Certain sirtuins such as SIRT1, SIRT7 and SIRT6 are nuclear residents, while others SIRT3–SIRT5 are spaced in mitochondria. SIRT2 although predominantly cytoplasmic shows nuclear shuttling (Parihar et al. 2015; Vaquero 2009). Sirtuins regulate senescence, inflammation and oxidative stress. Evidences suggest that sirtuin activation may prove fruitful both in metabolic and neurodegenerative diseases (Haigis and Sinclair 2010; Kupis et al. 2016; Sack and Finkel 2012; Santos et al. 2016). Class IV, the smallest class of HDACs has HDAC11 as a sole member. This HDAC encoded by chromosome 3 is predominantly localized in the nucleus (Gao et al. 2002). Thus only seven members of Class III HDACs are reliant on

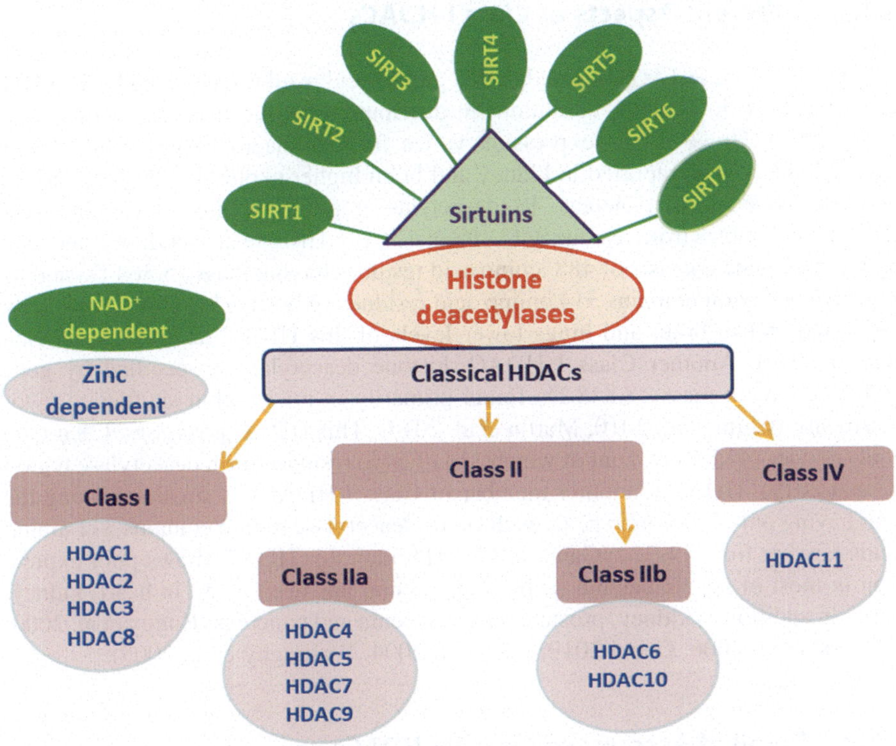

Fig. 3.1 Classification of 18 mammalian histone deacetylases identified from mammalian systems. HDACs may be zinc-dependent or may require NAD^+ for activity. The former are known as Classical HDACs, while latter as Sirtuins. Sirtuins are seven in number (SIRT1-SIRT7) and are mechanistically distinct from other HDACs. Classical HDACs include three classes including Class I in addition to Class II and IV. Class I has four members: HDAC1-HDAC3 and HDAC8. Class II is subdivided into two classes, namely Class IIa and Class IIb. Class IIa HDACs have a single catalytic domain and include HDAC4, HDAC5 and HDAC7 in addition to HDAC9. Class IIb has two HDACs: HDAC6 and HDAC10 under its umbrella. While HDAC6 has both the domains functional, HDAC10 has only one domain functional and other being partial is inactive in terms of deacetylation. Class IV HDACs are represented by lone and least studied HDAC, namely HDAC11. HDACs target not only histone substrates but also non-histone targets as well. On the whole they cause transcriptional silencing by histone hypoacetylation and subsequent chromatin compaction

NAD^+, while majority of HDACs (11) require zinc for their catalytic activity. Zinc reliant HDACs are also termed as classical HDACs. Classical HDACs include 4 members of Class I, 6 members of Class II and 1 lone member of Class IV (Fig. 3.1). These HDACs will be better understood shedding light on them individually.

3.1.1 Different Aspects of Class I HDACs

Histone deacetylase 1 is composed of 482 amino acid residues, encoded by HDAC1 gene, localized in nucleus, ubiquitous in distribution having mass 55,103 Da (Cai et al. 2000). However, its expression varies from tissue to tissue. While lower expression has been reported in kidney and brain higher expression has been noted in heart, gonads and pancreas. Its deacetylase region contains 313 amino acid residues and ranges from residue 9-321 (Ganai 2019). Histone deacetylase 2 encoded by HDAC2 gene consists of 488 amino acid residues having mass 55,364 Da and its deacetylase region contains 314 amino acid residues (9-322). This HDAC is widely expressed and in brain and lungs lower levels of this HDAC have been reported (Ganai 2019). Another Class I HDAC, histone deacetylase 3 specified by gene HDAC3, having mass 48,848 Da found primarily in nucleus but shuttles to cytoplasm also (Chini et al. 2010; Martin et al. 2014). This HDAC possesses 428 amino acid residues (41,758 Da) out of which 314 (3-316) residues form deacetylase region (Ganai 2019). HDAC8, another member of Class I HDACs is smallest among the four having only 377 amino acid residues. Its deacetylase region contains 311 amino acids ranging from 14-324 (Ganai 2019). Although this HDAC shows poor expression in most of the tissues but its high expression has been noted in heart, kidney, brain, in addition to kidney, prostate, lung, placenta and pancreas (Buggy et al. 2000; de Leval et al. 2006; Ganai 2019; Lee et al. 2004; Waltregny et al. 2005).

3.1.2 Detailed Account of Class IIa HDACs

HDAC4, a member of Class IIa, has 1084 amino acid residues (119,040 Da) out of which 430 form its deacetylase part. This HDAC interacts with myocyte enhancer factor MEF2A and the interaction region contains 196 residues (118-313). Moreover, this HDAC has been reported to undergo active nuclear export (Ganai 2019; Miska et al. 1999). HDAC5 is polymer of 1122 amino acid residues, its deacetylase regions include 345 residues ranging from 684–10,828. Its nuclear export signal lies towards C-terminal end (1081–1122) and the mass of this HDAC is 121,978 Da. During myocyte differentiation this HDAC translocates from nucleus to cytoplasm in a phosphorylation dependent manner (Ganai 2019; McKinsey et al. 2000). HDAC7, another Class IIa HDAC is condensation polymer of 952 amino acids and translocates from nucleus to cytoplasm. Calcium/calmodulin-dependent kinase I-dependent phosphorylation of this HDAC facilitates its interaction with 14-3-3 proteins thereby stabilizing it. (Li et al. 2004). Chromosomal Maintenance 1 (CRM1) or exportin 1 has been reported to mediate nuclear export of HDAC7 without phosphorylation and association of this HDAC with 14-3-3 proteins (Gao et al. 2006). HDAC9, one of the members of Class IIa is polymer of 1011 amino acid residues (111,297 Da). Its deacetylase region ranges from 631–978 and thus contains 348 amino acid residues. Although widely expressed, high levels of this HDAC occur in certain tissues including brain, heart in addition to muscle and testis (David et al. 2003; Ganai 2019; Petrie et al. 2003; Zhou et al. 2001, 2000).

3.1.3 Extensive Details of Class IIb HDACs and HDAC11

As mentioned above Class IIb has only two members, namely HDAC6 and HDAC10. HDAC6 also known as tubulin deacetylase is predominantly cytoplasmic and is composed of 1215 amino acid residues (131,419). On the other hand, HDAC10 actually known as polyamine deacetylase 10 has 669 amino acid residues. HDAC10 and HDAC6 differ from Class IIa HDACs have two putative catalytic domains. However, in case of HDAC10 only N-terminal catalytic domain is active while the C-terminal incomplete catalytic domain is inactive. Both the domains in case of HDAC6 are complete and studies with full length human HDAC6 have revealed that catalytic domain 1 exhibits narrow substrate specificity compared to catalytic domain 2 having broad substrate specificity (Hai and Christianson 2016). Tandem organization of catalytic domains in Class IIb HDACs makes them resistant to certain inhibitors including trapoxin B and sodium butyrate (Guardiola and Yao 2002). Based on these facts it won't be wrong to speculate that polyamine deacetylase and tubulin deacetylase share atypical structural and pharmacological features. HDAC6 deacetylates lysine residues of core histone located on the amino-terminal part (Grozinger et al. 1999). Through tubulin deacetylation, this HDAC regulates microtubule reliant cell motility (Hubbert et al. 2002). Moreover, HDAC6 plays a key role in misfolded protein degradation (Ganai 2017; Olzmann et al. 2007). HDAC10 has shown robust deacetylase activity against polyamines and further has been suggested to facilitate DNA mismatch repair (Radhakrishnan et al. 2015; Shinsky and Christianson 2018).

Histone deacetylase 11 encoded by HDAC11 gene has 347 amino acid residues (39,183 Da). Its histone deacetylase region has 313 amino acid residues ranging from 14–326. Expression of this HDAC appears to be tissue-restricted. Strong expression of this HDAC has been reported in brain, skeletal muscle, heart, kidney and gonads despite its weak expression in majority of tissues. Its tissue distribution and enzymatic activity have been well studied but meagre details regarding its functional role are available. This enzyme has been reported to silence the transcription of IL-10 gene (Villagra et al. 2009). Studies have shown that HDAC11 has growth inhibitory effect on influenza-A virus. Depletion of this enzyme in human lung epithelial cells facilitated the growth kinetics of this virus (Nutsford et al. 2019).

Till now I have discussed the different classes of HDACs based on the cofactor requirement and resemblance to yeast HDACs. Further I have given the layout of Classical HDACs and after that I have thoroughly discussed the individual members of various classes of Classical HDACs which will help in clear understanding of the forthcoming chapter regarding the implications of these HDACs in distinct cancers.

References

Allfrey VG, Faulkner R, Mirsky AE (1964) Acetylation and methylation of histones and their possible role in the regulation of RNA synthesis. Proc Natl Acad Sci U S A 51:786–794

Andreeva NB, Vishnevskaia T, Gazarian KG (1978) Role of serine-rich histone (H5) in bird erythrocyte genome inactivation. Mol Biol 12:123–134
Bannister AJ, Kouzarides T (2011) Regulation of chromatin by histone modifications. Cell Res 21:381–395
Barnes CE, English DM, Cowley SM (2019) Acetylation & Co: an expanding repertoire of histone acylations regulates chromatin and transcription. Essays Biochem 63:97–107
Bradner JE, West N, Grachan ML, Greenberg EF, Haggarty SJ, Warnow T, Mazitschek R (2010) Chemical phylogenetics of histone deacetylases. Nat Chem Biol 6:238–243
Buggy JJ, Sideris ML, Mak P, Lorimer DD, McIntosh B, Clark JM (2000) Cloning and characterization of a novel human histone deacetylase, HDAC8. Biochem J 350(Pt 1):199–205
Cai RL, Yan-Neale Y, Cueto MA, Xu H, Cohen D (2000) HDAC1, a histone deacetylase, forms a complex with Hus1 and Rad9, two G2/M checkpoint Rad proteins. J Biol Chem 275:27909–27916
Chini CC, Escande C, Nin V, Chini EN (2010) HDAC3 is negatively regulated by the nuclear protein DBC1. J Biol Chem 285:40830–40837
Davey CA, Sargent DF, Luger K, Maeder AW, Richmond TJ (2002) Solvent mediated interactions in the structure of the nucleosome core particle at 1.9Å resolution††we dedicate this paper to the memory of Max Perutz who was particularly inspirational and supportive to T.J.R. in the early stages of this study. J Mol Biol 319:1097–1113
David D, Cardoso J, Marques B, Marques R, Silva ED, Santos H, Boavida MG (2003) Molecular characterization of a familial translocation implicates disruption of HDAC9 and possible position effect on TGFbeta2 in the pathogenesis of Peters' anomaly. Genomics 81:489–503
de Leval L, Waltregny D, Boniver J, Young RH, Castronovo V, Oliva E (2006) Use of histone deacetylase 8 (HDAC8), a new marker of smooth muscle differentiation, in the classification of mesenchymal tumors of the uterus. Am J Surg Pathol 30:319–327
Denu JM (2005) The Sir 2 family of protein deacetylases. Curr Opin Chem Biol 9:431–440
Ganai S (2015) In silico approaches towards safe targeting of class I histone deacetylases. https://doi.org/10.1007/978-1-4614-6436-5_459-1, pp 1–9
Ganai SA (2016) Histone deacetylase inhibitor pracinostat in doublet therapy: a unique strategy to improve therapeutic efficacy and to tackle herculean cancer chemoresistance. Pharm Biol 54:1926–1935
Ganai SA (2017) Small-molecule modulation of HDAC6 activity: the propitious therapeutic strategy to vanquish neurodegenerative disorders. Curr Med Chem 24:4104–4120
Ganai SA (2019) HDACs and their distinct classes. In: Ganai SA (ed) Histone deacetylase inhibitors — epidrugs for neurological disorders. Springer, Singapore, pp 21–25
Gao L, Cueto MA, Asselbergs F, Atadja P (2002) Cloning and functional characterization of HDAC11, a novel member of the human histone deacetylase family. J Biol Chem 277:25748–25755
Gao C, Li X, Lam M, Liu Y, Chakraborty S, Kao H-Y (2006) CRM1 mediates nuclear export of HDAC7 independently of HDAC7 phosphorylation and association with 14-3-3s. FEBS Lett 580:5096–5104
Grozinger CM, Hassig CA, Schreiber SL (1999) Three proteins define a class of human histone deacetylases related to yeast Hda1p. Proc Natl Acad Sci U S A 96:4868–4873
Guardiola AR, Yao TP (2002) Molecular cloning and characterization of a novel histone deacetylase HDAC10. J Biol Chem 277:3350–3356
Hai Y, Christianson DW (2016) Histone deacetylase 6 structure and molecular basis of catalysis and inhibition. Nat Chem Biol 12:741–747
Haigis MC, Sinclair DA (2010) Mammalian sirtuins: biological insights and disease relevance. Annu Rev Pathol 5:253–295
Houtkooper RH, Pirinen E, Auwerx J (2012) Sirtuins as regulators of metabolism and healthspan. Nat Rev Mol Cell Biol 13:225–238
Hubbert C, Guardiola A, Shao R, Kawaguchi Y, Ito A, Nixon A, Yoshida M, Wang XF, Yao TP (2002) HDAC6 is a microtubule-associated deacetylase. Nature 417:455–458

Khan N, Jeffers M, Kumar S, Hackett C, Boldog F, Khramtsov N, Qian X, Mills E, Berghs SC, Carey N, Finn PW, Collins LS, Tumber A, Ritchie JW, Jensen PB, Lichenstein HS, Sehested M (2007) Determination of the class and isoform selectivity of small-molecule histone deacetylase inhibitors. Biochem J 409:581–589

Korolev N, Lyubartsev AP, Nordenskiöld L (2018) A systematic analysis of nucleosome core particle and nucleosome-nucleosome stacking structure. Sci Rep 8:1543

Kupis W, Pałyga J, Tomal E, Niewiadomska E (2016) The role of sirtuins in cellular homeostasis. J Physiol Biochem 72:371–380

Lahm A, Paolini C, Pallaoro M, Nardi MC, Jones P, Neddermann P, Sambucini S, Bottomley MJ, Lo Surdo P, Carfí A, Koch U, De Francesco R, Steinkühler C, Gallinari P (2007) Unraveling the hidden catalytic activity of vertebrate class IIa histone deacetylases. Proc Natl Acad Sci U S A 104:17335–17340

Lawlor L, Yang XB (2019) Harnessing the HDAC–histone deacetylase enzymes, inhibitors and how these can be utilised in tissue engineering. Int J Oral Sci 11:20

Lee H, Rezai-Zadeh N, Seto E (2004) Negative regulation of histone deacetylase 8 activity by cyclic AMP-dependent protein kinase A. Mol Cell Biol 24:765–773

Li X, Song S, Liu Y, Ko SH, Kao HY (2004) Phosphorylation of the histone deacetylase 7 modulates its stability and association with 14-3-3 proteins. J Biol Chem 279:34201–34208

Luger K, Mader AW, Richmond RK, Sargent DF, Richmond TJ (1997) Crystal structure of the nucleosome core particle at 2.8 a resolution. Nature 389:251–260

Martin D, Li Y, Yang J, Wang G, Margariti A, Jiang Z, Yu H, Zampetaki A, Hu Y, Xu Q, Zeng L (2014) Unspliced X-box-binding protein 1 (XBP1) protects endothelial cells from oxidative stress through interaction with histone deacetylase 3. J Biol Chem 289:30625–30634

McGinty RK, Tan S (2015) Nucleosome structure and function. Chem Rev 115:2255–2273

McKinsey TA, Zhang CL, Lu J, Olson EN (2000) Signal-dependent nuclear export of a histone deacetylase regulates muscle differentiation. Nature 408:106–111

Miska EA, Karlsson C, Langley E, Nielsen SJ, Pines J, Kouzarides T (1999) HDAC4 deacetylase associates with and represses the MEF2 transcription factor. EMBO J 18:5099–5107

Mottamal M, Zheng S, Huang TL, Wang G (2015) Histone deacetylase inhibitors in clinical studies as templates for new anticancer agents. Molecules (Basel, Switzerland) 20:3898–3941

Nutsford AN, Galvin HD, Ahmed F, Husain M (2019) The class IV human deacetylase, HDAC11, exhibits anti-influenza A virus properties via its involvement in host innate antiviral response. Cell Microbiol 21:e12989

Olins AL, Olins DE (1974) Spheroid chromatin units (ν bodies). Science (New York, N.Y.) 183:330–332

Olzmann JA, Li L, Chudaev MV, Chen J, Perez FA, Palmiter RD, Chin LS (2007) Parkin-mediated K63-linked polyubiquitination targets misfolded DJ-1 to aggresomes via binding to HDAC6. J Cell Biol 178:1025–1038

Parihar P, Solanki I, Mansuri ML, Parihar MS (2015) Mitochondrial sirtuins: emerging roles in metabolic regulations, energy homeostasis and diseases. Exp Gerontol 61:130–141

Petrie K, Guidez F, Howell L, Healy L, Waxman S, Greaves M, Zelent A (2003) The histone deacetylase 9 gene encodes multiple protein isoforms. J Biol Chem 278:16059–16072

Radhakrishnan R, Li Y, Xiang S, Yuan F, Yuan Z, Telles E, Fang J, Coppola D, Shibata D, Lane WS, Zhang Y, Zhang X, Seto E (2015) Histone deacetylase 10 regulates DNA mismatch repair and may involve the deacetylation of MutS homolog 2. J Biol Chem 290:22795–22804

Sack MN, Finkel T (2012) Mitochondrial metabolism, sirtuins, and aging. Cold Spring Harb Perspect Biol 4

Santos L, Escande C, Denicola A (2016) Potential modulation of sirtuins by oxidative stress. Oxidative Med Cell Longev 2016:9831825

Shinsky SA, Christianson DW (2018) Polyamine deacetylase structure and catalysis: prokaryotic acetylpolyamine amidohydrolase and eukaryotic HDAC10. Biochemistry 57:3105–3114

Simpson RT (1978) Structure of the chromatosome, a chromatin particle containing 160 base pairs of DNA and all the histones. Biochemistry 17:5524–5531

Tolsma TO, Hansen JC (2019) Post-translational modifications and chromatin dynamics. Essays Biochem 63:89–96

Tseng C, Yang X (2013) Packaging DNA into chromosomes: how do the long threads of DNA fit into the small interphase nucleus? pp 111–129

Vaquero A (2009) The conserved role of sirtuins in chromatin regulation. Int J Dev Biol 53:303–322

Verdin E, Dequiedt F, Kasler HG (2003) Class II histone deacetylases: versatile regulators. Trends Genet 19:286–293

Villagra A, Cheng F, Wang H-W, Suarez I, Glozak M, Maurin M, Nguyen D, Wright KL, Atadja PW, Bhalla K, Pinilla-Ibarz J, Seto E, Sotomayor EM (2009) The histone deacetylase HDAC11 regulates the expression of interleukin 10 and immune tolerance. Nat Immunol 10:92–100

Waltregny D, Glénisson W, Tran SL, North BJ, Verdin E, Colige A, Castronovo V (2005) Histone deacetylase HDAC8 associates with smooth muscle alpha-actin and is essential for smooth muscle cell contractility. FASEB J 19:966–968

Yang X-J, Grégoire S (2005) Class II histone deacetylases: from sequence to function, regulation, and clinical implication. Mol Cell Biol 25:2873–2884

Yang XJ, Seto E (2007) HATs and HDACs: from structure, function and regulation to novel strategies for therapy and prevention. Oncogene 26:5310–5318

Yang WM, Tsai SC, Wen YD, Fejer G, Seto E (2002) Functional domains of histone deacetylase-3. J Biol Chem 277:9447–9454

Zhou X, Richon VM, Rifkind RA, Marks PA (2000) Identification of a transcriptional repressor related to the noncatalytic domain of histone deacetylases 4 and 5. Proc Natl Acad Sci U S A 97:1056–1061

Zhou X, Marks PA, Rifkind RA, Richon VM (2001) Cloning and characterization of a histone deacetylase, HDAC9. Proc Natl Acad Sci U S A 98:10572–10577

Strong Involvement of Classical Histone Deacetylases and Mechanistically Distinct Sirtuins in Bellicose Cancers

4

The dynamic histone acetylation is precisely regulated by two antagonistic enzyme families. Histone acetyl transferases (HATs), the "writers" mediate this acetylation while histone deacetylases (HDACs), the "erasers" perform deacetylation (Yang and Seto 2007). Acetylation homeostasis regulated by these functionally opposing enzyme families plays a critical role in proper execution of gene expression programs. Anomalous activity of HDACs leads to transcriptional dysfunction by perturbing the finely tuned acetylation level (Ganai 2015; Saha and Pahan 2006). Plethora of proteins involved in tumour onset and progression are regulated by HDACs in one way or the other (Ropero and Esteller 2007). HDACs, by erasing acetyl moieties from nucleosomal histones create chromatin topology strongly unfavourable for transcription of tumorigenesis related genes (Parbin et al. 2014). Further HDACs modulate (deacetylate) a variety of protein targets, other than histones, having crucial links with transcription, cell growth control in addition to differentiation and programmed cell death (apoptosis) (Ganai 2018; Minucci and Pelicci 2006; Ropero and Esteller 2007; Singh et al. 2010). Thus it is quite obvious that aberrant HDAC activity may drive tumorigenesis by multiple ways. The contribution of HDACs in triggering different cancers will become clearer by discussing these HDACs class specific manner.

4.1 Class I HDACs in Fuelling Cancer

Class I HDACs have strong implications in distinct cancers. HDAC1 overexpression has been identified in liver cancer and cholangiocarcinoma (Mizuguchi et al. 2012; Xie et al. 2012). Overexpression of HDAC1 and HDAC2 has strong cross-talk with colon cancer progression (Yang et al. 2014a). In breast epithelial cells loss of oestrogen receptor α is crucial for transformation into breast cancer. HDAC1 interacts with this receptor and restrains its transcriptional activity. Thus HDAC1 facilitates breast cancer progressing by way of negatively regulating oestrogen receptor α (Kawai et al. 2003). This HDAC is potentially involved in cancer

S. A. Ganai, *Histone Deacetylase Inhibitors in Combinatorial Anticancer Therapy*,
https://doi.org/10.1007/978-981-15-8179-3_4

progression and development as post-transcriptional gene silencing of HDAC1 caused antiproliferative effect and apoptosis in colon and breast cancer cells (Weichert et al. 2008b). Enhanced expression of this HDAC has been identified in prostate cancer specimens (Halkidou et al. 2004). CBX4 protein has been reported to facilitate proliferation of clear renal cell carcinoma cells by recruiting HDAC1 to promoter region of tumour suppressor KLF6 gene. CBX4-induced cell growth and migration was rescued by disruption of its interaction with HDAC1 or ectopically expressing KLF6 (Jiang et al. 2020). It has been certified that this HDAC negatively regulates long non-coding RNA HRCEG and low expression of this RNA has been identified in gastric cancer tissues. HRCEG has strong antiproliferative effect and disfavours epithelial to mesenchymal transition. As expression of HRCEG is downregulated by HDAC1 this clearly suggests that HDAC1 overexpression may facilitate proliferation and epithelial to mesenchymal transition of gastric cancer cells by hampering HRCEG expression (Wu et al. 2020). Nuclear factor of activated T cell (NFAT), an oncogene, and HDAC1 have crucial implications in malignant glioblastoma phenotype (Song et al. 2020). Substantial upregulation of this HDAC has been seen in refractory AML patients and in doxorubicin-resistant AML cells (Lai et al. 2019). HDAC1 protein levels were found to be higher in squamous cell carcinoma compared to adenocarcinoma. Higher expression of this HDAC1 has negative correlation with overall lung cancer patient survival (Cao et al. 2017b). All members of Class I excluding HDAC8 showed differential expression in breast cancer. While hormone receptor possessing tumours showed higher expression of HDAC1, less differentiated tumours were associated with substantially escalated expression of HDAC2 and HDAC3 (Müller et al. 2013).

In lung cancer cells aberrant expression of HDAC2 has been noted and inactivation of this Class I HDAC hampered tumour cell growth and triggered apoptotic signalling in these cells (Jung et al. 2012). Through futuristic approaches it has been proved that HDAC2 and SIRT7 are highly expressed, whereas SIRT3 is most poorly expressed especially in aggressive breast cancer (basal-like). Enhanced expression of HDAC2 proved to be an indicator of high tumour grade besides poor prognosis and positive lymph node status (Shan et al. 2017). Higher expression of HDAC2 has been identified in hepatocellular carcinoma cells suggesting its role in tumour onset and development (Noh et al. 2011). Colon cancer development triggered by activation of Wnt signalling is mediated by HDAC2 induction (Zhu et al. 2004). This effect is not common to all cancers no cross-talk was seen between Wnt signalling and HDAC2 overexpression (Noh et al. 2011). Overexpression or mutation of this HDAC has been found in haematological malignancies. Studies involving transcriptome analyses have authenticated its unique role in leukaemogenesis (Conte et al. 2014). Immunohistochemistry study on 192 prostate carcinomas regarding the expression profile of Class I HDACs has been performed. While in 69.8% cases strong expression of HDAC1 was seen, higher expression of HDAC2 and HDAC3 was noted in 74% and 94.8% cases, respectively (Weichert et al. 2008a). Genomic amplification of nuclear-localized E3 ubiquitin ligase encoding gene (MDM2) is hallmark of dedifferentiated liposarcoma. Its expression seems to have cross-talk with HDAC2 activity as intervention against this enzyme attenuated expression of MDM2 (Seligson et al. 2019). HDAC2 and REST downregulation enhanced the aggressiveness of less aggressive breast cancer cells. This transformation was found to be mediated by escalated expression of Nav1.5 (sodium channel protein type 5 subunit alpha) and nNav1.5 (neonatal splice variant of Nav1.5) (Kamarulzaman et al. 2017).

Differential expression of HDAC3 has been reported in various colon cancer cell lines. Higher levels of this HDAC were found to be comparatively expressed by SW480 cell line. HDAC3 overexpression makes the colon cancer cells resistant to luminal butyrate suggesting a crucial role for this HDAC in proliferation and differentiation of these cells (Spurling et al. 2008). In hepatocellular carcinoma patients HDAC3 overrepresentation was directly proportional to tumour growth. Attenuation of xenograft tumour growth was seen on inhibiting the expression of this HDAC. Enhanced proliferation of cancer cells on HDAC3 overexpression has been attributed to signal transducer and activator of transcription 3 (STAT3) signalling (Lu et al. 2018). Upregulation of this HDAC has also been proved in cholangiocarcinoma tissues and cell lines. Therapeutic intervention with HDAC3 specific inhibitor triggered apoptosis in these cells (Yin et al. 2017). Enhanced expression of HDAC3 was also noted in gastric cancer tissues and cell lines. HDAC3 higher expression seems to be associated with enhanced proliferation as knockdown of this HDAC reduced cell viability besides colony formation and tumour weight. This effect of HDAC3 was found to be mediated by miR-454 (Xu et al. 2018). In human colon tumours as well as in duodenal adenomas (*Apc1638*$^{N/+}$ mice) enhanced protein levels of HDAC3 were seen. On silencing the expression of this HDAC in colon cancer cells, growth inhibition in addition to declined cell survival and enhanced programmed cell death (apoptosis) was recorded (Wilson et al. 2006). Thus HDAC3 being involved in multiple cancers may prove as a candidate target for small molecule based anticancer therapy.

Another HDAC of Class I, HDAC8 is multifaceted and its involvement has been proved in various cancers. Elevated expression of HDAC8 has been confirmed in oral squamous cell carcinoma cells and oral squamous cell carcinoma tissues through immunoblotting and immunohistochemistry method. Knockdown of HDAC8 showed substantial antiproliferative effect, induced programmed cell death in these cells (Ahn and Yoon 2017). This HDAC contributes significantly to neuroblastoma pathogenesis. Overexpression of HDAC8 has been observed in neuroblastoma cells and HDAC8 selective pharmacological intervention lessened proliferation, induced cell cycle arrest and mitigated clonogenic growth (Chakrabarti et al. 2015; Oehme et al. 2009). Further, in hepatocellular carcinoma cells and tissues, marked upregulation of HDAC8 has been seen through various state of the art techniques. Dramatic reduction in proliferation and more apoptosis was seen in hepatocellular cell carcinoma cells on knockdown of HDAC8 (Wu et al. 2013). Significantly increased levels of this HDAC and alleviated levels of miR-216b-5p have been proved in breast cancer cell lines and tissues. Lower levels of this microRNA were associated with elevated lymph node metastasis apart from tumour size progression. Attenuated breast cancer cell proliferation and advancement, mediated by downregulation of HDAC8, was seen on miR-216b-5p overexpression

Table 4.1 Class I HDACs and their aberrant expression in various cancers

Name of HDAC	Status in disease	Cancer type	Reference
HDAC1	Upregulated	Cholangiocarcinoma	Mizuguchi et al. (2012)
		Breast	Kawai et al. (2003)
		Colon and Breast	Weichert et al. (2008b)
		Prostate	Halkidou et al. (2004)
		Gastric	Wu et al. (2020)
		AML	Lai et al. (2019)
		Lung	Cao et al. (2017b)
		Breast	Müller et al. (2013)
HDAC2	Upregulated	Lung	Jung et al. (2012)
		Breast	Shan et al. (2017)
		Hepatocellular	Noh et al. (2011)
		Colon	Zhu et al. (2004)
		Prostate	Weichert et al. (2008a)
	Downregulated	Breast	Kamarulzaman et al. (2017)
HDAC3	Upregulated	Colon	Spurling et al. (2008)
		Hepatocellular	Lu et al. (2018)
		Cholangiocarcinoma	Yin et al. (2017)
		Gastric	Xu et al. (2018)
		Colon	Wilson et al. (2006)
HDAC8	Upregulated	Oral squamous cell carcinoma	Ahn and Yoon (2017)
		Neuroblastoma	Oehme et al. (2009)
		Hepatocellular	Wu et al. (2013)
		Breast	Menbari et al. (2019)
		Squamous cell carcinoma	Ahn (2018)
		Breast	Park et al. (2011)
		Squamous cell carcinoma	Wang et al. (2017)

(Menbari et al. 2019). Increased expression of HDAC8 contributes to squamous cell carcinoma. In murine squamous cell carcinoma cells enhanced expression of HDAC8 facilitated cell proliferation while inhibiting its expression attenuated the proliferation (Ahn 2018). Higher expression of several HDACs including HDAC8 has been reported in more invasive breast cancer cell line (MDA-MB-231) compared to lesser invasive one (MCF-7). Knockdown studies showed that for invasion and expression of MMP-9, HDAC1, HDAC6 and HDAC8 are crucial (Park et al. 2011). HDAC8 along with aryl hydrocarbon receptor were found to be markedly upregulated in both hepatocellular carcinoma tissues and cell models (Table 4.1). HDAC8 overexpression mediated its cancer promoting effect through downregulation of RB1 (retinoblastoma-associated protein encoding gene) (Wang et al. 2017).

4.2 Class IIa HDACs in Cancer Progression

Class IIa HDACs including HDAC4, 5, 7 and 9 have significant involvement in different cancers. In colon cancer cells overexpression of HDAC4 facilitated growth by way of downregulating p21 (Wilson et al. 2008). This HDAC was found to be highly expressed in both gastric cancer cell lines and tissues. HDAC4 inhibited p21 expression, facilitated proliferation and enhanced ATP levels. The HDAC4 induced effects were reversed by knockdown of p53 suggesting its key role in mediating HDAC4 signalling (Kang et al. 2014). Hypoxia-inducible factor 1 alpha, the vital component of HIF-1 transcriptional complex, modulates cancer development. Its stability is positively related to expression of HDAC4 as HDAC4 shRNA destabilizes hypoxia-inducible factor 1 alpha. Overexpression of this HDAC makes the cancer cells resistant to docetaxel chemotherapy (Geng et al. 2011). While inhibition of HDAC4 favoured cisplatin cytotoxicity its overexpression makes cells resistant to cisplatin. HDAC4 overexpression has been seen in gastric tumours in comparison to healthy tissues. Thus cisplatin resistance in gastric cancer cells may have strong link with HDAC4 expression (Spaety et al. 2019). In bladder cancer model enhanced expression of various HDACs (HDAC4, 7 and 9) has been observed. Similar situation has been noted in invasive clinical specimens (Buckwalter et al. 2019).

Frequent dysregulation of HDAC5 has been observed in human malignancies. Lung cancer cells and tissues are associated with escalated expression of HDAC5. While its overexpression facilitated proliferation of lung cancer cells, invasion and attenuated apoptosis, knockdown of this HDAC reversed these effects (Zhong et al. 2018). Cross-talk between HDACs and lysine specific demethylase 1 (LSD1) promotes proliferation of breast cancer cells (Vasilatos et al. 2013). Positive correlation between HDAC5 and LSD1 has been observed in human breast cancer cell lines. Primary breast cancer specimens showed substantial overexpression of these enzymes as compared to corresponding normal tissues (Cao et al. 2017a). These findings suggest that targeting the cross-talk between different epigenetic players may be an exciting therapeutic strategy. Indeed this strategy has been tested against breast cancer cells using Sulforaphane (HDAC inhibitor) either alone or in combination with LSD1 inhibitor (Cao et al. 2018; Ganai 2016). HDAC5 overexpression has been observed in breast cancers of very young women less than 35 years of age (Oltra et al. 2020).

HDAC7, another Class IIa HDAC has significant contribution in certain cancers. The expression profile of various HDACs belonging to different classes has been tested in surgically removed pancreatic tissues. Out of 11 pancreatic adenocarcinomas, 9 showed the marked increase in the expression of HDAC7 at message level. Further on western blotting similar trend was seen for protein levels of HDAC7 (Ouaïssi et al. 2008). Tumours are composed of heterogeneous cells among which only stem-like cells have regenerating capacity under in vivo conditions. These stem cells are not sensitive to conventional therapies and are thus crucial in metastasis and relapse. HDACs are involved in the cancer stem cell phenotype. In cancer stem cells specific overexpression of HDAC1 and HDAC7 has been observed in comparison to non-stem-tumour cells. Further it has been proved that only HDAC7 is enough for accentuating cancer stem cell phenotype (Witt et al. 2017). Expression of 12 HDAC genes has been checked in samples of childhood acute lymphoblastic leukaemia (ALL). Compared to normal bone marrow samples these samples showed escalated expression of five HDACs including HDAC6 and HDAC7 (Moreno et al. 2010). However, in certain malignancies like pro-B acute lymphoblastic leukaemia and Burkitt lymphoma alleviated expression of HDAC7 has been observed (Barneda-Zahonero et al. 2015). By way of obstructing Stat3 activation, HDAC7 facilitates lung tumorigenesis and may serve as a promising therapeutic target for effective lung cancer therapy (Lei et al. 2017).

Another member of Class IIa also shows aberrant expression in multiple cancers. Higher expression of HDAC9 has been observed in triple negative breast cancer cell lines as compared to non-triple negative breast cancer lines (Salgado et al. 2018). HDAC9 overexpression in pancreatic ductal adenocarcinoma cells facilitated their proliferation and migration. On HDAC9 silencing, these effects were reversed further certifying the involvement of HDAC9 in tumour progression (Li et al. 2020). Its overexpression has also been reported in breast cancer cells and tissues compared to controls. Lymph node metastasis and HDAC9 expression were found to be positively correlated. Further the expression of HDAC9 was found to be inversely proportional to overall survival (Huang et al. 2018). In breast cancer cells HDAC9 declines the expression of oestrogen receptor alpha and hampers its transcriptional activity. Higher messenger RNA levels of this HDAC were found in hydroxyl-tamoxifen resistant and in breast tumour cell lines devoid of oestrogen receptor alpha. Cells overexpressing this HDAC were found to be resistant to hydroxyl-tamoxifen induced antiproliferative effects comparative to parent breast cancer cells. These findings suggest HDAC9 as a promising target for sensitizing therapeutically resistant breast cancer cells to chemotherapeutic agents (Linares et al. 2019). Marked upregulation of HDAC9 has also been observed in gastric cancer cells and tissues. HDAC9 depletion mitigated cell growth, colony formation. In addition to this induction of apoptosis and cell cycle blockade was seen under HDAC9 depleted condition. Reduction in tumour growth and enhanced cisplatin sensitivity was seen on siRNA mediated knockdown of HDAC9 (Xiong et al. 2019). While HDAC9 elevation is associated with gastric cancer cells and tissues, downregulation of miR-383-5p has been noted. Studies have shown that miR-383-5p acts as post-transcriptional regulator of HDAC9 (Table 4.2). Its expression has been seen to be inversely proportional to HDAC9 expression, in gastric cancer tissues thereby suggesting HDAC9 as a propitious target for therapeutic intervention in gastric cancer (Xu et al. 2019).

4.3 Involvement of Class IIb HDACs in Cancer

Aberrant expression Class IIb HDACs (HDAC6/HDAC10) fuels cancer. In breast cancer patients having small tumours substantial overexpression of HDAC6 at messenger RNA level has been observed (Zhang et al. 2004). HDAC6 has been

Table 4.2 Aberrant expression of Class IIa, Class IIb and Class IV HDACs in distinct cancers

Name of HDAC	Expression status	Type of cancer	Literature evidence
HDAC4	Upregulated	Colon	Wilson et al. (2008)
		Gastric	Kang et al. (2014)
		Kidney	Geng et al. (2011)
		Gastric	Spaety et al. (2019)
		Bladder	Buckwalter et al. (2019)
HDAC5	Upregulated	Lung	Zhong et al. (2018)
		Breast	Cao et al. (2017a)
		Breast	Oltra et al. (2020)
HDAC7	Upregulated	Pancreatic	Ouaïssi et al. (2008)
		Childhood ALL	Moreno et al. (2010)
		Pro-B ALL/Burkitt lymphoma	Barneda-Zahonero et al. (2015)
		Lung	Lei et al. (2017)
HDAC9	Upregulated	Breast	Salgado et al. (2018)
		PDA	Li et al. (2020)
		Breast	Huang et al. (2018)
		Breast	Linares et al. (2019)
		Gastric	Xiong et al. (2019)
		Gastric	Xu et al. (2019)
HDAC6	Upregulated	Breast	Zhang et al. (2004)
		–	Lee et al. (2008)
		Ovarian carcinoma	Bazzaro et al. (2008)
		Oral squamous cell carcinoma	Sakuma et al. (2006)
		Endometrial	Zheng et al. (2020)
		Lung adenocarcinoma	Wang et al. (2016)
		Hepatocellular carcinoma	Kanno et al. (2012)
		Colon	Zhang et al. (2019b)
HDAC10	Upregulated	Lung	Yang et al. (2014b)
		Colon	Tao et al. (2017)
	Downregulated	Lung	Osada et al. (2004)
	Upregulated	MCL and CLL	Powers et al. (2016)
		Ovarian	Islam et al. (2017)
HDAC11	Upregulated	Basal-like breast cancer	Yi et al. (2019)
		NSCLC	Bora-Singhal et al. (2020)
		Liver cancer	Gong et al. (2019)
		Myeloproliferative neoplasm	Yue et al. (2020)

linked to oncogene-induced transformation and tumorigenesis. Cells like fibroblasts having deficiency of this enzyme are highly resistant to this transformation (Lee et al. 2008). In various cancer cell lines and other tumour models escalated expression of

HDAC6 has been seen. Its expression levels were found to be higher in ovarian carcinomas of both low- and high-grade ones as compared to benign epithelium (Aldana-Masangkay and Sakamoto 2011; Bazzaro et al. 2008). This HDAC was found to be upregulated both at mRNA and protein level in oral squamous cell carcinoma cell lines in comparison to normal oral keratinocytes (Sakuma et al. 2006). Experimental evidences suggest the involvement of this HDAC in endometrial cell proliferation in addition to metastasis and invasion. This effect of HDAC6 has been proved to be mediated by PTEN/AKT/mTOR pathway (Zheng et al. 2020). HDAC6 acts as a critical regulator of multiple signalling pathways related to cancer. Its increased expression has been found to be associated with lung adenocarcinoma as higher expression of this HDAC was found on molecular examination of adenocarcinoma cell lines. Further the expression of HDAC6 is inversely proportional to prognosis of patients having this cancer. This HDAC6 when present in enhanced levels not only facilitates the progression of lung adenocarcinoma cells but also makes them resistant to Gefitinib (epidermal growth factor receptor (EGFR) tyrosine kinase inhibitor) (Wang et al. 2016). HDAC6 has implications in invasion and metastasis of hepatocellular carcinoma. The expression of this HDAC in hepatocellular carcinoma cell lines was found to be higher in comparison to primary hepatocyte cultures (Kanno et al. 2012). This HDAC has been observed to be highly expressed in colon cancer where it facilitates colon cancer cell proliferation and migration. Its higher levels are associated with decreased overall survival. This effect of HDAC6 has been attributed to activation of MAPK/ERK pathway (Zhang et al. 2019b).

Anomalous expression of HDAC10 has also been observed in various cancers. This HDAC plays a critical role in restraining cervical cancer metastasis and its expression level was found to be substantially lesser in patients showing lymph node metastasis. In cervical cancer cells, forced expression of this HDAC lessened not only cell motility but also invasiveness (Song et al. 2013). Lung cancer growth has been found to be facilitated by HDAC10 as revealed by functional analysis. This effect of HDAC10 seems to be mediated by AKT phosphorylation. Its knockdown induced not only cell cycle arrest but also apoptosis by way of substantially alleviating the AKT phosphorylation (Yang et al. 2014b). While HDAC10 expression in colon cancer tissues has been linked to good prognosis, its expression in paracarcinoma tissues suggests poor prognosis (Tao et al. 2017). Patients having lung cancer showed decreased expression of this HDAC (Jin et al. 2014; Osada et al. 2004). Low levels of this HDAC maintain the viability of aggressive mantle cell lymphoma (MCL) in addition to chronic lymphocytic leukaemia (CLL)-cells and overexpression of HDAC10 culminated in induction of cell death (Powers et al. 2016). Higher levels of HDAC10 may make ovarian tumours resistant to cisplatin as primary ovarian carcinoma cells responded better to cisplatin on HDAC inhibitor treatment (Table 4.2) (Islam et al. 2017).

4.4 Class IV HDACs in Cancer Progression

This class includes the only HDAC known as HDAC11. Aberrant expression of this HDAC has strong cross-talk with a variety of cancers. Lower expression of this HDAC occurs in basal-like breast cancer cells. Under conditions of in vitro, HDAC11 overexpression attenuated their invasion and under in vivo conditions hampered metastasis (Yi et al. 2019). In comparison to corresponding healthy tissues this zinc-dependent enzyme (HDAC11) is overexpressed in a variety of carcinomas. Depleting the levels of this HDAC is enough to induce death in colon, prostate, ovarian and breast cancer cell lines. On the whole these findings suggest its crucial role in survival of cancer cells and thus may serve as a promising epi-target for therapeutic intervention (Deubzer et al. 2013). High levels of this HDAC have been related to poor patient outcome. Further it is quite well known that Sox2 expression is mandatory for upkeeping of cancer stem cells (CSCs). Depletion of HDAC11 substantially attenuated self-renewal of these cells (CSCs) from non-small cell lung cancer and reduced Sox2 expression. Hampering of Sox2 expression by HDAC11 has been found to be mediated by Gli1 (Bora-Singhal et al. 2020). High levels of HDAC11 are expressed by liver cancer cells and this has been found to have negative correlation with p53 expression (Table 4.2). This HDAC on forming the complex with transcription factor of p53 (Egr1) deacetylates it thereby preventing the transcription of p53. While overexpression of this HDAC suppresses apoptosis, inhibition of its expression confers reverse effect (Gong et al. 2019). HDAC11 has strong cross-talk with oncogenic JAK2-driven proliferation and survival of myeloproliferative neoplasm cells (Yue et al. 2020).

4.5 Role of Sirtuins in Cancer

Sirtuins may have contrary roles in cancer. This is because on one side they support certain cellular processes favourable for cancer onset while on the other hand they participate in processes which are meant for suppressing cancer. For instance, they maintain genomic stability, offer protection against oxidative stress and thus may have cancer preventive function (Mostoslavsky et al. 2006; Wang et al. 2008). On the flip side these NAD^+-dependent HDACs facilitate cell survival under conditions of stress and thus may promote tumorigenesis (Bosch-Presegué and Vaquero 2011; Imai et al. 2000). Sirtuins, the mechanistically distinct HDACs, on the whole are seven in number ranging from SIRT1-SIRT7 (Blander and Guarente 2004; Ganai 2014). SIRT1 (Silent information regulation factor 1) has both histone and a variety of non-histone targets which are linked to cancer. This HDAC can either act as tumour promoter or tumour suppressor based on its targets in particular signalling pathways. In breast cancer cells and tissues, upregulation of SIRT1 has been reported. This higher expression correlated with lymph node metastasis and tumour size also. Enhanced expression of this HDAC facilitated breast cancer growth while its depletion curbed these phenotypes. SIRT1 induced effects on cell proliferation were found to be partly mediated by AKT (Jin et al. 2018). In general SIRT1, in oral cancer has tumour-suppressive function thus agonists of this enzyme may prove as effective strategy for treating the precancerous oral lesions which later on transform into oral cancer (Islam et al. 2019). Upregulated levels of SIRT1 were noted in bladder cancer tissues. This HDAC seemed to maintain the proliferation and viability of bladder cancer cells and in SIRT1 knockdown condition both viability and proliferation was attenuated (Hu et al. 2017).

SIRT2, another member of Sirtuin family also contributes significantly to different cancers. In hepatocellular carcinoma and above 50% of the hepatocellular carcinoma tissues this HDAC has been observed to be upregulated. SIRT2 upregulation in primary hepatocellular tumours correlated with shorter overall survival and more advanced stage. Its suppression by short hairpin RNA substantially impeded motility and invasiveness (Chen et al. 2013). SIRT2 has shown tumour suppressor function in breast cancer cells. This HDAC by way of peroxiredoxin (an antioxidant protein) destabilization facilitates reactive oxygen species induced cell cytotoxicity in breast cancer cells (Fiskus et al. 2016). SIRT2 in serous ovarian carcinoma showed marked downregulation as compared to ovarian surface epithelium. This suggests that SIRT2 has tumour-attenuating function in ovarian cells thereby use of SIRT2 agonists may help in ameliorating serous ovarian carcinoma (Du et al. 2017). The expression level of this HDAC was relatively higher in hepatocellular carcinoma tissues in comparison to normal liver tissues. Its suppression alleviated proliferation and tumour cell growth (Xie et al. 2011).

Another member of Sirtuin family, SIRT3 also has implications in different cancers. Expression of this HDAC is low in breast cancers which favour the expression of HIF1α target genes by stabilizing this transcription factor (Finley et al. 2011). Its expression was also high in gastric tumour tissues especially of intestinal type gastric cancer (Cui et al. 2015). Downregulation of SIRT3 in these cells blocked oral squamous cell carcinoma cell growth, proliferation and sensitized them to cytotoxic effects of radio and chemotherapeutic agents. Thus it is quite evident that survival of such cells highly reliant on SIRT3 signalling (Alhazzazi et al. 2011). SIRT3 overexpression in human bladder cancer rescued cell growth arrest induced by p53 (Li et al. 2010). Its overexpression in breast cancer (lymph node-positive) suggests its role in advanced stages of this disease (Ashraf et al. 2006). Further in non-small cell lung cancer tissues high levels of SIRT3 were observed. Overall survival time of such patients was found to be shorter. Squamous cell carcinoma type non-small cell lung cancer tissues showed more elevated expression of SIRT3 as compared to others (Xiong et al. 2017).

Another member of Sirtuin family, namely SIRT4, is primarily involved in metabolism and genome stability. Unlike other Sirtuins, it has no NAD^+-dependent deacetylase function but does show ADP-ribosylase activity. This HDAC has tumour suppressor activity, restraining invasion, blocking cell cycle, attenuating proliferation and migration (Jeong et al. 2013). Although SIRT4 has been linked to tumorigenesis, its low expression is reported in various cancers including breast, colon, ovarian, bladder and stomach cancer (Carafa et al. 2019; Huang and Zhu 2018). SIRT4 enhanced the E-cadherin expression, lessened proliferation and invasion of colorectal cancer cells (Miyo et al. 2015). Breast cancer cells showed significantly enhanced protein levels of SIRT4 suggesting its implications in this cancer (Geng et al. 2017). In Burkitt's lymphoma, this HDAC impeded MYC-triggered lymphomagenesis, by impairing glutamine metabolism in mitochondria (Jeong et al. 2014). Proliferation of gastric cancer cells was hampered by enhanced expression of SIRT4 by way of cell cycle blockade (Hu et al. 2019).

SIRT5 like SIRT3 and SIRT4 is predominantly present in mitochondrial matrix (Kumar and Lombard 2015). It has been found that mitochondrial sirtuins in general and SIRT5 in particular are more expressed in hepatocellular carcinoma cell lines in comparison to normal liver cell lines. While knockdown of this HDAC attenuated hepatocellular carcinoma cell proliferation its overexpression showed contrary effect. Its knockdown induced apoptosis in hepatocellular carcinoma cells by way of mitochondrial pathway (Zhang et al. 2019a). Another study has also proved the role of SIRT5 in hepatocellular carcinoma signalling. Inhibition of this sirtuin substantially reduced proliferation and invasion of hepatocellular carcinoma cells whereas its (SIRT5) overexpression triggered proliferation and invasion in these cells. SIRT5 incited proliferation and invasion in hepatocellular carcinoma cells was found to be mediated by downregulation of E2F transcription 1 (Chang et al. 2018). Markedly increased expression of SIRT5, by way of enhancing the phosphorylation status of c-MET (hepatocyte growth factor receptor) in gastric cancer cells, facilitated their migration. Thus by altering SIRT5, the rate of migration of gastric cancer cells can be regulated thereby suggesting this sirtuin as a candidate target for vanquishing this cancer (Wu and Fang 2019). Further studies involving AML cell lines showed that SIRT5 promotes their growth and majority of these cell lines on SIRT5 knockdown showed attenuated growth, colony formation and enhanced apoptosis (Yan et al. 2018). Studies have proved that SIRT5 plays a crucial role in the regulation of cancer glutaminolysis by way of GLUD1 (glutamate dehydrogenase 1). By augmenting glutaminolysis, SIRT5 facilitates colorectal cancer carcinogenesis (Wang et al. 2018).

Several studies have related the depletion of SIRT6 with tumour progression in case of colorectal, ovarian, lung and hepatocellular cancer (Desantis et al. 2018). It has been demonstrated through in vivo study that deficiency of SIRT6 facilitates tumour growth and invasion as well. This sirtuin by way of repressing hypoxia-inducible factor 1-alpha (HIF-1α), in cancer cells, hampers glycolytic metabolism (Sebastián et al. 2012). Through upregulation of let-7 microRNA negative regulator Lin28b, inactivation of SIRT6 accentuated progression and metastasis of pancreatic ductal adenocarcinoma cells (Kugel et al. 2016). It has been proved in colon cancer model that SIRT6 degradation induced by dysregulated ubiquitin-specific peptidase USP10 function facilitates tumorigenesis (Lin et al. 2013). Significant reduction of SIRT6 has been noted in ovarian cancer cells and expression of this sirtuin, by way of downregulating Notch 3 expression attenuated proliferation of these cells (Zhang et al. 2015a). In cancer cells aerobic glycolysis is regulated by SIRT6. Loss of this HDAC results in tumour formation without awakening oncogenes. Further transformed SIRT6 deficient cells showed enhanced glycolysis and growth of tumour. This suggests the crucial role of SIRT6 not only in tumour establishment but also in upkeeping. Thus tumour reducing activity of SIRT6 is mediated by hampering cancer metabolism (Sebastián et al. 2012). Studies have shown that doxorubicin induced cytotoxicity in liver cancer cells is mediated by SIRT6 downregulation (Hu et al. 2018). Although SIRT6 knockdown markedly reduced migration and invasion of ovarian cancer cells (OVCAR3 and OVCAR5), no effect was seen on their proliferation. SIRT6 induced increase in invasiveness and triggering of epithelial to mesenchymal transition was mainly mediated by β-catenin. This conclusion has been drawn as knockdown of β-catenin impeded SIRT6 provoked signalling (Bae et al. 2018).

SIRT7, another member of Class III HDACs, plays a crucial role in metastasis. Inactivation of this sirtuin drastically impedes metastasis of cancer cells. SIRT7 gets recruited to E-cadherin (epithelial cadherin) promoter by SIRT1 and this cooperation leads to repression of epithelial cadherin (Malik et al. 2015). For upkeeping malignant phenotype SIRT7 mediated deacetylation of H3K18 seems to be critical (Malik et al. 2015). Overexpression of SIRT7 also facilitates colorectal cancer (Yu et al. 2014). In oral squamous cell carcinoma cells increased expression of this sirtuin hampers epithelial to mesenchymal transition. While this overexpression enhanced epithelial cadherin, downregulation of MMP7 and vimentin was noted. These effects have been ascribed to facilitation of SMAD4 deacetylation by the enhanced levels of SIRT7 (Carafa et al. 2019; Li et al. 2018b). Enhanced expression of SIRT7 occurs in human gastric cancers. Its expression significantly correlated with metastasis, disease stage and tumour size. Through downregulation of miR-34a by epigenetic mechanism (H3K18 deacetylation) SIRT7 inhibited cellular apoptosis. These findings suggest that reinstating expression of miR-34a or pharmacological intervention against SIRT7 may prove as an excellent therapeutic strategy for circumventing gastric cancer (Zhang et al. 2015b). Human prostate cancer progression has also been found to be reliant on SIRT7 expression (Ding et al. 2020). In human glioma tissues higher expression of this HDAC has been observed. Directly proportionality exists between glioma malignancy and SIRT7 expression. Knockdown of this enzyme decreased the phosphorylation of signal transducer and activator of transcription 3 (STAT3) in glioma cells (Mu et al. 2019). Oncogenic potential of SIRT7 has also been demonstrated in ovarian cancer cell lines. While its downregulation was associated with substantial reduction in ovarian cancer growth, enhanced apoptosis and supressed colony formation, its upregulation facilitated cancer cell migration (Wang et al. 2015). Higher expression of SIRT7 has been demonstrated in prostate cancer tumours compared to corresponding controls. Knockdown of SIRT7 in a couple of prostate cancer cell lines (androgen independent) hampered their migration. Further its overexpression in less aggressive prostate cancer cell line triggered invasion (Massa et al. 2017). In non-small cell lung cancer cell lines elevated expression of SIRT7 has proved through expression quantification methods including western blot analysis. Post-transcriptional silencing of SIRT7 markedly restrained the growth of these cells and resulted in induction of apoptosis. Further it has been proved that, by way of SIRT7 suppression, microRNA-3666 attenuates growth of non-small cell lung cancer cells (Shi et al. 2016). Again higher expression of this HDAC has been reported in cholangiocarcinoma tissues and cell lines (Table 4.3). On clinical analysis it has become quite evident that SIRT7

Table 4.3 Abnormal expression of sirtuins in different cancers

Sirtuin member	Expression profile	Name of cancer	Literature proof
SIRT1	Upregulated	Breast cancer	Jin et al. (2018)
	Downregulated	Oral cancer	Islam et al. (2019)
	Upregulated	Bladder cancer	Hu et al. (2017)
SIRT2	Upregulated	Hepatocellular carcinoma	Chen et al. (2013)
	Downregulated	Breast cancer	Fiskus et al. (2016)
		Serous ovarian carcinoma	Du et al. (2017)
	Upregulated	Hepatocellular carcinoma	Xie et al. (2011)
SIRT3	Downregulated	Breast cancer	Finley et al. (2011)
	Upregulated	Gastric cancer	Cui et al. (2015)
		Oral squamous cell carcinoma	Alhazzazi et al. (2011)
		Bladder cancer	Li et al. (2010)
		Breast cancer	Ashraf et al. (2006)
		NSCLC	Xiong et al. (2017)
SIRT4	Downregulated	Colorectal cancer	Miyo et al. (2015)
	Upregulated	Breast cancer	Geng et al. (2017)
	Downregulated	Burkitt's lymphoma	Jeong et al. (2014)
		Gastric cancer	Hu et al. (2019)
SIRT5	Upregulated	Hepatocellular carcinoma	Zhang et al. (2019a)
		Hepatocellular carcinoma	Chang et al. (2018)
		Gastric cancer	Wu and Fang (2019)
		AML	Yan et al. (2018)
		Colorectal cancer	Wang et al. (2018)
SIRT6	Downregulated	Pancreatic ductal adenocarcinoma	Kugel et al. (2016)
		Colon cancer	Lin et al. (2013)
		Ovarian cancer	Zhang et al. (2015a)
SIRT7	Upregulated	Epithelial prostate carcinomas	Malik et al. (2015)
		Colorectal cancer	Yu et al. (2014)
	Downregulated	Oral squamous cell carcinoma	Li et al. (2018b)
	Upregulated	Gastric cancer	Zhang et al. (2015b)
		Prostate cancer	Ding et al. (2020)
		Glioma	Mu et al. (2019)
		Ovarian cancer	Wang et al. (2015)
		Prostate cancer	Massa et al. (2017)
		NSCLC	Shi et al. (2016)
		Cholangiocarcinoma	Li et al. (2018a)

expression is directly proportional to tumour size and advanced stage (Li et al. 2018a).

Till here I have discussed thoroughly the involvement of Class I, Class IIa, Class IIb, Class III and Class IV HDACs in various monotonous cancers. Further, the expression profile of these HDAC classes in these cancers has been taken into account. The different signalling mechanisms provoked by these aberrantly expressed HDACs for promoting tumour onset and advancement has been elucidated. From this discussion it is clear that HDACs show differential expression in various cancers. Even a single HDAC, for instance, SIRT2 shows different functionality in two different cancers. Thus in the next chapter I will discuss the different mechanisms by way of which HDACs promote cancer formation.

References

Ahn MY (2018) HDAC inhibitor apicidin suppresses murine oral squamous cell carcinoma cell growth in vitro and in vivo via inhibiting HDAC8 expression. Oncol Lett 16:6552–6560

Ahn MY, Yoon JH (2017) Histone deacetylase 8 as a novel therapeutic target in oral squamous cell carcinoma. Oncol Rep 37:540–546

Aldana-Masangkay GI, Sakamoto KM (2011) The role of HDAC6 in cancer. J Biomed Biotechnol 2011:875824–875824

Alhazzazi TY, Kamarajan P, Joo N, Huang JY, Verdin E, D'Silva NJ, Kapila YL (2011) Sirtuin-3 (SIRT3), a novel potential therapeutic target for oral cancer. Cancer 117:1670–1678

Ashraf N, Zino S, MacIntyre A, Kingsmore D, Payne AP, George WD, Shiels PG (2006) Altered sirtuin expression is associated with node-positive breast cancer. Br J Cancer 95:1056–1061

Bae JS, Noh SJ, Kim KM, Park S-H, Hussein UK, Park HS, Park B-H, Ha SH, Lee H, Chung MJ, Moon WS, Cho DH, Jang KY (2018) SIRT6 is involved in the progression of ovarian carcinomas via β-catenin-mediated epithelial to mesenchymal transition. Front Oncol 8

Barneda-Zahonero B, Collazo O, Azagra A, Fernández-Duran I, Serra-Musach J, Islam AB, Vega-García N, Malatesta R, Camós M, Gómez A, Román-González L, Vidal A, López-Bigas N, Villanueva A, Esteller M, Parra M (2015) The transcriptional repressor HDAC7 promotes apoptosis and c-Myc downregulation in particular types of leukemia and lymphoma. Cell Death Dis 6:e1635

Bazzaro M, Lin Z, Santillan A, Lee MK, Wang MC, Chan KC, Bristow RE, Mazitschek R, Bradner J, Roden RB (2008) Ubiquitin proteasome system stress underlies synergistic killing of ovarian cancer cells by bortezomib and a novel HDAC6 inhibitor. Clin Cancer Res 14:7340–7347

Blander G, Guarente L (2004) The Sir2 family of protein deacetylases. Annu Rev Biochem 73:417–435

Bora-Singhal N, Mohankumar D, Saha B, Colin CM, Lee JY, Martin MW, Zheng X, Coppola D, Chellappan S (2020) Novel HDAC11 inhibitors suppress lung adenocarcinoma stem cell self-renewal and overcome drug resistance by suppressing Sox2. Sci Rep 10:4722

Bosch-Presegué L, Vaquero A (2011) The dual role of sirtuins in cancer. Genes Cancer 2:648–662

Buckwalter JM, Chan W, Shuman L, Wildermuth T, Ellis-Mohl J, Walter V, Warrick JI, Wu XR, Kaag M, Raman JD, DeGraff DJ (2019) Characterization of histone deacetylase expression within in vitro and in vivo bladder cancer model systems. Int J Mol Sci 20

Cao C, Vasilatos SN, Bhargava R, Fine JL, Oesterreich S, Davidson NE, Huang Y (2017a) Functional interaction of histone deacetylase 5 (HDAC5) and lysine-specific demethylase 1 (LSD1) promotes breast cancer progression. Oncogene 36:133–145

Cao L-L, Song X, Pei L, Liu L, Wang H, Jia M (2017b) Histone deacetylase HDAC1 expression correlates with the progression and prognosis of lung cancer: a meta-analysis. Medicine (Baltimore) 96:e7663–e7663

Cao C, Wu H, Vasilatos SN, Chandran U, Qin Y, Wan Y, Oesterreich S, Davidson NE, Huang Y (2018) HDAC5–LSD1 axis regulates antineoplastic effect of natural HDAC inhibitor sulforaphane in human breast cancer cells. Int J Cancer 143:1388–1401

Carafa V, Altucci L, Nebbioso A (2019) Dual tumor suppressor and tumor promoter action of sirtuins in determining malignant phenotype. Front Pharmacol 10

Chakrabarti A, Oehme I, Witt O, Oliveira G, Sippl W, Romier C, Pierce RJ, Jung M (2015) HDAC8: a multifaceted target for therapeutic interventions. Trends Pharmacol Sci 36:481–492

Chang L, Xi L, Liu Y, Liu R, Wu Z, Jian Z (2018) SIRT5 promotes cell proliferation and invasion in hepatocellular carcinoma by targeting E2F1. Mol Med Rep 17:342–349

Chen J, Chan AWH, To K-F, Chen W, Zhang Z, Ren J, Song C, Cheung Y-S, Lai PBS, Cheng S-H, Ng MHL, Huang A, Ko BCB (2013) SIRT2 overexpression in hepatocellular carcinoma mediates epithelial to mesenchymal transition by protein kinase B/glycogen synthase kinase-3β/β-catenin signaling. Hepatology 57:2287–2298

Conte M, Dell'Aversana C, Benedetti R, Petraglia F, Carissimo A, Petrizzi VB, Maria D'Arco A, Abbondanza C, Nebbioso A, Altucci L (2014) HDAC2 deregulation in tumorigenesis is causally connected to repression of immune modulation and defense escape. Oncotarget 6:886–901

Cui Y, Qin L, Wu J, Qu X, Hou C, Sun W, Li S, Vaughan ATM, Li JJ, Liu J (2015) SIRT3 enhances glycolysis and proliferation in SIRT3-expressing gastric cancer cells. PLoS One 10:e0129834

Desantis V, Lamanuzzi A, Vacca A (2018) The role of SIRT6 in tumors. Haematologica 103:1–4

Deubzer HE, Schier MC, Oehme I, Lodrini M, Haendler B, Sommer A, Witt O (2013) HDAC11 is a novel drug target in carcinomas. Int J Cancer 132:2200–2208

Ding M, Jiang C-Y, Zhang Y, Zhao J, Han B-M, Xia S-J (2020) SIRT7 depletion inhibits cell proliferation and androgen-induced autophagy by suppressing the AR signaling in prostate cancer. J Exp Clin Cancer Res 39:28

Du Y, Wu J, Zhang H, Li S, Sun H (2017) Reduced expression of SIRT2 in serous ovarian carcinoma promotes cell proliferation through disinhibition of CDK4 expression. Mol Med Rep 15:1638–1646

Finley LWS, Carracedo A, Lee J, Souza A, Egia A, Zhang J, Teruya-Feldstein J, Moreira PI, Cardoso SM, Clish CB, Pandolfi PP, Haigis MC (2011) SIRT3 opposes reprogramming of cancer cell metabolism through HIF1α destabilization. Cancer Cell 19:416–428

Fiskus W, Coothankandaswamy V, Chen J, Ma H, Ha K, Saenz DT, Krieger SS, Mill CP, Sun B, Huang P, Mumm JS, Melnick AM, Bhalla KN (2016) SIRT2 deacetylates and inhibits the peroxidase activity of peroxiredoxin-1 to sensitize breast cancer cells to oxidant stress-inducing agents. Cancer Res 76:5467–5478

Ganai SA (2014) HDAC inhibitors entinostat and suberoylanilide hydroxamic acid (SAHA): the ray of hope for cancer therapy. In: Wells RD, Bond JS, Klinman J, Masters BSS, Bell E (eds) Molecular life sciences: an encyclopedic reference. Springer, New York, pp 1–16

Ganai SA (2015) Strategy for enhancing the therapeutic efficacy of histone deacetylase inhibitor dacinostat: the novel paradigm to tackle monotonous cancer chemoresistance. Arch Pharm Res

Ganai SA (2016) Histone deacetylase inhibitor sulforaphane: the phytochemical with vibrant activity against prostate cancer. Biomedicine & pharmacotherapy 81:250–257

Ganai SA (2018) Histone deacetylase inhibitors modulating non-epigenetic players: the novel mechanism for small molecule based therapeutic intervention. Curr Drug Targets 19:593–601

Geng H, Harvey CT, Pittsenbarger J, Liu Q, Beer TM, Xue C, Qian DZ (2011) HDAC4 protein regulates HIF1alpha protein lysine acetylation and cancer cell response to hypoxia. J Biol Chem 286:38095–38102

Geng P, Zhang Y, Liu X, Zhang N, Liu Y, Liu X, Lin C, Yan X, Li Z, Wang G, Li Y, Tan J, Liu D-X, Huang B, Lu J (2017) Automethylation of protein arginine methyltransferase 7 and its impact on breast cancer progression. FASEB J 31:2287–2300

Gong D, Zeng Z, Yi F, Wu J (2019) Inhibition of histone deacetylase 11 promotes human liver cancer cell apoptosis. Am J Transl Res 11:983–990

Halkidou K, Gaughan L, Cook S, Leung HY, Neal DE, Robson CN (2004) Upregulation and nuclear recruitment of HDAC1 in hormone refractory prostate cancer. Prostate 59:177–189

Hu Q, Wang G, Peng J, Qian G, Jiang W, Xie C, Xiao Y, Wang X (2017) Knockdown of SIRT1 suppresses bladder cancer cell proliferation and migration and induces cell cycle arrest and antioxidant response through FOXO3a-mediated pathways. Biomed Res Int 2017:3781904
Hu JQ, Deng F, Hu XP, Zhang W, Zeng XC, Tian XF (2018) Histone deacetylase SIRT6 regulates chemosensitivity in liver cancer cells via modulation of FOXO3 activity. Oncol Rep 40:3635–3644
Hu Y, Lin J, Lin Y, Chen X, Zhu G, Huang G (2019) Overexpression of SIRT4 inhibits the proliferation of gastric cancer cells through cell cycle arrest. Oncol Lett 17:2171–2176
Huang G, Zhu G (2018) Sirtuin-4 (SIRT4), a therapeutic target with oncogenic and tumor-suppressive activity in cancer. Onco Targets Ther 11:3395–3400
Huang Y, Jian W, Zhao J, Wang G (2018) Overexpression of HDAC9 is associated with poor prognosis and tumor progression of breast cancer in Chinese females. Onco Targets Ther 11:2177–2184
Imai S, Armstrong CM, Kaeberlein M, Guarente L (2000) Transcriptional silencing and longevity protein Sir2 is an NAD-dependent histone deacetylase. Nature 403:795–800
Islam MM, Banerjee T, Packard CZ, Kotian S, Selvendiran K, Cohn DE, Parvin JD (2017) HDAC10 as a potential therapeutic target in ovarian cancer. Gynecol Oncol 144:613–620
Islam S, Abiko Y, Uehara O, Chiba I (2019) Sirtuin 1 and oral cancer. Oncol Lett 17:729–738
Jeong SM, Xiao C, Finley LW, Lahusen T, Souza AL, Pierce K, Li YH, Wang X, Laurent G, German NJ, Xu X, Li C, Wang RH, Lee J, Csibi A, Cerione R, Blenis J, Clish CB, Kimmelman A, Deng CX, Haigis MC (2013) SIRT4 has tumor-suppressive activity and regulates the cellular metabolic response to DNA damage by inhibiting mitochondrial glutamine metabolism. Cancer Cell 23:450–463
Jeong SM, Lee A, Lee J, Haigis MC (2014) SIRT4 protein suppresses tumor formation in genetic models of Myc-induced B cell lymphoma. J Biol Chem 289:4135–4144
Jiang N, Niu G, Pan YH, Pan W, Zhang MF, Zhang CZ, Shen H (2020) CBX4 transcriptionally suppresses KLF6 via interaction with HDAC1 to exert oncogenic activities in clear cell renal cell carcinoma. EBioMedicine 53:102692
Jin Z, Jiang W, Jiao F, Guo Z, Hu H, Wang L, Wang L (2014) Decreased expression of histone deacetylase 10 predicts poor prognosis of gastric cancer patients. Int J Clin Exp Pathol 7:5872–5879
Jin X, Wei Y, Xu F, Zhao M, Dai K, Shen R, Yang S, Zhang N (2018) SIRT1 promotes formation of breast cancer through modulating Akt activity. J Cancer 9:2012–2023
Jung KH, Noh JH, Kim JK, Eun JW, Bae HJ, Xie HJ, Chang YG, Kim MG, Park H, Lee JY, Nam SW (2012) HDAC2 overexpression confers oncogenic potential to human lung cancer cells by deregulating expression of apoptosis and cell cycle proteins. J Cell Biochem 113:2167–2177
Kamarulzaman NS, Dewadas HD, Leow CY, Yaacob NS, Mokhtar NF (2017) The role of REST and HDAC2 in epigenetic dysregulation of Nav1.5 and nNav1.5 expression in breast cancer. Cancer Cell Int 17:74
Kang ZH, Wang CY, Zhang WL, Zhang JT, Yuan CH, Zhao PW, Lin YY, Hong S, Li CY, Wang L (2014) Histone deacetylase HDAC4 promotes gastric cancer SGC-7901 cells progression via p21 repression. PLoS One 9:e98894
Kanno K, Kanno S, Nitta H, Uesugi N, Sugai T, Masuda T, Wakabayashi G, Maesawa C (2012) Overexpression of histone deacetylase 6 contributes to accelerated migration and invasion activity of hepatocellular carcinoma cells. Oncol Rep 28:867–873
Kawai H, Li H, Avraham S, Jiang S, Avraham HK (2003) Overexpression of histone deacetylase HDAC1 modulates breast cancer progression by negative regulation of estrogen receptor α. Int J Cancer 107:353–358
Kugel S, Sebastián C, Fitamant J, Ross KN, Saha SK, Jain E, Gladden A, Arora KS, Kato Y, Rivera MN, Ramaswamy S, Sadreyev RI, Goren A, Deshpande V, Bardeesy N, Mostoslavsky R (2016) SIRT6 suppresses pancreatic cancer through control of Lin28b. Cell 165:1401–1415
Kumar S, Lombard DB (2015) Mitochondrial sirtuins and their relationships with metabolic disease and cancer. Antioxid Redox Signal 22:1060–1077

Lai QY, He YZ, Peng XW, Zhou X, Liang D, Wang L (2019) Histone deacetylase 1 induced by neddylation inhibition contributes to drug resistance in acute myelogenous leukemia. Cell Commun Signal 17:86

Lee YS, Lim KH, Guo X, Kawaguchi Y, Gao Y, Barrientos T, Ordentlich P, Wang XF, Counter CM, Yao TP (2008) The cytoplasmic deacetylase HDAC6 is required for efficient oncogenic tumorigenesis. Cancer Res 68:7561–7569

Lei Y, Liu L, Zhang S, Guo S, Li X, Wang J, Su B, Fang Y, Chen X, Ke H, Tao W (2017) Hdac7 promotes lung tumorigenesis by inhibiting Stat3 activation. Mol Cancer 16:170

Li S, Banck M, Mujtaba S, Zhou M-M, Sugrue MM, Walsh MJ (2010) p53-induced growth arrest is regulated by the mitochondrial SirT3 deacetylase. PLoS One 5:e10486

Li W, Sun Z, Chen C, Wang L, Geng Z, Tao J (2018a) Sirtuin7 has an oncogenic potential via promoting the growth of cholangiocarcinoma cells. Biomed Pharmacother 100:257–266

Li W, Zhu D, Qin S (2018b) SIRT7 suppresses the epithelial-to-mesenchymal transition in oral squamous cell carcinoma metastasis by promoting SMAD4 deacetylation. J Exp Clin Cancer Res 37:148

Li H, Li X, Lin H, Gong J (2020) High HDAC9 is associated with poor prognosis and promotes malignant progression in pancreatic ductal adenocarcinoma. Mol Med Rep 21:822–832

Lin Z, Yang H, Tan C, Li J, Liu Z, Quan Q, Kong S, Ye J, Gao B, Fang D (2013) USP10 antagonizes c-Myc transcriptional activation through SIRT6 stabilization to suppress tumor formation. Cell Rep 5:1639–1649

Linares A, Assou S, Lapierre M, Thouennon E, Duraffourd C, Fromaget C, Boulahtouf A, Tian G, Ji J, Sahin O, Badia E, Boulle N, Cavaillès V (2019) Increased expression of the HDAC9 gene is associated with antiestrogen resistance of breast cancers. Mol Oncol 13:1534–1547

Lu X-F, Cao X-Y, Zhu Y-J, Wu Z-R, Zhuang X, Shao M-Y, Xu Q, Zhou Y-J, Ji H-J, Lu Q-R, Shi Y-J, Zeng Y, Bu H (2018) Histone deacetylase 3 promotes liver regeneration and liver cancer cells proliferation through signal transducer and activator of transcription 3 signaling pathway. Cell Death Dis 9:398

Malik S, Villanova L, Tanaka S, Aonuma M, Roy N, Berber E, Pollack J, Michishita-Kioi E, Chua K (2015) SIRT7 inactivation reverses metastatic phenotypes in epithelial and mesenchymal tumors. Sci Rep 5:9841

Massa F, Kaminski L, Clavel S, Djabari Z, Robert G, Laurent K, Michiels J-F, Durand M, Ricci J-E, Tanti J-F, Bost F, Ambrosetti D (2017) Sirtuin 7: a new marker of aggressiveness in prostate cancer. Oncotarget 8

Menbari MN, Rahimi K, Ahmadi A, Elyasi A, Darvishi N, Hosseini V, Mohammadi-Yeganeh S, Abdi M (2019) MiR-216b-5p inhibits cell proliferation in human breast cancer by down-regulating HDAC8 expression. Life Sci 237:116945

Minucci S, Pelicci PG (2006) Histone deacetylase inhibitors and the promise of epigenetic (and more) treatments for cancer. Nat Rev Cancer 6:38–51

Miyo M, Yamamoto H, Konno M, Colvin H, Nishida N, Koseki J, Kawamoto K, Ogawa H, Hamabe A, Uemura M, Nishimura J, Hata T, Takemasa I, Mizushima T, Doki Y, Mori M, Ishii H (2015) Tumour-suppressive function of SIRT4 in human colorectal cancer. Br J Cancer 113:492–499

Mizuguchi Y, Specht S, Lunz JG 3rd, Isse K, Corbitt N, Takizawa T, Demetris AJ (2012) SPRR2A enhances p53 deacetylation through HDAC1 and down regulates p21 promoter activity. BMC Mol Biol 13:20

Moreno DA, Scrideli CA, Cortez MA, de Paula Queiroz R, Valera ET, da Silva Silveira V, Yunes JA, Brandalise SR, Tone LG (2010) Differential expression of HDAC3, HDAC7 and HDAC9 is associated with prognosis and survival in childhood acute lymphoblastic leukaemia. Br J Haematol 150:665–673

Mostoslavsky R, Chua KF, Lombard DB, Pang WW, Fischer MR, Gellon L, Liu P, Mostoslavsky G, Franco S, Murphy MM, Mills KD, Patel P, Hsu JT, Hong AL, Ford E, Cheng HL, Kennedy C, Nunez N, Bronson R, Frendewey D, Auerbach W, Valenzuela D, Karow M, Hottiger MO, Hursting S, Barrett JC, Guarente L, Mulligan R, Demple B,

Yancopoulos GD, Alt FW (2006) Genomic instability and aging-like phenotype in the absence of mammalian SIRT6. Cell 124:315–329

Mu P, Liu K, Lin Q, Yang W, Liu D, Lin Z, Shao W, Ji T (2019) Sirtuin 7 promotes glioma proliferation and invasion through activation of the ERK/STAT3 signaling pathway. Oncol Lett 17:1445–1452

Müller BM, Jana L, Kasajima A, Lehmann A, Prinzler J, Budczies J, Winzer K-J, Dietel M, Weichert W, Denkert C (2013) Differential expression of histone deacetylases HDAC1, 2 and 3 in human breast cancer - overexpression of HDAC2 and HDAC3 is associated with clinicopathological indicators of disease progression. BMC Cancer 13:215

Noh JH, Jung KH, Kim JK, Eun JW, Bae HJ, Xie HJ, Chang YG, Kim MG, Park WS, Lee JY, Nam SW (2011) Aberrant regulation of HDAC2 mediates proliferation of hepatocellular carcinoma cells by deregulating expression of G1/S cell cycle proteins. PLoS One 6:e28103

Oehme I, Deubzer HE, Wegener D, Pickert D, Linke J-P, Hero B, Kopp-Schneider A, Westermann F, Ulrich SM, von Deimling A, Fischer M, Witt O (2009) Histone deacetylase 8 in neuroblastoma tumorigenesis. Clin Cancer Res 15:91–99

Oltra SS, Cejalvo JM, Tormo E, Albanell M, Ferrer A, Nacher M, Bermejo B, Hernando C, Chirivella I, Alonso E, Burgués O, Peña-Chilet M, Eroles P, Lluch A, Ribas G, Martinez MT (2020) HDAC5 inhibitors as a potential treatment in breast cancer affecting very Young women. Cancers 12:412

Osada H, Tatematsu Y, Saito H, Yatabe Y, Mitsudomi T, Takahashi T (2004) Reduced expression of class II histone deacetylase genes is associated with poor prognosis in lung cancer patients. Int J Cancer 112:26–32

Ouaïssi M, Sielezneff I, Silvestre R, Sastre B, Bernard JP, Lafontaine JS, Payan MJ, Dahan L, Pirrò N, Seitz JF, Mas E, Lombardo D, Ouaissi A (2008) High histone deacetylase 7 (HDAC7) expression is significantly associated with adenocarcinomas of the pancreas. Ann Surg Oncol 15:2318–2328

Parbin S, Kar S, Shilpi A, Sengupta D, Deb M, Rath SK, Patra SK (2014) Histone deacetylases: a saga of perturbed acetylation homeostasis in cancer. J Histochem Cytochem 62:11–33

Park SY, Jun JA, Jeong KJ, Heo HJ, Sohn JS, Lee HY, Park CG, Kang J (2011) Histone deacetylases 1, 6 and 8 are critical for invasion in breast cancer. Oncol Rep 25:1677–1681

Powers J, Lienlaf M, Perez-Villarroel P, Deng S, Knox T, Villagra A, Sahakian E (2016) Expression and function of histone deacetylase 10 (HDAC10) in B cell malignancies. Methods Mol Biol 1436:129–145

Ropero S, Esteller M (2007) The role of histone deacetylases (HDACs) in human cancer. Mol Oncol 1:19–25

Saha RN, Pahan K (2006) HATs and HDACs in neurodegeneration: a tale of disconcerted acetylation homeostasis. Cell Death Differ 13:539–550

Sakuma T, Uzawa K, Onda T, Shiiba M, Yokoe H, Shibahara T, Tanzawa H (2006) Aberrant expression of histone deacetylase 6 in oral squamous cell carcinoma. Int J Oncol 29:117–124

Salgado E, Bian X, Feng A, Shim H, Liang Z (2018) HDAC9 overexpression confers invasive and angiogenic potential to triple negative breast cancer cells via modulating microRNA-206. Biochem Biophys Res Commun 503:1087–1091

Sebastián C, Zwaans BM, Silberman DM, Gymrek M, Goren A, Zhong L, Ram O, Truelove J, Guimaraes AR, Toiber D, Cosentino C, Greenson JK, MacDonald AI, McGlynn L, Maxwell F, Edwards J, Giacosa S, Guccione E, Weissleder R, Bernstein BE, Regev A, Shiels PG, Lombard DB, Mostoslavsky R (2012) The histone deacetylase SIRT6 is a tumor suppressor that controls cancer metabolism. Cell 151:1185–1199

Seligson ND, Stets CW, Demoret BW, Awasthi A, Grosenbacher N, Shakya R, Hays JL, Chen JL (2019) Inhibition of histone deacetylase 2 reduces MDM2 expression and reduces tumor growth in dedifferentiated liposarcoma. Oncotarget 10

Shan W, Jiang Y, Yu H, Huang Q, Liu L, Guo X, Li L, Mi Q, Zhang K, Yang Z (2017) HDAC2 overexpression correlates with aggressive clinicopathological features and DNA-damage response pathway of breast cancer. Am J Cancer Res 7:1213–1226

Shi H, Ji Y, Zhang D, Liu Y, Fang P (2016) MicroRNA-3666-induced suppression of SIRT7 inhibits the growth of non-small cell lung cancer cells. Oncol Rep 36:3051–3057

Singh BN, Zhang G, Hwa YL, Li J, Dowdy SC, Jiang S-W (2010) Nonhistone protein acetylation as cancer therapy targets. Expert Rev Anticancer Ther 10:935–954

Song C, Zhu S, Wu C, Kang J (2013) Histone deacetylase (HDAC) 10 suppresses cervical cancer metastasis through inhibition of matrix metalloproteinase (MMP) 2 and 9 expression. J Biol Chem 288:28021–28033

Song Y, Jiang Y, Tao D, Wang Z, Wang R, Wang M, Han S (2020) NFAT2-HDAC1 signaling contributes to the malignant phenotype of glioblastoma. Neuro-Oncology 22:46–57

Spaety ME, Gries A, Badie A, Venkatasamy A, Romain B, Orvain C, Yanagihara K, Okamoto K, Jung AC, Mellitzer G, Pfeffer S, Gaiddon C (2019) HDAC4 levels control sensibility toward cisplatin in gastric cancer via the p53-p73/BIK pathway. Cancers 11

Spurling CC, Godman CA, Noonan EJ, Rasmussen TP, Rosenberg DW, Giardina C (2008) HDAC3 overexpression and colon cancer cell proliferation and differentiation. Mol Carcinog 47:137–147

Tao X, Yan Y, Lu L, Chen B (2017) HDAC10 expression is associated with DNA mismatch repair gene and is a predictor of good prognosis in colon carcinoma. Oncol Lett 14:4923–4929

Vasilatos SN, Katz TA, Oesterreich S, Wan Y, Davidson NE, Huang Y (2013) Crosstalk between lysine-specific demethylase 1 (LSD1) and histone deacetylases mediates antineoplastic efficacy of HDAC inhibitors in human breast cancer cells. Carcinogenesis 34:1196–1207

Wang RH, Sengupta K, Li C, Kim HS, Cao L, Xiao C, Kim S, Xu X, Zheng Y, Chilton B, Jia R, Zheng ZM, Appella E, Wang XW, Ried T, Deng CX (2008) Impaired DNA damage response, genome instability, and tumorigenesis in SIRT1 mutant mice. Cancer Cell 14:312–323

Wang HL, Lu RQ, Xie SH, Zheng H, Wen XM, Gao X, Guo L (2015) SIRT7 exhibits oncogenic potential in human ovarian cancer cells. Asian Pac J Cancer Prev 16:3573–3577

Wang Z, Tang F, Hu P, Wang Y, Gong J, Sun S, Xie C (2016) HDAC6 promotes cell proliferation and confers resistance to gefitinib in lung adenocarcinoma. Oncol Rep 36

Wang LT, Chiou SS, Chai CY, Hsi E, Wang SN, Huang SK, Hsu SH (2017) Aryl hydrocarbon receptor regulates histone deacetylase 8 expression to repress tumor suppressive activity in hepatocellular carcinoma. Oncotarget 8:7489–7501

Wang Y-Q, Wang H-L, Xu J, Tan J, Fu L-N, Wang J-L, Zou T-H, Sun D-F, Gao Q-Y, Chen Y-X, Fang J-Y (2018) Sirtuin5 contributes to colorectal carcinogenesis by enhancing glutaminolysis in a deglutarylation-dependent manner. Nat Commun 9:545

Weichert W, Röske A, Gekeler V, Beckers T, Stephan C, Jung K, Fritzsche FR, Niesporek S, Denkert C, Dietel M, Kristiansen G (2008a) Histone deacetylases 1, 2 and 3 are highly expressed in prostate cancer and HDAC2 expression is associated with shorter PSA relapse time after radical prostatectomy. Br J Cancer 98:604–610

Weichert W, Roske A, Niesporek S, Noske A, Buckendahl AC, Dietel M, Gekeler V, Boehm M, Beckers T, Denkert C (2008b) Class I histone deacetylase expression has independent prognostic impact in human colorectal cancer: specific role of class I histone deacetylases in vitro and in vivo. Clin Cancer Res 14:1669–1677

Wilson AJ, Byun DS, Popova N, Murray LB, L'Italien K, Sowa Y, Arango D, Velcich A, Augenlicht LH, Mariadason JM (2006) Histone deacetylase 3 (HDAC3) and other class I HDACs regulate colon cell maturation and p21 expression and are deregulated in human colon cancer. J Biol Chem 281:13548–13558

Wilson AJ, Byun DS, Nasser S, Murray LB, Ayyanar K, Arango D, Figueroa M, Melnick A, Kao GD, Augenlicht LH, Mariadason JM (2008) HDAC4 promotes growth of colon cancer cells via repression of p21. Mol Biol Cell 19:4062–4075

Witt AE, Lee CW, Lee TI, Azzam DJ, Wang B, Caslini C, Petrocca F, Grosso J, Jones M, Cohick EB, Gropper AB, Wahlestedt C, Richardson AL, Shiekhattar R, Young RA, Ince TA (2017) Identification of a cancer stem cell-specific function for the histone deacetylases, HDAC1 and HDAC7, in breast and ovarian cancer. Oncogene 36:1707–1720

Wu Z, Fang H (2019) Expression of SIRT5 protein in gastric cancer cells. J Biol Regul Homeost Agents 33:1675–1683
Wu J, Du C, Lv Z, Ding C, Cheng J, Xie H, Zhou L, Zheng S (2013) The up-regulation of histone deacetylase 8 promotes proliferation and inhibits apoptosis in hepatocellular carcinoma. Dig Dis Sci 58:3545–3553
Wu S, Wu E, Wang D, Niu Y, Yue H, Zhang D, Luo J, Chen R (2020) LncRNA HRCEG, regulated by HDAC1, inhibits cells proliferation and epithelial-mesenchymal-transition in gastric cancer. Cancer Genet 241:25–33
Xie HJ, Jung KH, Nam SW (2011) Overexpression of SIRT2 contributes tumor cell growth in hepatocellular carcinomas. Mol Cell Toxicol 7:367–374
Xie HJ, Noh JH, Kim JK, Jung KH, Eun JW, Bae HJ, Kim MG, Chang YG, Lee JY, Park H, Nam SW (2012) HDAC1 inactivation induces mitotic defect and caspase-independent autophagic cell death in liver cancer. PLoS One 7:e34265
Xiong Y, Wang M, Zhao J, Wang L, Li X, Zhang Z, Jia L, Han Y (2017) SIRT3 is correlated with the malignancy of non-small cell lung cancer. Int J Oncol 50:903–910
Xiong K, Zhang H, Du Y, Tian J, Ding S (2019) Identification of HDAC9 as a viable therapeutic target for the treatment of gastric cancer. Exp Mol Med 51:1–15
Xu G, Zhu H, Zhang M, Xu J (2018) Histone deacetylase 3 is associated with gastric cancer cell growth via the miR-454-mediated targeting of CHD5. Int J Mol Med 41:155–163
Xu G, Li N, Zhang Y, Zhang J, Xu R, Wu Y (2019) MicroRNA-383-5p inhibits the progression of gastric carcinoma via targeting HDAC9 expression. Brazilian J Med Biol Res 52:e8341
Yan D, Pomicter AD, Heaton WL, Mason CC, Ahmann J, Senina A, Franzini A, Halverson B, Than H, Clair PM, Khorashad JS, O'Hare T, Deininger MW (2018) SIRT5 as a therapeutic target in acute myeloid leukemia. Blood 132:907–907
Yang XJ, Seto E (2007) HATs and HDACs: from structure, function and regulation to novel strategies for therapy and prevention. Oncogene 26:5310–5318
Yang H, Salz T, Zajac-Kaye M, Liao D, Huang S, Qiu Y (2014a) Overexpression of histone deacetylases in cancer cells is controlled by interplay of transcription factors and epigenetic modulators. FASEB J 28:4265–4279
Yang Y, Huang Y, Wang Z, Wang H-T, Duan B, Ye D, Wang C, Jing R, Leng Y, Xi J, Chen W, Wang G, Jia W, Zhu S, Kang J (2014b) HDAC10 promotes lung cancer proliferation via AKT phosphorylation. Oncotarget 7
Yi Z, Wenwen L, Kun W, Jian S (2019) Overexpression of histone deacetylase 11 suppresses basal-like breast cancer cell invasion and metastasis. J South Med Univ 39:751–759
Yin Y, Zhang M, Dorfman RG, Li Y, Zhao Z, Pan Y, Zhou Q, Huang S, Zhao S, Yao Y, Zou X (2017) Histone deacetylase 3 overexpression in human cholangiocarcinoma and promotion of cell growth via apoptosis inhibition. Cell Death Dis 8:e2856–e2856
Yu H, Ye W, Wu J, Meng X, Liu RY, Ying X, Zhou Y, Wang H, Pan C, Huang W (2014) Overexpression of sirt7 exhibits oncogenic property and serves as a prognostic factor in colorectal cancer. Clin Cancer Res 20:3434–3445
Yue L, Sharma V, Horvat NP, Akuffo AA, Beatty MS, Murdun C, Colin C, Billington JMR, Goodheart WE, Sahakian E, Zhang L, Powers JJ, Amin NE, Lambert-Showers QT, Darville LN, Pinilla-Ibarz J, Reuther GW, Wright KL, Conti C, Lee JY, Zheng X, Ng PY, Martin MW, Marshall CG, Koomen JM, Levine RL, Verma A, Grimes HL, Sotomayor EM, Shao Z, Epling-Burnette PK (2020) HDAC11 deficiency disrupts oncogene-induced hematopoiesis in myelo-proliferative neoplasms. Blood 135:191–207
Zhang Z, Yamashita H, Toyama T, Sugiura H, Omoto Y, Ando Y, Mita K, Hamaguchi M, Hayashi S-I, Iwase H (2004) HDAC6 expression is correlated with better survival in breast Cancer. Clin Cancer Res 10:6962–6968
Zhang J, Yin XJ, Xu CJ, Ning YX, Chen M, Zhang H, Chen SF, Yao LQ (2015a) The histone deacetylase SIRT6 inhibits ovarian cancer cell proliferation via down-regulation of notch 3 expression. Eur Rev Med Pharmacol Sci 19:818–824

Zhang S, Chen P, Huang Z, Hu X, Chen M, Hu S, Hu Y, Cai T (2015b) Sirt7 promotes gastric cancer growth and inhibits apoptosis by epigenetically inhibiting miR-34a. Sci Rep 5:9787

Zhang R, Wang C, Tian Y, Yao Y, Mao J, Wang H, Li Z, Xu Y, Ye M, Wang L (2019a) SIRT5 promotes hepatocellular carcinoma progression by regulating mitochondrial apoptosis. J Cancer 10:3871–3882

Zhang S-L, Zhu H-Y, Zhou B-Y, Chu Y, Huo J-R, Tan Y-Y, Liu D-L (2019b) Histone deacetylase 6 is overexpressed and promotes tumor growth of colon cancer through regulation of the MAPK/ERK signal pathway. Onco Targets Ther 12:2409–2419

Zheng Y, Yang X, Wang C, Zhang S, Wang Z, Li M, Wang Y, Wang X, Yang X (2020) HDAC6, modulated by miR-206, promotes endometrial cancer progression through the PTEN/AKT/mTOR pathway. Sci Rep 10:3576

Zhong L, Sun S, Yao S, Han X, Gu M, Shi J (2018) Histone deacetylase 5 promotes the proliferation and invasion of lung cancer cells. Oncol Rep 40:2224–2232

Zhu P, Martin E, Mengwasser J, Schlag P, Janssen K-P, Göttlicher M (2004) Induction of HDAC2 expression upon loss of APC in colorectal tumorigenesis. Cancer Cell 5:455–463

Recap of Distinct Molecular Signalling Mechanisms Modulated by Histone Deacetylases for Cancer Genesis and Progression

5

Various post-translational modifications occurring mainly on N-terminal tails of histone proteins modulate their function. Histone acetylation being one of these modifications, regulated by histone acetyltransferases (HATs) and histone deacetylases (HDACs), plays a critical role in governing transcriptional events precisely (Seto and Yoshida 2014; Yang and Seto 2007). In fact the opposite activities of these enzymes upkeep acetylation/deacetylation equilibrium, the critical condition for preventing transcriptional dysregulation (Fig. 5.1) (Fraga et al. 2005; Ganai 2019). Thus not surprisingly, anomalous alterations in calibrated acetylation may lead to cancer development through transcriptional deregulation (Ganai 2015; Parbin et al. 2014; Saha and Pahan 2006). In a variety of cancers aberrant expression of HDACs is often noticed. Further, HDACs in addition to histone substrates modulate several non-histone targets also which may also incite cancer signalling (Ganai 2018; Singh et al. 2010). Moreover, onset and advancement of certain cancers has been attributed to mutations in HDACs (Ropero et al. 2006). HDACs by supressing apoptosis and cell cycle kinase inhibitors may fuel tumour onset. Moreover, overactivity of HDACs through downregulation of cell adhesion molecules and extracellular matrix-related genes triggers invasion (Onder et al. 2008). Further HDAC overexpression may accentuate tumour progression through downregulation of differentiation related transcription factors and by induction of certain factors such as vascular endothelial growth factor favourable for angiogenesis (Glozak and Seto 2007; Li and Seto 2016). Thus HDACs facilitate tumour initiation and progression on one hand by supressing certain molecular players and on the other hand by inducing others.

5.1 Aberrant HDAC Activity in Tumorigenesis

Abnormal expression of HDACs has been demonstrated in a variety of malignancies ranging from solid to haematological ones. For instance, overexpression of HDAC1 has been observed in colon, prostate, gastric and breast carcinomas (Choi et al. 2001;

S. A. Ganai, *Histone Deacetylase Inhibitors in Combinatorial Anticancer Therapy*,
https://doi.org/10.1007/978-981-15-8179-3_5

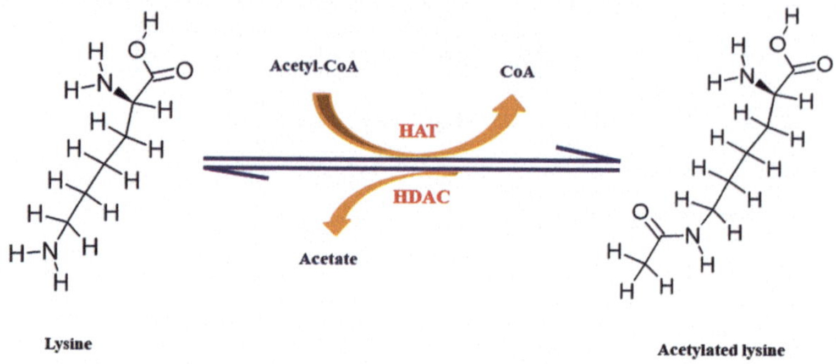

Fig. 5.1 Chemical mechanisms showing how histone acetyltransferase (HAT) transfers acetyl moiety from acetyl-CoA to ε-amino group of lysine. On receiving acetyl group lysine transforms to acetylated lysine. Histone deacetylase (HDAC) removes the acetyl moiety from acetylated lysine to convert it into deacetylated lysine or simply lysine. Lysine residues occurring in histone proteins similarly get acetylated by histone acetyltransferases (HATs) and deacetylated by HDACs. The antagonistic functions of HATs and HDACs upkeeps acetylation/deacetylation balance crucial for normal cell growth

Halkidou et al. 2004; Wilson et al. 2006; Zhang et al. 2005). Enhanced expression of HDAC2 has been linked to gastric and cervical cancers (Huang et al. 2005; Song et al. 2005). While high expression of HDAC3 has been observed in colon cancer (Wilson et al. 2006), escalated expression of tubulin deacetylase HDA6 has been proved in breast cancer (Zhang et al. 2004). In pancreatic tumour samples special overexpression of HDAC1, HDAC7 and HDAC8 has been proved (Cai et al. 2018). By way of inhibiting p21 (cyclin-dependent kinase inhibitor)-expression, HDAC4 overexpression facilitates gastric cancer progression (Kang et al. 2014). Further studies have shown that this HDAC on interacting with Sp1 (*specificity protein 1*) impedes the expression of p21 through epigenetic mechanism (H3 deacetylation) (Mottet et al. 2009).

Apart from overexpression, reduced expression of Class II HDACs has been observed to facilitate lung cancer suggesting the possible repression of crucial genes by them (Osada et al. 2004). Similarly lower SIRT2 expression has been noted in serous ovarian carcinoma. SIRT2 in underexpressed state is not able to silence the expression of cyclin-dependent kinase 4, the regulator of G1/S transition (Du et al. 2017). SIRT3, another member of sirtuin family shows lower expression in breast cancer cells. Downregulated levels of this HDAC enhances ROS (reactive oxygen species) generation, stabilizes HIF1α, the regulator of glycolytic gene transcription (Finley et al. 2011). Downregulation of SIRT4 has been proved in gastric, colorectal cancer and Burkitt's lymphoma (Hu et al. 2019; Jeong et al. 2014; Miyo et al. 2015). This HDAC restrains gastric cancer by upregulating the expression of epithelial cadherin (E-cadherin). Low levels of SIRT4 promote epithelial to mesenchymal transition by downregulating the defined cadherin (Sun et al. 2018). Moreover downregulation of SIRT6 has strong link with a variety of cancers including ovarian, colon and pancreatic ductal adenocarcinoma (Kugel et al. 2016; Lin et al. 2013;

Zhang et al. 2015). In gastric cancer tissues low expression of SIRT6 showed negative correlation with phosphorylated STAT3 (p-STAT3). Activation of JAK2/STAT3 by downregulated levels of SIRT6 then activates the cyclin D1 and Bcl2 genes thereby facilitating tumorigenesis (Zhou et al. 2017). Further reduced expression of SIRT6 has been seen in colon cancer tissues and cell lines suggesting its role as a tumour suppressor. SIRT6, by way of modulating PTEN (phosphatase and tensin homologue deleted on chromosome 10)/AKT (serine/threonine kinase) signalling attenuates colon cancer progression (Tian and Yuan 2018). In ovarian serous carcinoma lower expression of SIRT2 facilitates proliferation. This effect of reduced SIRT2 levels on proliferation has been ascribed to enhanced protein levels of an established carcinogen CDK4 (Fig. 5.2) (Du et al. 2017).

5.2 HDAC Mutations and Cancer

Mounting evidences suggest that genetic inactivation of various HDACs may also promote tumorigenesis. While 8.3% of dedifferentiated liposarcoma (DLPS) showed somatic mutations in HDAC1 (Taylor et al. 2011), homozygous deletions of Class II member (HDAC4) were detected in 4% of melanomas (Stark and Hayward 2007). Truncating mutations have been reported in HDAC2, a Class I HDAC, in sporadic carcinomas having microsatellite instability. In two colorectal and two endometrial cell lines, frameshift mutation has seen identified in this classical HDAC. This mutation results in functional abrogation of this HDAC and mitigates the sensitivity of these cells to trichostatin A, a hydroxamates group HDAC inhibitor (Li and Seto 2016; Ropero et al. 2006).

5.3 Deacetylation of Non-histone Substrates by HDACs Facilitates Cancer

The name of histone deacetylases changed to protein deacetylases after experimental evidences suggested plethora of non-histone targets modulated by HDACs (Ganai 2018; Singh et al. 2010). Acetylation status of non-histone substrates has multiple effects on protein function. In fact acetylation level of non-histone proteins governs stability, protein–protein interactions and subcellular localization (Ganai 2018; Glozak et al. 2005). For instance, HDAC6 upregulation deacetylates heat shock protein 90 (HSP90) and thus facilitates its association with androgen receptor (AR) thereby stabilizing the latter and triggering prostate cancer (Gibbs et al. 2009; Kovacs et al. 2005). Sulforaphane, a glucoraphanin derived HDAC inhibitor (Ganai 2016), hyperacetylates HSP90, through inhibition of HDAC6 enzymatic activity (Gibbs et al. 2009). This results in disruption of interaction between HSP90 and AR and subsequent destabilization of the latter culminating in attenuation of prostate cancer signalling (Fig. 5.2) (Ganai 2018; Gibbs et al. 2009). There is also possibility that acetylation and deacetylation may encourage or discourage

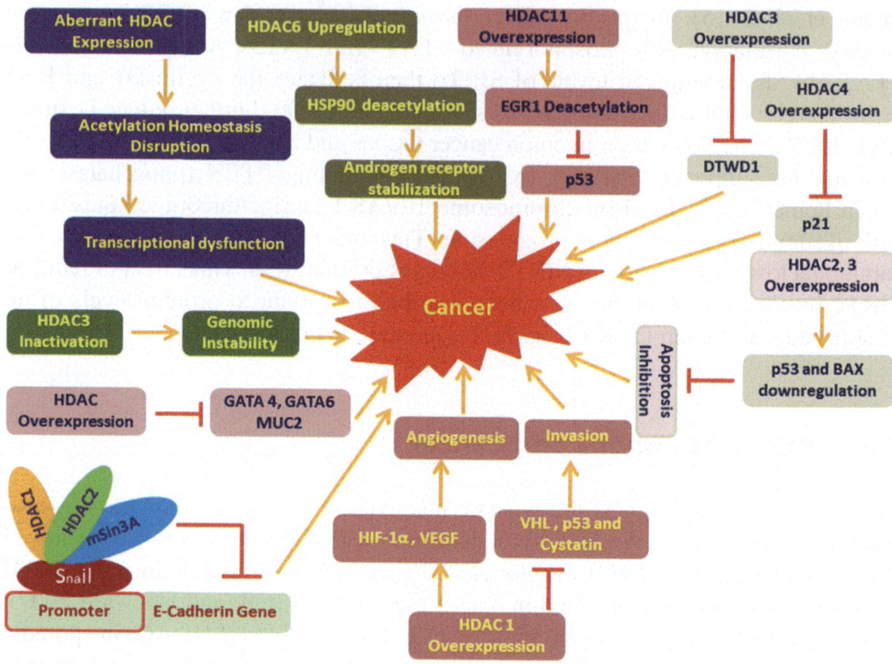

Fig. 5.2 Compendium of various signalling mechanisms through which histone deacetylases promote tumour initiation and advancement. Aberrant expression of HDACs disrupts acetylation homeostasis which results in transcriptional dysfunction and subsequent cancer. HDACs by modulating non-histone proteins also trigger cancer signalling. Heat shock protein 90 (HSP90) is deacetylated by upregulated HDAC6 which facilitates its association with androgen receptor. This stabilizes the androgen receptor thereby promoting cancer. Egr1, a transcription factor regulates p53 expression positively. HDAC11 overexpression through deacetylation of this transcription factor represses p53 leading to cancer. DTW domain-containing protein 1 (DTWD1), a tumour suppressor is regulated positively by p53. HDAC3 overexpression through chromatin compaction hampers binding of transcription factors and thus facilitates cancer. HDAC4 overexpression by supressing cyclin-dependent kinase inhibitor/tumour suppressor (p53) also fuels cancer. Enhanced expressions of HDAC2 and HDAC3 favour cancer growth by restraining apoptosis via downregulation/inactivation of p53 and BAX. Inactivation of HDAC3 causes cancer through genomic instability. HDAC overexpression augments cancer by supressing differentiation. HDAC overexpression represses differentiation factors such as GATA4, GAT6 and their loss in differentiated cells triggers de-differentiation and this may lead to cancer. Further HDAC overexpression via downregulation of MUC2 expression, a critical protein for gastrointestinal cell differentiation may promote cancer. Recruited by Snail (repressor), HDAC1 and HDAC2 in association with corepressor mSin3A mediate the downregulation of epithelial cadherin (E-cadherin). E-cadherin being cell–cell adhesion protein prevents epithelial invasion, the primary step in metastasis. Thus downregulation of this cadherin by HDACs makes the conditions conducive for cancer progression by way of facilitating invasion. Further enhanced expression of HDAC1 promotes cancer progression by supporting angiogenesis through down regulation of p53 and von Hippel Lindau and via induction of HIF-1α and VEGF (hypoxia-responsive genes)

phosphorylation of non-histone targets that may impact their activation/deactivation (Glozak et al. 2005; Kouzarides 2000).

5.4 Downregulation of Tumour Suppressor Genes/ Cyclin-Dependent Kinase Inhibitors by HDACs Fuels Cancer

Aberrant activity of various HDACs also promotes tumour onset and progression through downregulation of tumour suppressor genes. It is well established that HDACs function as transcriptional corepressors. In liver cancer overexpressed HDAC11 forms a complex with Egr1, the transcription factor positively regulating tumour suppressor p53. Deacetylation of Egr1 mediated by HDAC11 supresses p53 expression which in turn prevents apoptosis of liver cancer cells (Gong et al. 2019). Escalated expression of HDAC3 promotes gastric cancer by inhibiting the expression of tumour suppressor DTWD1 (DTW domain-containing protein 1). This effect of HDAC3 is mediated by p53, the positive regulator of DTWD1 transcription. However, no effect on p53 stability occurs but HDAC3 creates chromatin environment less favourable for binding of transcription factors thereby hampering DTWD1 expression (Ma et al. 2015). Overexpression of Class IIa HDAC (HDAC4) promotes growth of colon cancer cells and tumour by way of repressing p21 (cyclin-dependent kinase inhibitor/tumour suppressor) (Wang et al. 2001; Wilson et al. 2008). HDAC4 as component of corepressor complex (HDAC4-HDAC3-N-CoR/SMRT) is recruited to p21 promoter through Sp1 resulting in inhibition of p21 expression (Wilson et al. 2008). Most of the breast and ovarian cancers do not express ARHI (maternally expressed tumour suppressor) unlike normal breast epithelial and ovarian cells (Yu et al. 1999). Its downregulation in breast cancer cells is mediated by E2F1 and E2F2 (transcription factors) and their complexes with certain classical HDACs. Finer details provided by futuristic techniques suggest that only some zinc-dependent HDACs were able to substantially lower the promoter activity of this tumour suppressor. Among the effective HDACs, HDAC1, HDAC3 and HDAC11 are more noteworthy. Thus by attenuating the expression of tumour suppressor gene, HDAC expression results in tumour onset and development (Fig. 5.2) (Feng et al. 2005).

5.5 Suppression of Apoptosis by HDACs Promote Cancer

In a variety of cancers HDACs regulate apoptosis by modulating the expression of antiapoptotic and pro-apoptotic proteins. Pharmacological intervention of cancer cells with HDAC inhibitors trigger apoptosis by way of intrinsic/extrinsic pathway or by escalating the sensitivity of these cells to apoptosis (Zhang and Zhong 2014). Hyperexpression of HDAC2 suppresses apoptosis of human lung cancer cells through p53 and Bax deactivation. Inactivation of this HDAC promotes apoptosis in these cells by activation of p53 and Bax in addition to suppression of Bcl2 (Jung et al. 2012). Upregulation of HDAC8 in gastric cancers cells facilitates their proliferation but inhibits apoptosis. Depletion of this HDAC attenuates proliferation and enhances apoptosis. This effect is mediated by elevated expression of various apoptosis facilitating proteins including Bmf, caspase-6 and activated caspase-3 following HDAC8 decline. Thus HDAC8 being overexpressed in gastric cancer cells protects them from apoptosis by suppressing the expression of apoptosis favouring proteins (Song et al. 2015). HDAC3 overexpression prevents the apoptosis but enhances the proliferation of cholangiocarcinoma cells. HDAC3 overexpression, through downregulation of p53 and BAX in these cells blocks apoptosis (Fig. 5.2). Thus intervention against HDAC3 may prove as effective strategy for tackling cholangiocarcinoma (Yin et al. 2017).

5.6 HDACs and DNA Damage Repair

In DNA damage response HDACs play a crucial role as they modulate chromatin topology (Li and Zhu 2014). Class I HDACs (HDAC1 and HDAC2) on recruitment to DNA damage sites promote non-homologous end joining by way of hypoacetylating histones (H3K56 and H4K16) (Miller et al. 2010). This reflects the critical role of these two Class I HDACs in double strand break repair. HDAC3, another Class I HDAC is related to control of DNA damage despite being not localized to damaged sites (Miller et al. 2010). Its inactivation leads to genomic instability and deletion of this HDAC in liver results in hepatocellular carcinoma (Bhaskara et al. 2010). Class I HDACs influence DNA repair not only by hypoacetylating histones but also by regulating other proteins having implications in DNA damage response. Among these proteins ATM, BRCA1, ATR and FUS are important to mention (Thurn et al. 2013). It has been proved that HDAC9 and HDAC10 depletion obstruct homologous recombination and make the cells vulnerable to antitumour antibiotic treatment (Mitomycin-C) (Kotian et al. 2011). Multiple steps of response pathway (DNA damage) are regulated by SIRT1 (Gorospe and de Cabo 2008). This HDAC interacts with a variety of DNA damage response proteins and deacetylates them. These proteins include NBS1, Ku70, APE1, PARP-1 in addition to KAP1 TopBP1 (Li and Seto 2016; Lin et al. 2015; Luna et al. 2013; Wang et al. 2014). Another sirtuin, SIRT6, impedes genomic instability by influencing base excision repair of DNA (Mostoslavsky et al. 2006). Through deacetylation of carboxyl-terminal binding protein apart from carboxyl-terminal interacting protein this HDAC participates in homologous recombination (Kaidi et al. 2010). Further on recruitment to DNA damage sites, SIRT6 facilitates double strand break repair through PARP1 mono-ADP-ribosylation (Mao et al. 2011).

5.7 HDACs Through Differentiation Deregulation Cause Cancer

Genes having the proliferation restraining ability are often silenced in cancer cells. Unsuitable proliferation due to differentiation inhibition results in cancer initiation. In a variety of cancers, repression of differentiation factors belonging to GATA family has been identified. The mechanism behind the silencing of these crucial transcription factors varies among cell types (Glozak and Seto 2007). Transcription factors such as GATA4 and GATA6 show higher expression in ovarian epithelial cells but this is not true in case of ovarian cancer cells. Loss of GATA6 is associated with morphological transformation and de-differentiation of ovarian epithelial cells. Inhibition of GATA4 and GATA6 expression has been attributed to histone H3 and H4 hypoacetylation apart from loss of H3–K4 methylation within their promoter regions (Cai et al. 2009; Caslini et al. 2006). Pharmacological intervention with trichostatin A reinstates the expression of GATA factors and their target genes among which Dab2, the tumour suppressor is prominent (Caslini et al. 2006).

MUC2 encoding mucin-2 protein has crucial role in gastrointestinal cell differentiation. Inhibition of MUC2 expression has involvement in colorectal cancer. Moreover, escalated adenoma formation has been observed in mice MUC2 null mice (Velcich et al. 2002). Enhanced acetylation (H3K9 and H3K27) in the MUC2 promoter region promotes its expression. In addition the promoter region, in cells of MUC2 expressing gene, displayed lessened CpG island methylation. Trichostatin A based intervention in pancreatic cancer cells induced MUC2 expression through epigenetic mechanism (histone acetylation) (Yamada et al. 2006). MUC2 is differentially regulated by HDACs depending on cell type. This statement is based on the finding where undifferentiated adenocarcinoma cells showed repression of MUC2 on therapeutic intervention with short chain fatty acid group HDAC inhibitor sodium butyrate (Fig. 5.2) (Augenlicht et al. 2003).

It is well known that haematopoiesis involves step-wise differentiation programs for which precise expression of differentiation factors is critical. Formation of mature haematopoietic cell from pluripotent stem cells needs a cooperation of various specific molecular players. Obstruction of differentiation pathway at any of the step may culminate in the burgeoning of leukaemic cells (Glozak and Seto 2007). Leukaemias and lymphomas are associated with chromosomal translocations. These translocations result in inappropriate gene expression by mediating aberrant HDAC recruitment to promoter regions. RUNX1 gene encoding runt-related transcription factor 1, the key regulator of absolute haematopoiesis is quite often interrupted in leukaemia. Runx1-ETO (t (8; 21) translocation product) commonly seen in AML forms contacts with various HDACs and also binds to corepressors (mSin3A and N-CoR) (Amann et al. 2001). Besides it has been observed that Runx1 binds HDAC5–6 and HDAC9 with different propensities (Durst et al. 2003). Unlike ETO which lacks DNA binding ability, the chimeric protein possesses this ability due to Runx1-DNA binding domain. This chimeric protein by way of repressing p14 (ARF) tumour suppressor may lengthen the myeloid progenitor cells lifespan (Linggi et al. 2002). This fusion protein also silences the expression of c-fms, the gene encoding macrophage colony-stimulating factor receptor 1 through HDAC1 mediated deacetylation of H3K9 and H3K14 (Follows et al. 2003). Transcriptional silencing of genes involved in differentiation, by ETO and PLZF interplay also involved HDAC-reliant mechanism (Melnick et al. 2000). Other studies have also shown that ETO forms contact with BCL6 (repressor) and facilitates silencing through recruitment of HDACs (Fig. 5.2) (Chevallier et al. 2004; Lemercier et al. 2002). Thus from the above findings, it is obvious that inappropriate recruitment of HDACs to promoters of genes related to differentiation restrains this process resulting in perpetuation of undifferentiated progenitor cell proliferation and subsequent lymphoma or leukaemia.

5.8 HDACs Promote Angiogenesis and Metastasis for Cancer Progression

Histone acetylation and deacetylation regulate not only the genes meant for cancer genesis but also modulate the genes implicated in cancer progression. For progression of cancer angiogenesis, invasion and migration supporting molecular players should increase while molecules related to adhesion should get downregulated. Hypoxia frequently encountered in solid malignancies triggers angiogenesis and results in induction HDAC expression and function. Overexpression of HDAC1 enhances angiogenesis through induction of HIF-1α and VEGF (vascular endothelial growth factor), the hypoxia-responsive genes. This effect is also favoured by HDAC1 mediated repression of p53 and VHL (von Hippel Lindau), the established tumour suppressors (Deroanne et al. 2002; Kim et al. 2001). Studies in triple negative breast cancer cells have proved that Class I HDACs excluding HDAC8 contribute to vasculogenic mimicry, having survival role in these cells (Maiti et al. 2019). Studies on endothelial cells have shown that VE-cadherin silences the expression of VEGFR-2 through Class I (HDAC1) and Class II HDACs (HDAC4–6) (Hrgovic et al. 2017).

HDACs facilitate cellular invasion by regulating various genes (extracellular matrix-related). Especially Class I HDACs are strongly involved for controlling their expression. Cystatin a peptide inhibitor restraining invasion of tumour is negatively regulated by HDAC1.This conclusion has been drawn from the experimental finding where either knockdown of this HDAC or cystatin overexpression attenuated cellular invasion (Whetstine et al. 2005). The involvement of HDACs is also involved in invasion of v-Fos-transformed fibroblasts. Therapeutic intervention involving low doses of these inhibitors, in this model, alleviated the invasion through RYBP, protocadherin and STAT6 derepression. Further the invasion is impeded on overexpression of any one of the above-mentioned genes (McGarry et al. 2004). Expression of vital gene encoding a protein that mediates cell–cell adhesion (E-cadherin) is directly governed by Class I HDACs of Classical HDACs. Epithelial invasion the premier and critical step for metastasis is promoted by the downregulation of this gene. HDAC1 and HDAC2 apart from mSin3A (corepressor) are recruited by Snail (repressor) to the promoter of E-cadherin thereby silencing its expression through epigenetic mechanism (H3/H4 deacetylation and increased H3K9 methylation) (Peinado et al. 2004). Another mechanistically different HDAC, SIRT4, in colorectal cancer cells hampered invasion by escalating E-cadherin expression (Miyo et al. 2015). SIRT1 and SIRT7 interplay silences the expression of E-cadherin. SIRT7 inactivation drastically attenuates metastasis via upregulation of this cadherin (Malik et al. 2015). HDACs in acute lymphoblastic leukaemia cells regulate the expression of CXCR4. This chemokine receptor facilitates migration of ALL cells to different regions including liver, brain and spleen. HDAC inhibitors attenuated migration of these cells through downregulation of CXCR4 (Crazzolara et al. 2002).

On the whole, I have discussed the various mechanisms through which HDACs promote cancer initiation and progression. From the above evidences it is clear that HDAC have crucial implications in both initiation and progression of cancer. HDACs facilitate cancer initiation by downregulating/inactivating the molecular players involved in processes supporting continuous proliferation. For instance, HDACs downregulate cell cycle inhibitors and differentiation inducing factors leading to continuous proliferation. HDACs supress apoptosis by enhancing pro-apoptotic factors such as Bax and by escalating antiapoptotic Bcl2 thereby favouring cell survival. Further for promoting progressing HDAC supress molecular players involved in cell adhesion like epithelial cadherin. Besides HDAC by way of accentuating angiogenesis, invasion and migration augment cancer progression.

References

Amann JM, Nip J, Strom DK, Lutterbach B, Harada H, Lenny N, Downing JR, Meyers S, Hiebert SW (2001) ETO, a target of t(8;21) in acute leukemia, makes distinct contacts with multiple histone deacetylases and binds mSin3A through its oligomerization domain. Mol Cell Biol 21:6470–6483

Augenlicht L, Shi L, Mariadason J, Laboisse C, Velcich A (2003) Repression of MUC2 gene expression by butyrate, a physiological regulator of intestinal cell maturation. Oncogene 22:4983–4992

Bhaskara S, Knutson SK, Jiang G, Chandrasekharan MB, Wilson AJ, Zheng S, Yenamandra A, Locke K, Yuan JL, Bonine-Summers AR, Wells CE, Kaiser JF, Washington MK, Zhao Z, Wagner FF, Sun ZW, Xia F, Holson EB, Khabele D, Hiebert SW (2010) Hdac3 is essential for the maintenance of chromatin structure and genome stability. Cancer Cell 18:436–447

Cai KQ, Caslini C, Capo-chichi CD, Slater C, Smith ER, Wu H, Klein-Szanto AJ, Godwin AK, Xu X-X (2009) Loss of GATA4 and GATA6 expression specifies ovarian cancer histological subtypes and precedes neoplastic transformation of ovarian surface epithelia. PLoS One 4: e6454–e6454

Cai M-H, Xu X-G, Yan S-L, Sun Z, Ying Y, Wang B-K, Tu Y-X (2018) Depletion of HDAC1, 7 and 8 by histone deacetylase inhibition confers elimination of pancreatic cancer stem cells in combination with gemcitabine. Sci Rep 8:1621

Caslini C, Capo-chichi CD, Roland IH, Nicolas E, Yeung AT, Xu XX (2006) Histone modifications silence the GATA transcription factor genes in ovarian cancer. Oncogene 25:5446–5461

Chevallier N, Corcoran C, Lennon C, Hyjek E, Chadburn A, Bardwell V, Licht J, Melnick A (2004) ETO protein of t(8;21) AML is a corepressor for Bcl6 B-cell lymphoma oncoprotein. Blood 103:1454–1463

Choi JH, Kwon HJ, Yoon BI, Kim JH, Han SU, Joo HJ, Kim DY (2001) Expression profile of histone deacetylase 1 in gastric cancer tissues. Jpn J Cancer Res 92:1300–1304

Crazzolara R, Jöhrer K, Johnstone RW, Greil R, Ofler RK, Meister B, Bernhard D (2002) Histone deacetylase inhibitors potently repress CXCR4 chemokine receptor expression and function in acute lymphoblastic leukaemia. Br J Haematol 119:965–969

Deroanne CF, Bonjean K, Servotte S, Devy L, Colige A, Clausse N, Blacher S, Verdin E, Foidart J-M, Nusgens BV, Castronovo V (2002) Histone deacetylases inhibitors as anti-angiogenic agents altering vascular endothelial growth factor signaling. Oncogene 21:427–436

Du Y, Wu J, Zhang H, Li S, Sun H (2017) Reduced expression of SIRT2 in serous ovarian carcinoma promotes cell proliferation through disinhibition of CDK4 expression. Mol Med Rep 15:1638–1646

Durst KL, Lutterbach B, Kummalue T, Friedman AD, Hiebert SW (2003) The inv(16) fusion protein associates with corepressors via a smooth muscle myosin heavy-chain domain. Mol Cell Biol 23:607–619

Feng W, Lu Z, Luo RZ, Zhang X, Seto E, Bast RC, Yu Y (2005) The ARHI tumor suppressor gene is repressed by multiple histone deacetylases (HDACs) in breast cancer cells. Cancer Res 65:643–643

Finley LWS, Carracedo A, Lee J, Souza A, Egia A, Zhang J, Teruya-Feldstein J, Moreira PI, Cardoso SM, Clish CB, Pandolfi PP, Haigis MC (2011) SIRT3 opposes reprogramming of cancer cell metabolism through HIF1α destabilization. Cancer Cell 19:416–428

Follows GA, Tagoh H, Lefevre P, Hodge D, Morgan GJ, Bonifer C (2003) Epigenetic consequences of AML1-ETO action at the human c-FMS locus. EMBO J 22:2798–2809

Fraga MF, Ballestar E, Villar-Garea A, Boix-Chornet M, Espada J, Schotta G, Bonaldi T, Haydon C, Ropero S, Petrie K, Iyer NG, Pérez-Rosado A, Calvo E, Lopez JA, Cano A, Calasanz MJ, Colomer D, Piris MA, Ahn N, Imhof A, Caldas C, Jenuwein T, Esteller M (2005) Loss of acetylation at Lys16 and trimethylation at Lys20 of histone H4 is a common hallmark of human cancer. Nat Genet 37:391–400

Ganai SA (2015) In silico approaches towards safe targeting of class I histone deacetylases, pp 1–9. https://doi.org/10.1007/978-1-4614-6436-5_459-1

Ganai SA (2016) Histone deacetylase inhibitor sulforaphane: the phytochemical with vibrant activity against prostate cancer. Biomed Pharmacother 81:250–257

Ganai SA (2018) Histone deacetylase inhibitors modulating non-epigenetic players: the novel mechanism for small molecule based therapeutic intervention. Curr Drug Targets 19:593–601

Ganai SA (2019) HDACs and their distinct classes. In: Ganai SA (ed) Histone deacetylase inhibitors — epidrugs for neurological disorders. Springer, Singapore, pp 21–25

Gibbs A, Schwartzman J, Deng V, Alumkal J (2009) Sulforaphane destabilizes the androgen receptor in prostate cancer cells by inactivating histone deacetylase 6. Proc Natl Acad Sci 106:16663–16668

Glozak MA, Seto E (2007) Histone deacetylases and cancer. Oncogene 26:5420–5432

Glozak MA, Sengupta N, Zhang X, Seto E (2005) Acetylation and deacetylation of non-histone proteins. Gene 363:15–23

Gong D, Zeng Z, Yi F, Wu J (2019) Inhibition of histone deacetylase 11 promotes human liver cancer cell apoptosis. Am J Transl Res 11:983–990

Gorospe M, de Cabo R (2008) AsSIRTing the DNA damage response. Trends Cell Biol 18:77–83

Halkidou K, Gaughan L, Cook S, Leung HY, Neal DE, Robson CN (2004) Upregulation and nuclear recruitment of HDAC1 in hormone refractory prostate cancer. Prostate 59:177–189

Hrgovic I, Doll M, Pinter A, Kaufmann R, Kippenberger S, Meissner M (2017) Histone deacetylase inhibitors interfere with angiogenesis by decreasing endothelial VEGFR-2 protein half-life in part via a VE-cadherin-dependent mechanism. Exp Dermatol 26:194–201

Hu Y, Lin J, Lin Y, Chen X, Zhu G, Huang G (2019) Overexpression of SIRT4 inhibits the proliferation of gastric cancer cells through cell cycle arrest. Oncol Lett 17:2171–2176

Huang BH, Laban M, Leung CH, Lee L, Lee CK, Salto-Tellez M, Raju GC, Hooi SC (2005) Inhibition of histone deacetylase 2 increases apoptosis and p21Cip1/WAF1 expression, independent of histone deacetylase 1. Cell Death Differ 12:395–404

Jeong SM, Lee A, Lee J, Haigis MC (2014) SIRT4 protein suppresses tumor formation in genetic models of Myc-induced B cell lymphoma. J Biol Chem 289:4135–4144

Jung KH, Noh JH, Kim JK, Eun JW, Bae HJ, Xie HJ, Chang YG, Kim MG, Park H, Lee JY, Nam SW (2012) HDAC2 overexpression confers oncogenic potential to human lung cancer cells by deregulating expression of apoptosis and cell cycle proteins. J Cell Biochem 113:2167–2177

Kaidi A, Weinert BT, Choudhary C, Jackson SP (2010) Human SIRT6 promotes DNA end resection through CtIP deacetylation. Science (New York, N.Y.) 329:1348–1353

Kang ZH, Wang CY, Zhang WL, Zhang JT, Yuan CH, Zhao PW, Lin YY, Hong S, Li CY, Wang L (2014) Histone deacetylase HDAC4 promotes gastric cancer SGC-7901 cells progression via p21 repression. PLoS One 9:e98894

Kim MS, Kwon HJ, Lee YM, Baek JH, Jang JE, Lee SW, Moon EJ, Kim HS, Lee SK, Chung HY, Kim CW, Kim KW (2001) Histone deacetylases induce angiogenesis by negative regulation of tumor suppressor genes. Nat Med 7:437–443

Kotian S, Liyanarachchi S, Zelent A, Parvin JD (2011) Histone deacetylases 9 and 10 are required for homologous recombination. J Biol Chem 286:7722–7726

Kouzarides T (2000) Acetylation: a regulatory modification to rival phosphorylation? EMBO J 19:1176–1179

Kovacs JJ, Murphy PJ, Gaillard S, Zhao X, Wu JT, Nicchitta CV, Yoshida M, Toft DO, Pratt WB, Yao TP (2005) HDAC6 regulates Hsp90 acetylation and chaperone-dependent activation of glucocorticoid receptor. Mol Cell 18:601–607

Kugel S, Sebastián C, Fitamant J, Ross KN, Saha SK, Jain E, Gladden A, Arora KS, Kato Y, Rivera MN, Ramaswamy S, Sadreyev RI, Goren A, Deshpande V, Bardeesy N, Mostoslavsky R (2016) SIRT6 suppresses pancreatic cancer through control of Lin28b. Cell 165:1401–1415

Lemercier C, Brocard MP, Puvion-Dutilleul F, Kao HY, Albagli O, Khochbin S (2002) Class II histone deacetylases are directly recruited by BCL6 transcriptional repressor. J Biol Chem 277:22045–22052

Li Y, Seto E (2016) HDACs and HDAC inhibitors in cancer development and therapy. Cold Spring Harb Perspect Med 6:a026831

Li Z, Zhu W-G (2014) Targeting histone deacetylases for cancer therapy: from molecular mechanisms to clinical implications. Int J Biol Sci 10:757–770

Lin Z, Yang H, Tan C, Li J, Liu Z, Quan Q, Kong S, Ye J, Gao B, Fang D (2013) USP10 antagonizes c-Myc transcriptional activation through SIRT6 stabilization to suppress tumor formation. Cell Rep 5:1639–1649

Lin Y-H, Yuan J, Pei H, Liu T, Ann DK, Lou Z (2015) KAP1 deacetylation by SIRT1 promotes non-homologous end-joining repair. PLoS One 10:e0123935

Linggi B, Müller-Tidow C, van de Locht L, Hu M, Nip J, Serve H, Berdel WE, van der Reijden B, Quelle DE, Rowley JD, Cleveland J, Jansen JH, Pandolfi PP, Hiebert SW (2002) The t(8;21) fusion protein, AML1–ETO, specifically represses the transcription of the p14ARF tumor suppressor in acute myeloid leukemia. Nat Med 8:743–750

Luna A, Aladjem MI, Kohn KW (2013) SIRT1/PARP1 crosstalk: connecting DNA damage and metabolism. Genome Integr 4:6–6

Ma Y, Yue Y, Pan M, Sun J, Chu J, Lin X, Xu W, Feng L, Chen Y, Chen D, Shin VY, Wang X, Jin H (2015) Histone deacetylase 3 inhibits new tumor suppressor gene DTWD1 in gastric cancer. Am J Cancer Res 5:663–673

Maiti A, Qi Q, Peng X, Yan L, Takabe K, Hait NC (2019) Class I histone deacetylase inhibitor suppresses vasculogenic mimicry by enhancing the expression of tumor suppressor and anti-angiogenesis genes in aggressive human TNBC cells. Int J Oncol 55:116–130

Malik S, Villanova L, Tanaka S, Aonuma M, Roy N, Berber E, Pollack J, Michishita-Kioi E, Chua K (2015) SIRT7 inactivation reverses metastatic phenotypes in epithelial and mesenchymal tumors. Sci Rep 5:9841

Mao Z, Hine C, Tian X, Van Meter M, Au M, Vaidya A, Seluanov A, Gorbunova V (2011) SIRT6 promotes DNA repair under stress by activating PARP1. Science (New York, N.Y.) 332:1443–1446

McGarry LC, Winnie JN, Ozanne BW (2004) Invasion of v-Fos(FBR)-transformed cells is dependent upon histone deacetylase activity and suppression of histone deacetylase regulated genes. Oncogene 23:5284–5292

Melnick AM, Westendorf JJ, Polinger A, Carlile GW, Arai S, Ball HJ, Lutterbach B, Hiebert SW, Licht JD (2000) The ETO protein disrupted in t(8;21)-associated acute myeloid leukemia is a corepressor for the promyelocytic leukemia zinc finger protein. Mol Cell Biol 20:2075–2086

Miller KM, Tjeertes JV, Coates J, Legube G, Polo SE, Britton S, Jackson SP (2010) Human HDAC1 and HDAC2 function in the DNA-damage response to promote DNA nonhomologous end-joining. Nat Struct Mol Biol 17:1144–1151

Miyo M, Yamamoto H, Konno M, Colvin H, Nishida N, Koseki J, Kawamoto K, Ogawa H, Hamabe A, Uemura M, Nishimura J, Hata T, Takemasa I, Mizushima T, Doki Y, Mori M, Ishii H (2015) Tumour-suppressive function of SIRT4 in human colorectal cancer. Br J Cancer 113:492–499

Mostoslavsky R, Chua KF, Lombard DB, Pang WW, Fischer MR, Gellon L, Liu P, Mostoslavsky G, Franco S, Murphy MM, Mills KD, Patel P, Hsu JT, Hong AL, Ford E, Cheng HL, Kennedy C, Nunez N, Bronson R, Frendewey D, Auerbach W, Valenzuela D, Karow M, Hottiger MO, Hursting S, Barrett JC, Guarente L, Mulligan R, Demple B, Yancopoulos GD, Alt FW (2006) Genomic instability and aging-like phenotype in the absence of mammalian SIRT6. Cell 124:315–329

Mottet D, Pirotte S, Lamour V, Hagedorn M, Javerzat S, Bikfalvi A, Bellahcène A, Verdin E, Castronovo V (2009) HDAC4 represses p21(WAF1/Cip1) expression in human cancer cells through a Sp1-dependent, p53-independent mechanism. Oncogene 28:243–256

Onder TT, Gupta PB, Mani SA, Yang J, Lander ES, Weinberg RA (2008) Loss of E-cadherin promotes metastasis via multiple downstream transcriptional pathways. Cancer Res 68:3645–3654

Osada H, Tatematsu Y, Saito H, Yatabe Y, Mitsudomi T, Takahashi T (2004) Reduced expression of class II histone deacetylase genes is associated with poor prognosis in lung cancer patients. Int J Cancer 112:26–32

Parbin S, Kar S, Shilpi A, Sengupta D, Deb M, Rath SK, Patra SK (2014) Histone deacetylases: a saga of perturbed acetylation homeostasis in cancer. J Histochem Cytochem 62:11–33

Peinado H, Ballestar E, Esteller M, Cano A (2004) Snail mediates E-cadherin repression by the recruitment of the Sin3A/histone deacetylase 1 (HDAC1)/HDAC2 complex. Mol Cell Biol 24:306–319

Ropero S, Fraga MF, Ballestar E, Hamelin R, Yamamoto H, Boix-Chornet M, Caballero R, Alaminos M, Setien F, Paz MF, Herranz M, Palacios J, Arango D, Orntoft TF, Aaltonen LA, Schwartz S Jr, Esteller M (2006) A truncating mutation of HDAC2 in human cancers confers resistance to histone deacetylase inhibition. Nat Genet 38:566–569

Saha RN, Pahan K (2006) HATs and HDACs in neurodegeneration: a tale of disconcerted acetylation homeostasis. Cell Death Differ 13:539–550

Seto E, Yoshida M (2014) Erasers of histone acetylation: the histone deacetylase enzymes. Cold Spring Harb Perspect Biol 6:a018713

Singh BN, Zhang G, Hwa YL, Li J, Dowdy SC, Jiang S-W (2010) Nonhistone protein acetylation as cancer therapy targets. Expert Rev Anticancer Ther 10:935–954

Song J, Noh JH, Lee JH, Eun JW, Ahn YM, Kim SY, Lee SH, Park WS, Yoo NJ, Lee JY, Nam SW (2005) Increased expression of histone deacetylase 2 is found in human gastric cancer. Acta Pathol Microbiol Immunol Scand 113:264–268

Song S, Wang Y, Xu P, Yang R, Ma Z, Liang S, Zhang G (2015) The inhibition of histone deacetylase 8 suppresses proliferation and inhibits apoptosis in gastric adenocarcinoma. Int J Oncol 47:1819–1828

Stark M, Hayward N (2007) Genome-wide loss of heterozygosity and copy number analysis in melanoma using high-density single-nucleotide polymorphism arrays. Cancer Res 67:2632–2642

Sun H, Huang D, Liu G, Jian F, Zhu J, Zhang L (2018) SIRT4 acts as a tumor suppressor in gastric cancer by inhibiting cell proliferation, migration, and invasion. Onco Targets Ther 11:3959–3968

Taylor BS, DeCarolis PL, Angeles CV, Brenet F, Schultz N, Antonescu CR, Scandura JM, Sander C, Viale AJ, Socci ND, Singer S (2011) Frequent alterations and epigenetic silencing of differentiation pathway genes in structurally rearranged liposarcomas. Cancer Discov 1:587–597

Thurn KT, Thomas S, Raha P, Qureshi I, Munster PN (2013) Histone deacetylase regulation of ATM-mediated DNA damage signaling. Mol Cancer Ther 12:2078–2087

Tian J, Yuan L (2018) Sirtuin 6 inhibits colon cancer progression by modulating PTEN/AKT signaling. Biomed Pharmacother 106:109–116

Velcich A, Yang W, Heyer J, Fragale A, Nicholas C, Viani S, Kucherlapati R, Lipkin M, Yang K, Augenlicht L (2002) Colorectal cancer in mice genetically deficient in the mucin Muc2. Science (New York, N.Y.) 295:1726–1729

Wang T-J, Huang M-S, Hong C-Y, Tse V, Silverberg GD, Hsiao M (2001) Comparisons of tumor suppressor p53, p21, and p16 gene therapy effects on glioblastoma tumorigenicity in situ. Biochem Biophys Res Commun 287:173–180

Wang RH, Lahusen TJ, Chen Q, Xu X, Jenkins LM, Leo E, Fu H, Aladjem M, Pommier Y, Appella E, Deng CX (2014) SIRT1 deacetylates TopBP1 and modulates intra-S-phase checkpoint and DNA replication origin firing. Int J Biol Sci 10:1193–1202

Whetstine JR, Ceron J, Ladd B, Dufourcq P, Reinke V, Shi Y (2005) Regulation of tissue-specific and extracellular matrix-related genes by a class I histone deacetylase. Mol Cell 18:483–490

Wilson AJ, Byun DS, Popova N, Murray LB, L'Italien K, Sowa Y, Arango D, Velcich A, Augenlicht LH, Mariadason JM (2006) Histone deacetylase 3 (HDAC3) and other class I HDACs regulate colon cell maturation and p21 expression and are deregulated in human colon cancer. J Biol Chem 281:13548–13558

Wilson AJ, Byun D-S, Nasser S, Murray LB, Ayyanar K, Arango D, Figueroa M, Melnick A, Kao GD, Augenlicht LH, Mariadason JM (2008) HDAC4 promotes growth of colon cancer cells via repression of p21. Mol Biol Cell 19:4062–4075

Yamada N, Hamada T, Goto M, Tsutsumida H, Higashi M, Nomoto M, Yonezawa S (2006) MUC2 expression is regulated by histone H3 modification and DNA methylation in pancreatic cancer. Int J Cancer 119:1850–1857

Yang XJ, Seto E (2007) HATs and HDACs: from structure, function and regulation to novel strategies for therapy and prevention. Oncogene 26:5310–5318

Yin Y, Zhang M, Dorfman RG, Li Y, Zhao Z, Pan Y, Zhou Q, Huang S, Zhao S, Yao Y, Zou X (2017) Histone deacetylase 3 overexpression in human cholangiocarcinoma and promotion of cell growth via apoptosis inhibition. Cell Death Dis 8:e2856–e2856

Yu Y, Xu F, Peng H, Fang X, Zhao S, Li Y, Cuevas B, Kuo WL, Gray JW, Siciliano M, Mills GB, Bast RC Jr (1999) NOEY2 (ARHI), an imprinted putative tumor suppressor gene in ovarian and breast carcinomas. Proc Natl Acad Sci U S A 96:214–219

Zhang J, Zhong Q (2014) Histone deacetylase inhibitors and cell death. Cell Mol Life Sci 71:3885–3901

Zhang Z, Yamashita H, Toyama T, Sugiura H, Omoto Y, Ando Y, Mita K, Hamaguchi M, Hayashi S, Iwase H (2004) HDAC6 expression is correlated with better survival in breast cancer. Clin Cancer Res 10:6962–6968

Zhang Z, Yamashita H, Toyama T, Sugiura H, Ando Y, Mita K, Hamaguchi M, Hara Y, Kobayashi S, Iwase H (2005) Quantitation of HDAC1 mRNA expression in invasive carcinoma of the breast*. Breast Cancer Res Treat 94:11–16

Zhang J, Yin XJ, Xu CJ, Ning YX, Chen M, Zhang H, Chen SF, Yao LQ (2015) The histone deacetylase SIRT6 inhibits ovarian cancer cell proliferation via down-regulation of notch 3 expression. Eur Rev Med Pharmacol Sci 19:818–824

Zhou J, Wu A, Yu X, Zhu J, Dai H (2017) SIRT6 inhibits growth of gastric cancer by inhibiting JAK2/STAT3 pathway. Oncol Rep 38:1059–1066

Compendium of Mechanistic Insights of Distinct Conventional Anticancer Therapies and Their Grievous Toxicities

6

Omitting cardiovascular diseases, cancer is the first chief death cause globally (Nagai and Kim 2017; Siegel et al. 2020). The term cancer includes over two hundred seventy-seven (277) diseases all specified by uncurbed cell proliferation and this evidence is sufficient for saying that it is not a single disease (Hassanpour and Dehghani 2017). Hippocrates, the father of contemporary medicine was the first physician who used the "carcinos" and "carcinoma" words to describe tumours. The word cancer has been derived from the Greek term "karkinos" or crab. Actually it was Celsus, a Roman physician who translated this Greek word into cancer which is the Latin version for crab (Bynum 2012; Di Lonardo et al. 2015; Grant 2001; Hajdu 2011). Galen, a Greek physician used the term oncos (swelling) in order to describe tumours (Di Lonardo et al. 2015). Majority of cancers form a lump termed as tumour and not all lumps should be considered cancer. While cancerous lumps are known as malignant tumours, the non-cancerous lumps are termed as benign tumours. Certain cancers like leukaemia do not form tumours. Thus all tumours cannot be labelled cancerous, only malignant ones are cancerous. While certain cancers show fast growth and spreading, others show reverse trend. Cancers respond to various treatments differentially.

6.1 Overview of Different Cancer Types

Cancers are classified on the basis of the tissue type in which the cancer sprouts (histological type) and by the primary site of cancer development. From histological point of view, hundreds of distinct cancers exist. These different cancers have been placed under the confines of six major categories. These categories based on histological type will be discussed one after the other. Carcinoma designates the malignant neoplasm having epithelial origin or in other words carcinomas are the malignancies of epithelial tissue (Coleman 2018). It has been estimated that 80–90% of all cancer cases belong to this type. Epithelial tissue occurs all through the body. Carcinoma has two major subtypes, namely adenocarcinoma and squamous cell

S. A. Ganai, *Histone Deacetylase Inhibitors in Combinatorial Anticancer Therapy*,
https://doi.org/10.1007/978-981-15-8179-3_6

carcinoma. While adenomacarcinoma originates from an organ or gland, squamous cell carcinoma develops from the squamous epithelium (Neville et al. 2019). Unlike squamous cell carcinomas occurring in different body areas, adenocarcinomas predominantly appear in mucus membranes. Most of the carcinomas affect glands or organs having secretory function including breasts and lungs producing milk and secreting mucus, respectively. Among all cases of cancer 80–90% are carcinomas. While sarcoma designates the cancer of connective tissues, leukaemia represents the cancer of bone marrow, creating blood cells (Nenot and Stather 1979). In general sarcoma occurs in young adults and it has been seen that the most frequently occurring sarcoma develops on bone as hurting mass (Morrison 2003). The Greek meaning of leukaemia is white blood and this cancer usually results in excessive production of white blood cells which are immature (Piller 2001). As the immature cells are functionally compromised, thus patients with leukaemia are oftentimes vulnerable to infections (Piller 2001). Lymphoma, the most frequent form of haematological malignancy represents the lymphatic system cancer. This system forms the component of body's defence system and it has been reported that the incidence of non-Hodgkin lymphoma is relatively higher than Hodgkin lymphoma (Ganai 2016a). The hallmark of Hodgkin's lymphoma is the presence of HRS (Hodgkin and Reed/Sternberg) cells (Kuppers and Hansmann 2005). Plasma cells are present in the bone marrow and the cancer of these cells is termed as myeloma (Collins 2004). Myeloma cells which are malignant plasma cells assemble in the bone marrow. Localized tumours of plasma cells known as plasmacytomas grow inside or outside the bone. When multiple plasmacytomas develop inside or outside the bone this condition is known as multiple myeloma (Di Micco and Di Micco 2005). Certain cancers are of mixed type consisting of two or more types of tissue. Teratocarcinoma, carcinosarcoma and adenosquamous carcinoma form the examples of such tumours (Bastide et al. 2010; Feng et al. 2015; Malavalli et al. 2013). Based on the site of development, cancers have many types. These types are prostate, skin, pancreatic, colorectal, breast, lung, ovarian and cervical cancer (Apalla et al. 2017; Cohen et al. 2019; Ganai 2016b, c; Granados-Romero et al. 2017; Kim and Andriole 2018; Yabar and Winter 2016).

6.2 Conventional Chemotherapy

As per national cancer institute dictionary, treatment that has widespread acceptance and is followed by majority of health professionals is known as conventional treatment. The conventional therapy designates the traditional anticancer therapy procedures used nowadays in clinic. Conventional anticancer therapies encompass fractionated/conventional radiation therapy, surgery and traditional chemotherapy (Fig. 6.1) (Ece et al. 2014). Many cancer types respond to traditional anticancer agents such as alkylating agents, antimetabolites, intercalating agents/drugs, antimitotic drugs and topoisomerase inhibitors (Meegan and O'Boyle 2019).

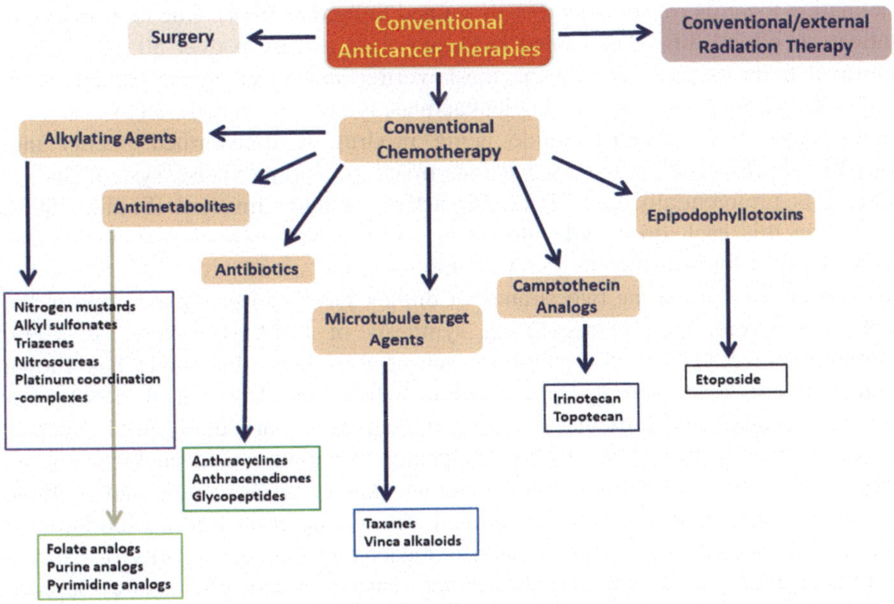

Fig. 6.1 An overview of conventional therapies used against cancer. These therapies include traditional chemotherapy, external beam radiation therapy and surgery. Surgery is mainly used before metastasis. Conventional radiation therapy is also used for cancer treatment. Conventional chemotherapy involves certain drugs such as those belonging to alkylating agents, antibiotics have antitumour action, agents interfering microtubules, enzymes like camptothecin analogues and epipodophyllotoxins such as etoposide

6.2.1 Alkylating Agents in Anticancer Therapy

The premier non-hormonal drugs used for anticancer therapy were alkylating agents (Hall and Tilby 1992). These agents being genotoxic directly bind to DNA and as such interfere with replication and transcriptional events leading to mutations. Thus alkylating agents cause severe DNA damage sufficient for inducing apoptosis in cancer cells (Guimaraes et al. 2013). Mechanistically, alkylating agents cause DNA duplex cross-linking especially at guanine (N-7 position) (Ralhan and Kaur 2007). These agents have various subgroups including nitrogen mustards, nitrosoureas, triazenes, alkyl sulfonates, ethylenamine/methylenamine derivatives and platinum coordination complexes.

6.2.1.1 Nitrogen Mustards

Among the alkylating agents nitrogen mustards are the predominantly used antineoplastic agents (More et al. 2019). Only five among thousands of synthesized and tested nitrogen mustards are frequently used for anticancer therapy nowadays. These include the original nitrogen mustard mechlorethamine, melphalan, cyclophosphamide, chlorambucil and ifosfamide (Ralhan and Kaur 2007; Sreerama 2014). Nitrogen mustard derivative cyclophosphamide due to its modified structure has higher inclination towards cancer cells (Friedman and Seligman 1954). Due to its effective anticancer effect, this drug gained FDA approval and even over 50 years of its approval finds its place among the most worthy anticancer agents (Emadi et al. 2009). This drug is used for treating leukaemias, lymphoma in addition to ovary and breast cancers. Cyclophosphamide being prodrug is transformed mainly into 4-hydroxycyclophosphamide by hepatic mixed function oxidase system and in tumour is preferentially converted into active nitrogen mustard (Siddik 2005). Following this each drug molecule forms adducts with two distinct nucleotides sequentially which results in the formation of interstrand cross-link. Once the cross-links are formed the two strands of duplex DNA fail to separate during the replication event, thereby restraining synthesis of DNA (Colvin et al. 1999; Guimaraes et al. 2013). Structural isomer of this drug, ifosfamide has shown comparatively lesser toxicity in clinical investigations (Corsi et al. 1978). For treating sarcomas and testicular tumours, ifosfamide is particularly used (Loehrer Sr. et al. 1988; Pratt et al. 1989). Melphalan, an amino acid analogue, as an alkylating agent is used for treating ovarian cancer, breast cancer and multiple myeloma (Costa et al. 1973; Rivkin et al. 1989; Young et al. 1990). Chlorambucil unlike cyclophosphamide or melphalan is nicely tolerated by majority of patients and thus can be used as substitute in patients sensitive to side effects of cyclophosphamide or melphalan. For treating lymphoma, ovarian carcinoma and chronic lymphocytic leukaemia this alkylating agent is used (Harding et al. 1988; Portlock et al. 1987; Rundles et al. 1959; Wiltshaw 1965).

6.2.1.2 Alkyl Sulfonates

Busulfan, an alkyl alkane sulfonate with selective toxicity towards early myeloid precursors is among the oldest alkylating agents (Elson 1958; Fried et al. 1977; Haddow and Timmis 1953). Activity of busulfan against chronic myelocytic leukaemia has been attributed to the above-mentioned selectivity (Galton 1953; Galton et al. 1958). Alkyl sulfamate analogue of this drug, hepsulfam showed no superiority over busulfan in clinical trials (Pacheco et al. 1989).

6.2.1.3 Triazenes

Temozolomide and dacarbazine are the triazene compounds having clinical significance. These agents cause methylation of *O*6-guanine by way of methyldiazonium ion. Triazene compounds have efficient pharmacokinetic properties and their toxicity is also limited (Marchesi et al. 2007). For treating malignant melanoma and Hodgkin disease dacarbazine is used (Advani 2011; Flaherty 2006). Oral alkylating agent temozolomide is effective against melanoma and due to its efficient bioavailability especially in the central nervous system is also useful for overcoming primary brain tumours (Neyns et al. 2010).

6.2.1.4 Nitrosoureas

The name of nitrosoureas comes among the list of most active anticancer agents. Anticancer activity of these drugs has been noted not only on oral administration but also through parenteral route. These agents undergo decomposition

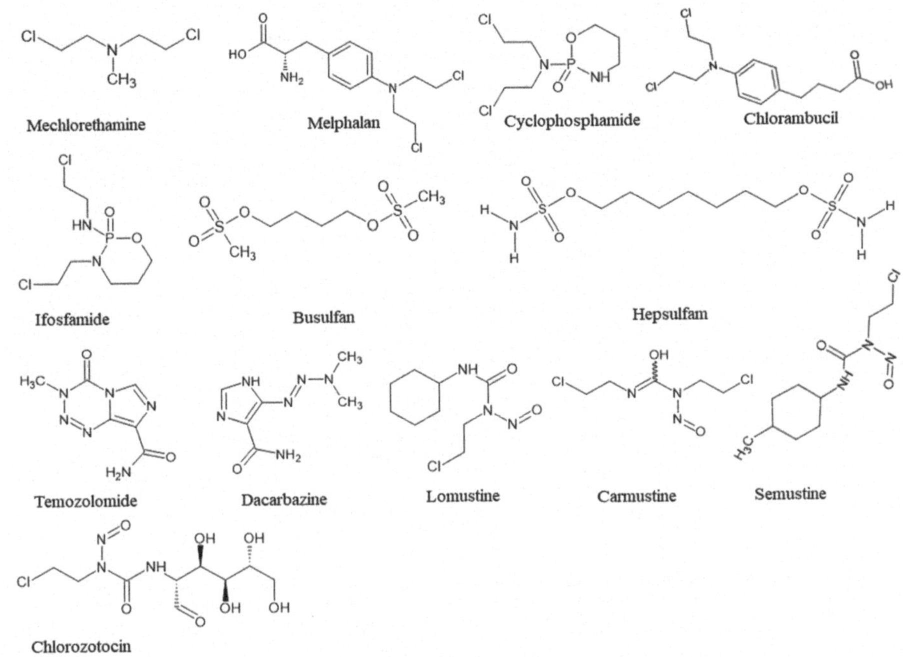

Fig. 6.2 Structural details of various alkylating anticancer agents belonging to nitrogen mustards, alkyl sulfonates, triazenes and nitrosoureas. Chemical structures from mechlorethamine to ifosfamide come under nitrogen mustards. Busulfan and hepsulfam are alkyl sulfonates, while temozolomide and dacarbazine are triazenes. The remaining structures from lomustine to chlorozotocin are nitrosoureas. For generation of chemical structures ACD/ChemSketch (Freeware) was used

non-enzymatically to gain the alkylating and carbamylating functions (Lemoine et al. 1991). Decomposition yields 2-chloroethyl carbonium ion, a strong electrophile having the ability to alkylate adenine, cytidine and guanine. A variety of compounds including 1-(2-chloroethyl)-3-cyclohexyl-1-nitrosourea (CCNU/lomustine), bis(chloroethyl) nitrosourea (BCNU/carmustine), 2-chloroethylnitrosoureas (CENUs), 1-(2-chloroethyl)-3-(4-methylcyclohexyl)-1-nitrosourea (methyl-CCNU/semustine) and chlorozotocin belong to this category of drugs (Fig. 6.2) (Newton 2006). CCNU and BCNU being lipid soluble are the most frequently used nitrosoureas for cancer chemotherapy. Due to hydrophobicity feature, nitrosoureas can cross the blood–brain barrier making them suitable candidates for treating brain tumours (Krouwer et al. 1990; Lefkowitz et al. 1990; Schabel Jr. et al. 1963; Walker et al. 1978).

6.2.1.5 Platinum Coordination Complexes

The efficacy of platinum compounds in cancer patients was recognized in seventies (Desoize 2004). Due to severe side effects of cisplatin, the analogues of this compound were developed with the aim to mitigate its toxicity and to circumvent

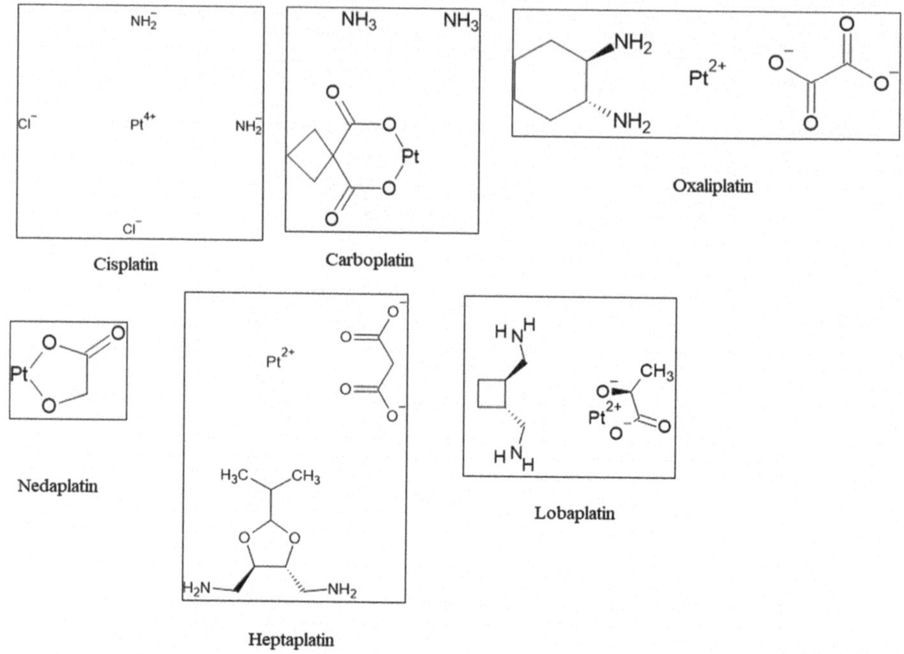

Fig. 6.3 Here the chemical structures of six platinum coordination complexes which are currently in use are shown. While carboplatin to nedaplatin are second generation analogues of cisplatin, lobaplatin and heptaplatin are third generation analogues

the platinum resistance issue (Boulikas et al. 2007; Gore et al. 1989). Six of the platinum compounds are in clinical use nowadays. These compounds include cisplatin; second generation analogues such as carboplatin, oxaliplatin and nedaplatin; third generation analogues including heptaplatin and lobaplatin (Fig. 6.3) (Johnstone et al. 2016). Certain platinum complexes are undergoing clinical evaluation for their possible use in anticancer therapy. Platinum compounds being prodrugs undergo aquation to form diaquo-platinum compound which is active. The active aquated platinum specifically reacts with guanine and adenine (N-7 position) and may result in formation of intrastrand and interstrand cross-links (Desoize 2004; Hall and Hambley 2002). Thus the cytotoxic effects of these compounds may be ascribed to their ability of forming platinum-DNA adducts (Saris et al. 1996). Cisplatin is used either as a single agent or in conjunction with other anticancer agents. Cisplatin cures 90% of the testicular cancers when used in combinatorial therapy with etoposide, vinblastine or bleomycin (Devanandan and Chowdary 2013). In most of the ovary carcinoma patients, cisplatin/carboplatin in combination with other anticancer agent paclitaxel induces full response (Bicaku et al. 2012). Compared to cisplatin, carboplatin causes less ototoxicity, nephrotoxicity, neurotoxicity, nausea and is clinically better tolerated (Karasawa and Steyger 2015). Anticancer activity of oxaliplatin has been reported against gastric and colorectal cancers (Comella et al. 2009; Zhang et al. 2019). Further the clinical trial studies have proved enhanced therapeutic effect of oxaliplatin in combination with other therapeutic agents as compared to its use as single agent. Unlike cisplatin, this anticancer agent has not shown nephrotoxic effects (Ludwig et al. 2004; Raymond et al. 1998).

6.2.2 Antimetabolites

Certain cytotoxic agents are structurally similar to substances utilized in normal biochemical pathways (Lansiaux 2011). They block DNA synthesis on competing with natural substrate for the enzyme active site (Carver 2011). These agents known as antimetabolites act by way of mimicking pyrimidines and purines crucial for DNA synthesis or by preventing native synthesis (Lansiaux 2011; Szucs and Jones 2016). Thus antimetabolites are the earliest anticancer agents which are rationally designed and targeted against DNA and RNA (Peters 2014). Due to faster cell division in cancer cells compared to normal cells antimetabolites show selectivity to former cells to certain extent (Avendaño and Menéndez 2008). Antimetabolites, the normal metabolites analogues restrain cell division and growth. Majority of the antimetabolites act during the synthetic phase of cell cycle (Freres et al. 2017). Antimetabolites include folic acid analogues, cytidine and pyrimidine and purine analogues.

6.2.2.1 Analogues of Folic Acid and Their Mechanism of Action

An antifolate drug aminopterin became famous after it was found to induce temporary remissions in acute lymphoblastic leukaemia children (Farber and Diamond 1948). As usual, this finding triggered the search for other metabolites with lesser toxicity profile compared to this anticancer agent. Following aminopterin, methotrexate (folate analogue) demonstrated anticancer activity. Folic acid analogues exhibit anticancer activity in various ways. Mainly these analogues show competition with folates for cellular uptake (Wosikowski et al. 2003). Moreover these analogues by way of restraining dihdrofolate reductase prevent its conversion to active tetrahydofolate (Osborn et al. 1958). Being an essential cofactor in biosynthesis of certain DNA and RNA precursors, the depletion of active tetrahydofolate may obviously hamper DNA synthesis and subsequent replication.

Despite the several choices of antifolate drugs methotrexate is frequently used. In osteosarcoma, acute lymphoblastic leukaemia and lymphoma this therapeutic agent is largely used (Hagner and Joerger 2010). On entering into cells through reduced folate carrier or with the assistance of folate binding protein, methotrexate undergoes polyglutamation, the reaction being catalysed by folylpolyglutamate synthetase. This modified methotrexate has longer retention time in cells compared to normal methotrexate (Chabner et al. 1985; Cho et al. 2007; Mikkelsen et al. 2011). The prime target of methotrexate and its cell modified version methotrexate-polyglutamate is the dihydrofolate reductase (Mikkelsen et al. 2011). By inhibiting this enzyme, methotrexate and methotrexate-polyglutamate alleviate the tetrahydrofolate cofactors. This in turn attenuates the synthesis of new purine and thymidylate nucleotides and consequently the nucleic acid biosynthesis. Moreover, methotrexate-polyglutamates have also been reported to interfere with three other folate-dependent enzymes including thymidylate synthase (Chabner et al. 1985; Hsiao et al. 2014). Selectivity of methotrexate towards cancerous tissues is to a certain extent explained by the fact that normal cells exhibit relatively lower polyglutamation (Guimaraes et al. 2013).

Pemetrexed, a folate analogue has already been approved by FDA for malignant pleural mesothelioma treatment and for advanced or metastatic lung cancer (non-small cell type) (Buqué et al. 2012; Hazarika et al. 2004). The efficacy of this anticancer agent is being evaluated against other cancers including gastric, cervical and breast (Goedhals et al. 2006; Martin 2006; Zhang et al. 2015). As a polyglutamate this anticancer agent primarily inhibits thymidylate synthase and glycinamide ribonucleotide transformylase, an enzyme involved in de novo purine formation. Pemetrexed like methotrexate also obstructs dihydrofolate reductase but this inhibition does not seen to be potent or primary as no drastic reduction occurs in the level of reduced folates (Guimaraes et al. 2013). This anticancer agent is used in both forms as a single agent and in combination with other agents like cisplatin (Vogelzang et al. 2003).

6.2.2.2 Purine Analogues and Their Targets

Among the purine nucleoside analogues, 6-mercaptopurine is the earliest that has been approved for treating acute leukaemias (Guimaraes et al. 2013; Stanczyk et al. 2012). Cladribine, fludarabine and pentostatin, the next generation purine nucleoside analogues have strong importance as therapeutic agents against haematological malignancies (Robak et al. 2009). 6-mercaptopurine, an analogue of hypoxanthine gained FDA approval in 1953 and is still among the most important drugs employed for the treatment of acute leukaemias (Bhagavan and Ha 2015). 6-mercaptopurine being prodrug undergoes intracellular transformation to methylated TIMP (6-thioinosine-5′-monophosphate). First with the help of hypoxanthine guanine phosphoribosyltransferase, 6-mercaptopurine is converted into TIMP. Thiopurine S-methyltransferase (TPMT) then converts TIMP into methylated TIMP, the potential inhibitor of purine biosynthesis (de novo) (Dubinsky et al. 2000).

For chronic lymphocytic leukaemia treatment, fludarabine phosphate (FAMP) got approval in 1991 (Guimaraes et al. 2013). In order to induce cytotoxicity this prodrug needs metabolic conversion. It undergoes quick dephosphorylation to 9-β-D-arabinosyl-2-fluoroadenine (F-ara-A) following which cellular uptake occurs where it is converted into active form 2-fluoro-ara-ATP (F-ara-ATP) by deoxycytidine kinase (Chun et al. 1991; Plunkett et al. 1990). It affects various enzymes involved in DNA synthesis in one way or the other way. These enzymes include ribonucleotide reductase, DNA ligase and DNA primase. F-ara-ATP also restrains DNA polymerization on getting incorporated into DNA at the terminus where new nucleotides are joined (Catapano et al. 1993; Yang et al. 1992).

Another next generation analogue cladribine was approved by FDA for treatment of hairy cell leukaemia (Ogura 2003). Cladribine on getting phosphorylated accumulates in cells as 2-chlorodeoxyadenosine triphosphate (2-CdA-TP) (Arner 1996). 2-CdA-TP gets incorporated into DNA, thereby obstructing its synthesis and repair. Besides, this metabolite exerts potent inhibitory activity against ribonucleotide reductase (Beutler 1992). Pentostatin (deoxycoformycin), a natural product from *Streptomyces antibioticus* was the premier promising agent for hairy cell leukaemia treatment (Kane et al. 1992; Klohs and Kraker 1992). Nowadays its use has been replaced by cladribine discussed above. The prime target of this anticancer agent is adenosine deaminase, an enzyme has a role in purine salvage pathway (Kane et al. 1992). In cells accumulation of adenosine and deoxyadenosine nucleotides as a consequence of adenosine deaminase inhibition inhibits DNA synthesis by way of obstructing ribonucleotide reductase (Guimaraes et al. 2013).

6.2.2.3 Pyrimidine Analogues as Anticancer Agents

After it became evident that malignant tissues use uracil comparatively faster than normal tissues, the idea of possible use of pyrimidine analogues in anticancer therapy gained interest. The first pyrimidine analogue, 5-fluorouracil (5-FU) was synthesized by Charles Heidelberger and his associates (Lee 2005; Longley et al. 2003). With the advent of time other pyrimidine analogues such as gemcitabine, cytosine arabinoside and capecitabine were also developed. At present pyrimidine analogues are extensively used in anticancer therapy.

5-fluorouracil (5-FU) is used against a variety of cancers including colorectal, pancreatic and breast cancer (Zhang et al. 2008). On entering into cells this pyrimidine analogue undergoes metabolic conversion to 5-fluoro-2′-deoxyuridine-5-′-monophosphate (FdUMP). This active metabolite promotes incorporation of uracil into DNA by depleting thymine levels via inhibition of thymidylate synthase culminating in DNA breaks. Fluorouridine triphosphate (FUTP), another metabolite of 5-FU impairs RNA processing by getting incorporated into RNA extensively (Miura et al. 2010). Gemcitabine, an analogue of deoxycytidine, in 1996 was approved by FDA for pancreatic cancer and non-small cell lung cancer treatment. Its approval was extended to metastatic breast cancer treatment in 2004 (Toschi et al. 2005). Following this gemcitabine approval was further extended in 2006 for treatment of advanced ovarian cancer (Guimaraes et al. 2013). Gemcitabine undergoes metabolic conversion by several enzymes to 5′-triphosphate derivatives (dFdCTP) which on incorporation into DNA restrains replication and triggers programmed cell death (Wong et al. 2009).

Capecitabine, an oral prodrug of 5-fluorouracil, gained FDA approval (1998) to treat metastatic breast cancer patients which was later extended to metastatic colorectal carcinoma. By thymidine phosphorylase, the elevated levels of which are comparatively seen in a variety of tumours, capecitabine undergoes metabolic conversion to active fluorouracil (Walko and Lindley 2005). 1-β-D-Arabinofuranosylcytosine also known as cytarabine after being integrated into DNA as a sham nucleotide interferes DNA polymerase activity which in turns impairs the synthesis of DNA (Fig. 6.4). Being cell cycle specific, cytarabine is S-phase restricted antimetabolite. This anticancer agent has FDA approval for a variety of malignancies. These include leukaemia that has reached to meninges and in combination with other drugs for chronic myeloid leukaemia, acute lymphoblastic leukaemia and acute myeloid leukaemia (Guimaraes et al. 2013). Besides liposomal

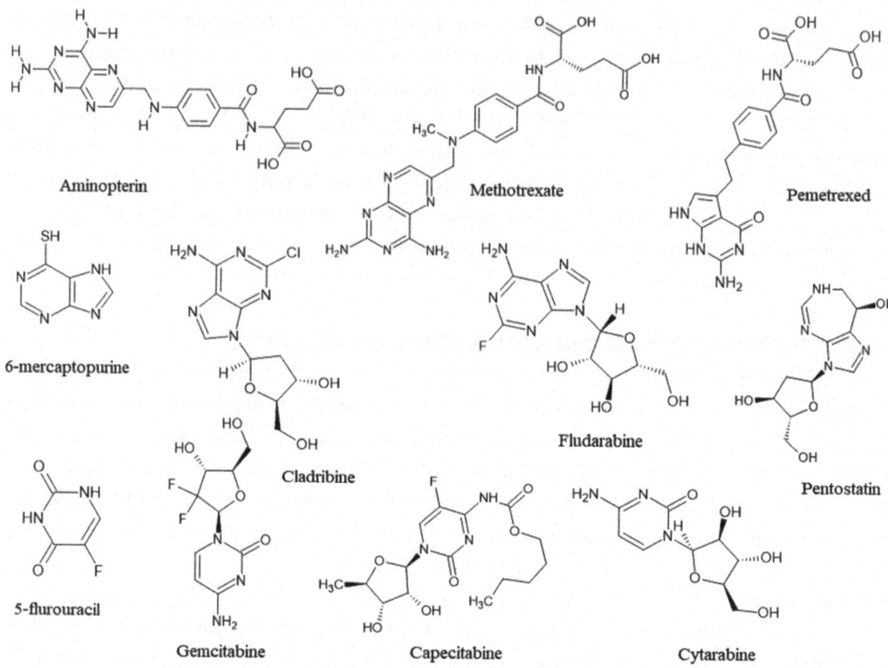

Fig. 6.4 Detailed chemical structure of various antimetabolites having antineoplastic properties. Structures from aminopterin to pemetrexed clockwise come under the confines of folate analogues. While structures from 6-mercaptopurine to pentostatin (clockwise) belong to purine analogues, the remaining chemical structures including 5-fluorouracil and onwards are pyrimidine analogues

cytarabine through intrathecal administration has FDA approval to treat lymphoma that has reached to meninges (Kripp and Hofheinz 2008). Cytarabine, a prodrug, inside cells gets activated by multiple enzymatic phosphorylation steps to ara-C 5′-triphosphate (Ara-CTP) (Kim et al. 2013).

6.2.3 Anticancer Antibiotics

Certain antibiotics have emerged as promising anticancer agents. Antibiotics belonging to anthracyclines (aromatic polyketides), glycopeptides and indocarbazoles class have proved effective against cancer (Saeidnia 2015). Anthracyclines such as epirubicin, daunorubicin, idarubicin and doxorubicin are often used for treating various cancers. While epirubicin and doxorubicin are effective against solid tumours of humans, idarubicin and daunorubicin are more promising against acute leukaemias. Doxorubicin is a key anticancer antibiotic used against a wide range of cancers including gastric, lung, breast, paediatric cancers and sarcomas. Its epimer, namely epirubicin is used to tackle breast cancer (Cortes-Funes and Coronado 2007; Thorn et al. 2011). Anthracyclines show anticancer effect by multiple mechanisms. Topoisomerase II prevents DNA from becoming tangled during replication by

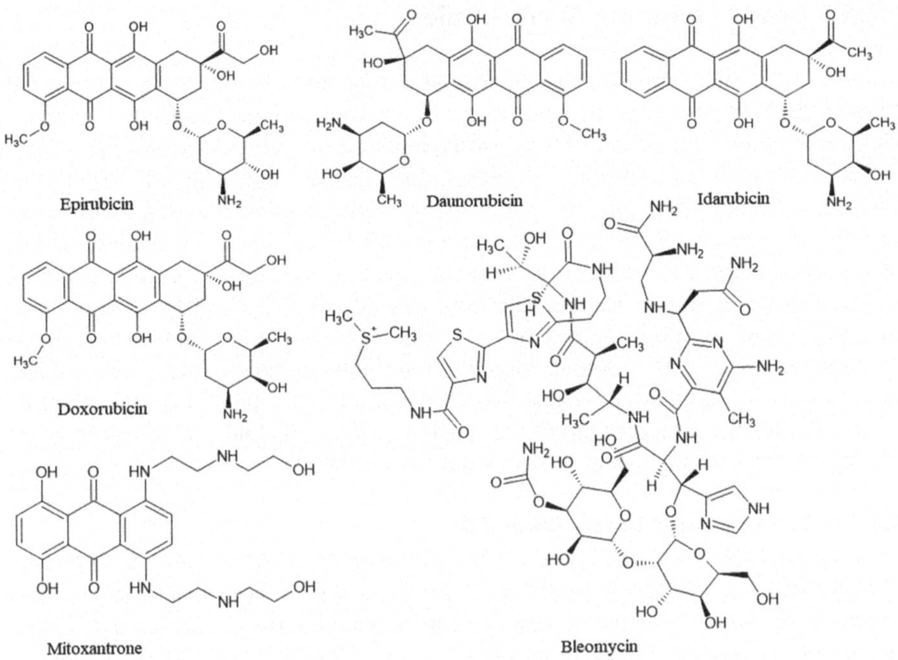

Fig. 6.5 Various anticancer antibiotics and their chemical structures. The first four structures clockwise from top most row are anthracyclines. Anthracenedione derivative mitoxantrone also acts as anticancer agent. The last structure bleomycin is glycopeptide antibiotic functioning as antitumour agent

nicking and resealing. Anthracyclines inhibit the DNA-topoisomerase II prior to resealing and after nicking phase. As a result of this large number of DNA fragments get produced facilitating programmed cell death. Anthracyclines by producing free radicals result in damage of proteins, lipids and cell membranes. Further anthracyclines being DNA intercalators get incorporated into DNA and inhibit DNA-templated reactions such as replication and transcription (Bardal et al. 2011; Marinello et al. 2018). Compared to free/conventional doxorubicin, liposomal formulations of this anticancer agent (doxorubicin) showed mitigated toxicity without alteration of efficacy against metastatic breast cancer (Batist et al. 2001).

Anthracenediones class of anticancer antibiotics unlike anthracyclines exert less cardiotoxicity. Mitoxantrone, the doxorubicin analogue and top active member of this class acquired FDA approval to treat AML in 1987 and for prostate cancer in 1996 (Fox 2004; Koutinos et al. 2002). This anticancer agent not only intercalates into DNA but also obstructs the topoisomerase II (Yoneda and Cross 2010). Another antitumour antibiotic bleomycin belonging to glycopeptides has also shown significant anticancer effect (Fig. 6.5) (Galm et al. 2005). Bleomycin exerts cytotoxic effect through free radical generation culminating in DNA breaks (Tounekti et al. 2001).

6.2.4 Drugs Targeting Microtubules

Among the major components of cytoskeleton, microtubules are formed of heterodimers of α–β-tubulin and in addition microtubule associated proteins (Schwarzerová et al. 2019). They participate in a variety of cellular processes including mitosis and protein transportation (Janke and Magiera 2020). As microtubules are central players in mitosis, thus therapeutics affecting them serves as propitious anticancer agents (Karahalil et al. 2019; Pasquier and Kavallaris 2008). Drugs influencing microtubules are currently used in the clinics for circumventing haematological malignancies and solid tumours (Mukhtar et al. 2014). These drugs are either used as single agents or in conjunction with other anticancer agents (Smorenburg et al. 2001). Agents targeting microtubules function as strong mitotic poisons may be microtubule-destabilizing or microtubule-stabilizing (Fanale et al. 2015). While the examples of former include vinca alkaloids, taxanes form the example of latter (Fanale et al. 2015; Moudi et al. 2013).

6.2.4.1 Taxanes and Vinca Alkaloids

As mentioned above taxanes come under anticancer agents stabilizing microtubules. Chemically they are diterpenoid natural products with inherent anticancer effect (Wani et al. 1971). Paclitaxel and docetaxel are two taxane anticancer agents effectively used against a broad range of cancers (Ojima et al. 2016). Paclitaxel was isolated from the bark of *Taxus brevifolia* by Wani et al. (1971). USFDA approved injectable paclitaxel (Taxol) in 1992 to treat refractory ovarian cancer. This approval in 1994 was extended to breast cancer (refractory or anthracyclines resistant) and then further broadened to Kaposi's sarcoma and non-small cell lung cancer in 1997 and 1998, respectively (Ojima et al. 2016). Paclitaxel's semi-synthetic analogue docetaxel (taxotere) showed comparatively superior efficacy in certain cases. Docetaxel received first FDA approval for advanced breast cancer treatment (Bissery et al. 1995). Later on this approval was expanded to non-small cell lung cancer, metastatic HRPC (hormone refractory prostate cancer) in addition to head and neck cancer (Gueritte 2001; Ojima et al. 2016). Another taxane anticancer drug cabazitaxel (Jevtana), acquainted with better anticancer activity than the above-mentioned taxane agents, has been approved by FDA. Due to its poor inclination towards P-glycoprotein cabazitaxel has proved promising even against docetaxel-resistant tumours (Kartner et al. 1983). Paclitaxel and docetaxel cause mitotic arrest and subsequent programmed cell death by interacting with β-tubulin and stabilizing microtubules (Horwitz et al. 1993; Snyder et al. 2001). Moreover, taxanes may promote apoptosis in cancer cells by mediating the inactivation (phosphorylation) of certain proteins having key role in restraining apoptosis (Haldar et al. 1996).

Vinca alkaloids are the oldest microtubule target agents extracted from periwinkle plant. Vinca alkaloids vinblastine and vincristine in USA have been approved for clinical use (Fig. 6.6) (Martino et al. 2018; Tagliamento et al. 2019). For treating urothelium carcinoma in Europe, synthetic vinca alkaloid vinflunine has been approved (Moudi et al. 2013). Speaking mechanistically, these alkaloids cause

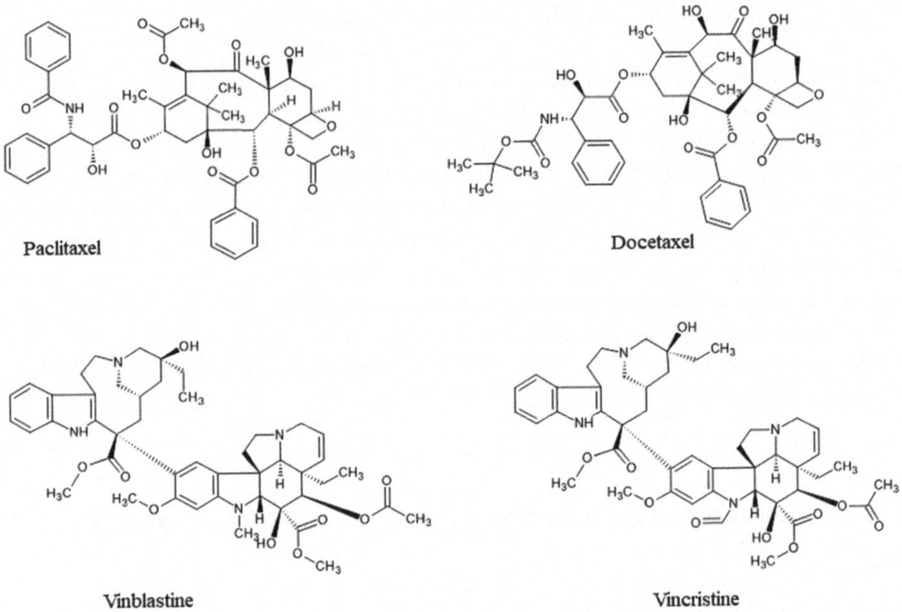

Fig. 6.6 Conventional antineoplastic drugs targeting microtubules. Paclitaxel and docetaxel are taxanes. The remaining drugs from vinblastine to vincristine are vinca alkaloids

depolymerization of microtubules and thus obstruct mitotic advancement culminating in cell death through apoptotic mechanism (Jordan et al. 1991; Jordan and Wilson 2004). In combination with other agents vincristine is used to treat neuroblastoma, Wilms tumour and certain haematological malignancies including lymphoma and paediatric leukaemias (Groninger et al. 2002; Kingston 2009).

6.2.5 Analogues of Camptothecin in Cancer Treatment

Camptothecin present in various parts of *Camptotheca acuminate* targets topoisomerase I specifically (Pommier 2006). Topoisomerase I plays a key role in DNA replication and during RNA formation. This pentacyclic alkaloid occurs in bark, wood and fruits of the above-mentioned Asian tree (Wall et al. 1966). Topoisomerase I performs the function of relaxing the supercoiled DNA by inducing a single strand break followed by relegation. Camptothecin inhibits the rejoining step of cleavage/religation process, thereby interrupting the process of cell division (Kacprzak 2013; Liu et al. 2000). Two camptothecin analogues irinotecan and topotecan have gained FDA approval for clinical usage (Li et al. 2017). These analogues having less toxicity but more solubility are nowadays used for intervention against various cancers. While irinotecan is used against colorectal cancer (metastatic), topotecan has its use against cervical, lung and ovarian cancer treatment (Guimaraes et al. 2013).

6.2.6 Epipodophyllotoxins for Tackling Cancer

Though podophyllotoxin, a lignan was isolated first from *Podophyllum peltatum* in 1880, its structure was elucidated by Hartwell and Schrecker in 1951 (Canel et al. 2000; Hartwell and Schrecker 1951). Due to unfavourable toxicity profile, this toxin in spite of having anticancer activity was not clinically used (Lucas et al. 2010). Among the various less toxic podophyllotoxin analogues, etoposide has been approved to be used in clinic (Fig. 6.7). Etoposide although mainly used for lung and testicular cancer treatment has proven to be effective against gastric cancer, lymphomas and acute nonlymphocytic leukaemia as well (Cragg et al. 2005; Hande 1998; Lucas et al. 2010; van Maanen et al. 1988). These agents by way of inhibiting topoisomerase II result in the accumulation of DNA breaks, thereby facilitating caspase driven apoptosis (Hande 1998).

6.3 Conventional Radiation Therapy and Its Mechanism of Action

For damaging cancer cells ionizing radiations (physical agents) are used. By way of damaging DNA of the cells the high energy radiations exert antiproliferative effect (Jackson and Bartek 2009). As radiations do not differentiate among cancer and normal cells, thus during radiotherapy extreme care is taken to give maximum exposure to cancer cells and to minimize the same towards normal cells (Emami

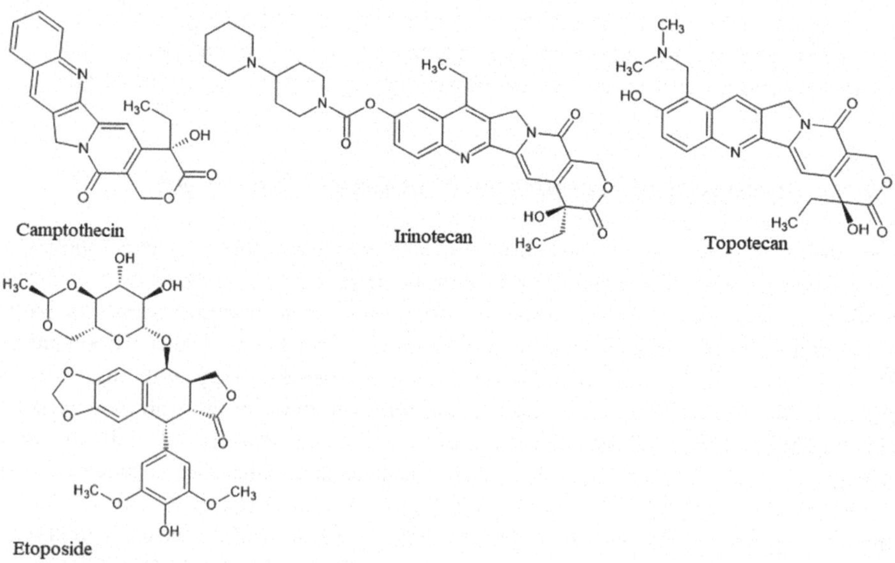

Fig. 6.7 Chemical structures of camptothecin, its two analogues (irinotecan and topotecan) and two podophyllotoxin analogue, namely etoposide

et al. 1991). However, radiation causes differential killing of cancer cells as these cells unlike normal cells are not effective in repairing radiation-induced damage (Begg et al. 2011). X-rays are most frequently used in radiation therapy, the other options being gamma radiations and photons. Radiations are delivered to tumour location mainly as external beam and thus this type of therapy is termed as external beam radiation therapy. Radiation induced killing of cancer cells depends on various aspects like targeted tissues/cells radiosensitivity, linear energy transfer, fractionation in addition to total dose (Baskar 2010; Hall 2007). Radiation therapy induces cell killing through a variety of mechanisms. By direct way radiation damages cellular DNA by interacting with it. Excitation/ionization of cellular water by radiations results in the production of free radicals which in turn cause DNA damage. Radiations induce double and single strand DNA breaks the former being more crucial in killing of cancer and nearby normal cells (Baskar et al. 2012). The predominant pathways of cell death induced by radiations are apoptosis and mitotic cell death also known as mitotic catastrophe (Verheij 2008). Radiotherapy is often used in conjunction with chemotherapy to achieve maximum therapeutic benefit.

6.4 Surgery

Another conventional method of treating cancer is surgery (Guimaraes et al. 2013). This method is the highest effective method for vanquishing more cancer types prior to metastasis. Surgery is used either alone or in combination with chemotherapy and radiotherapy (Shewach and Kuchta 2009). Prior to surgery neoadjuvant therapy is given to shrink the tumour dimensions and following surgery adjuvant therapy is provided to eliminate the cancer cells up to the utmost extent (Delaney et al. 2005; Glynne-Jones and Chau 2013; Sun et al. 2018). This traditional method may cure the person before the onset of metastasis. Before surgery it is not always possible to diagnose whether the tumour has undergone metastasis or not. While performing surgery, doctors remove sentinel nodes (lymph nodes near to tumour) to confirm whether the metastasis has occurred or yet to occur (Bouquet de Jolinière et al. 2018; Chéreau et al. 2013). If metastasis has occurred, then surgery needs to be followed by chemotherapy/radiation therapy to restrain the recurrence. Surgery was regarded as the only liked treatment for cancer cure prior to 1950 (Abbas and Rehman 2018).

6.5 Dreadful Toxicities of Conventional Anticancer Drugs

The main toxicities associated with alkylating agents are related to gastrointestinal tract and bone marrow. Transient escalation of serum aminotransferase has been noted in a portion of patients. While alkylating agents when administered in high doses cause sinusoidal obstruction syndrome, the prolonged use results in nodular regenerative hyperplasia. Chlorambucil on clinical use produces certain undesirable effects like vomiting, nausea, anaemia, neurotoxicity and bone marrow suppression (Nicolle et al. 2004; Springer et al. 1990). Certain toxicities like ototoxicity, neurotoxicity and nephrotoxicity are associated with the use of cisplatin (Brabec et al. 2017). The other platinum compound carboplatin also shows these toxicities but with lesser intensity. Carboplatin shows primary thrombocytopenia and myelosuppression as dose-limiting toxicities (Guimaraes et al. 2013).

Methotrexate also exerts certain toxicities to patients. These include liver damage, cytopenia, alopecia, serious infections, mucocutaneous complications and allergic interstitial pneumonitis (Salliot and van der Heijde 2009). Adverse reactions including anorexia and neutropenia were reported with the therapeutic use of pemetrexed drug (Adjei 2004). Purine analogues such as fludarabine have also shown certain toxicities like lymphocytopenia, myelosuppression and severe neurotoxicity (Cheson et al. 1994). Moreover bone marrow suppression has also been noted in case of cladribine and pentostatin. A variety of toxicities were found on use of pyrimidine analogues. Fluorouracil induces stomatitis, vomiting, bone marrow suppression, cardiotoxicity and thrombocytopenia (Han et al. 2008; Stewart et al. 2010). While hand-foot syndrome, alopecia and cardiotoxicity are the side effects recorded with capecitabine, gemcitabine has implications in myelosuppression, flu-like symptoms and rash (Hui and Reitz 1997; Stewart et al. 2010). Also unpleasant effects have been observed with cytarabine. Among the incited effects were acute pulmonary syndrome, mucositis, thrombocytopenia, neurotoxicity, severe anaemia and myelosuppression (Baker et al. 1991; Stentoft 1990). The major serious concern with the use of anthracyclines such as doxorubicin is cardiotoxicity (Pai and Nahata 2000). Mitoxantrone, an anthracenedione (synthetic) also induces cardiotoxicity but to a lesser extent than doxorubicin. Moreover, this anticancer agent causes unwanted effects like acute myelogenous leukaemia and severe leukopenia (Lv et al. 2019; Shaikh et al. 2016).

Neuropathy and neutropenia are among the predominant effects of taxanes. The intensity of docetaxel related neutropenia is higher than that of paclitaxel (Gradishar et al. 2012; Rowinsky et al. 1993). Pulmonary oedema and peripheral oedema are also triggered by docetaxel due to its fluid retention effect (Dunsford et al. 1999; Read et al. 2002). Moreover the cancer cells become resistant to taxanes as these drugs have high binding affinity for P-glycoprotein due to which they are effluxed from the cells. Neurotoxicity is often noticed on use of vincristine (Moudi et al. 2013; Quasthoff and Hartung 2002). Myelosuppression is the prime toxicity related to the latter two vinca alkaloids (Goa and Faulds 1994; Moudi et al. 2013). Irinotecan, a camptothecin analogue often causes diarrhoea and other side effects like vomiting, alopecia, fatigue, neutropenia. Neutropenia has been observed as the prime concern associated with topotecan administration (Grochow et al. 1992; Ormrod and Spencer 1999).

Thrombosis, pain, join disorders and hot flashes are the frequent unwanted effects noticed with the use of antiestrogen fulvestrant (Valachis et al. 2010). Radiation therapy also kills the normal cells lying in the close vicinity of tumour. Conventional anticancer agents strongly affect normal swiftly dividing cells such as that of bone marrow, spermatogenic cells, hair follicles, gut and lymphoid tissue (Kakde et al. 2011; Links and Brown 1999).

Hitherto, I have rigorously discussed what actually the conventional therapy means. Following this the various types of conventional anticancer agents and their therapeutic role in tackling various malignancies have been explained deeply. Importantly, strong emphasis was given to nitrogen mustards, microtubule stabilizing and destabilizing agents, folate analogues, pyrimidine and purine analogues and conventional radiation therapy. Most importantly, the serious toxicities associated with the use of conventional therapeutic agents against a variety of cancers have been explained to a detailed degree. As a whole, conventional anticancer agents are devoid of selectivity and get distributed in the body arbitrarily culminating in serious off-target effects. Moreover cancer cells by effluxing the conventional agents (docetaxel and paclitaxel) easily develop resistance against them. Based on these reasons it is highly advisable to opt for other anticancer therapies which are selective and impart least or no significant toxicity to normal cells.

References

Abbas Z and Rehman S (2018) An overview of Cancer treatment modalities. Neoplasm Book, 1524

Adjei AA (2004) Pharmacology and mechanism of action of pemetrexed. Clin Lung Cancer 5 (Suppl 2):S51–S55

Advani R (2011) Optimal therapy of advanced Hodgkin lymphoma. Hematology 2011:310–316

Apalla Z, Nashan D, Weller RB, Castellsagué X (2017) Skin cancer: epidemiology, disease burden, pathophysiology, diagnosis, and therapeutic approaches. Dermatol Ther (Heidelb) 7:5–19

Arner ES (1996) On the phosphorylation of 2-chlorodeoxyadenosine (CdA) and its correlation with clinical response in leukemia treatment. Leuk Lymphoma 21:225–231

Avendaño C, Menéndez JC (2008) Chapter 2 - antimetabolites. In: Avendaño C, Menéndez JC (eds) Medicinal chemistry of anticancer drugs. Elsevier, Amsterdam, pp 9–52

Baker WJ, Royer GL Jr, Weiss RB (1991) Cytarabine and neurologic toxicity. J Clin Oncol 9:679–693

Bardal SK, Waechter JE, Martin DS (2011) Chapter 20 - Neoplasia. In: Bardal SK, Waechter JE, Martin DS (eds) Applied pharmacology. Content repository only! Elsevier, Philadelphia, pp 305–324

Baskar R (2010) Emerging role of radiation induced bystander effects: cell communications and carcinogenesis. Genome Integr 1:13

Baskar R, Lee KA, Yeo R, Yeoh K-W (2012) Cancer and radiation therapy: current advances and future directions. Int J Med Sci 9:193–199

Bastide K, Ugolin N, Levalois C, Bernaudin JF, Chevillard S (2010) Are adenosquamous lung carcinomas a simple mix of adenocarcinomas and squamous cell carcinomas, or more complex at the molecular level? Lung Cancer (Amsterdam, Netherlands) 68:1–9

Batist G, Ramakrishnan G, Rao CS, Chandrasekharan A, Gutheil J, Guthrie T, Shah P, Khojasteh A, Nair MK, Hoelzer K, Tkaczuk K, Park YC, Lee LW (2001) Reduced cardiotoxicity and preserved antitumor efficacy of liposome-encapsulated doxorubicin and cyclophosphamide compared with conventional doxorubicin and cyclophosphamide in a randomized, multicenter trial of metastatic breast cancer. J Clin Oncol 19:1444–1454

Begg AC, Stewart FA, Vens C (2011) Strategies to improve radiotherapy with targeted drugs. Nat Rev Cancer 11:239–253

Beutler E (1992) Cladribine (2-chlorodeoxyadenosine). Lancet (London, England) 340:952–956

Bhagavan NV, Ha C-E (2015) Chapter 25 - Nucleotide metabolism. In: Bhagavan NV, Ha C-E (eds) Essentials of medical biochemistry, 2nd edn. Academic Press, San Diego, pp 465–487

Bicaku E, Xiong Y, Marchion DC, Chon HS, Stickles XB, Chen N, Judson PL, Hakam A, Gonzalez-Bosquet J, Wenham RM, Apte SM, Fulp W, Cubitt CL, Chen DT, Lancaster JM (2012) In vitro analysis of ovarian cancer response to cisplatin, carboplatin, and paclitaxel identifies common pathways that are also associated with overall patient survival. Br J Cancer 106:1967–1975

Bissery MC, Nohynek G, Sanderink GJ, Lavelle F (1995) Docetaxel (Taxotere): a review of preclinical and clinical experience. Part I: Preclinical experience. Anticancer Drugs 6 (339–355):363–338

Boulikas T, Pantos A, Bellis E, Christofis P (2007) Designing platinum compounds in cancer: structures and mechanisms. Cancer Ther 5:537–583

Bouquet de Jolinière J, Major A, Khomsi F, Ben Ali N, Guillou L, Feki A (2018) The sentinel lymph node in breast cancer: problems posed by examination during surgery. A review of current literature and management. Front Surg 5:56

Brabec V, Hrabina O, Kasparkova J (2017) Cytotoxic platinum coordination compounds. DNA binding agents. Coord Chem Rev 351:2–31

Buqué A, Muhialdin JS, Muñoz A, Calvo B, Carrera S, Aresti U, Sancho A, Rubio I, López-Vivanco G (2012) Molecular mechanism implicated in pemetrexed-induced apoptosis in human melanoma cells. Mol Cancer 11:25

Bynum WF (2012) Leslie Morton and Robert J Moore, A chronology of medicine and related sciences, Aldershot, Scolar Press, 1997, pp. 784, £75.00 (1-85928-215-6). Med Hist 42:541–542

Canel C, Moraes RM, Dayan FE, Ferreira D (2000) Podophyllotoxin. Phytochemistry 54:115–120

Carver JR (2011) 9 - Management of cardiac and pulmonary treatment–related side effects. In: Davis MP, Feyer PC, Ortner P, Zimmermann C (eds) Supportive oncology. W.B. Saunders, Saint Louis, pp 67–94

Catapano CV, Perrino FW, Fernandes DJ (1993) Primer RNA chain termination induced by 9-beta-D-arabinofuranosyl-2-fluoroadenine 5′-triphosphate. A mechanism of DNA synthesis inhibition. J Biol Chem 268:7179–7185

Chabner BA, Allegra CJ, Curt GA, Clendeninn NJ, Baram J, Koizumi S, Drake JC, Jolivet J (1985) Polyglutamation of methotrexate. Is methotrexate a prodrug? J Clin Invest 76:907–912

Chéreau E, Hudry D, Lambaudie E, Cohen M, Houvenaeghel G, Coutant C (2013) Modèles prédictifs d'envahissement du ganglion non sentinelle en cas de ganglion sentinelle positif dans le cancer du sein. Oncologie 15:304–310

Cheson BD, Vena DA, Foss FM, Sorensen JM (1994) Neurotoxicity of purine analogs: a review. J Clin Oncol 12:2216–2228

Cho R, Cole P, Sohn K-J, Gaisano G, Croxford R, Kamen B, Kim Y-I (2007) Effects of folate and folyipolyglutamyl synthase modulation on chemosensitivity of breast cancer cells. Mol Cancer Ther 6:2909–2920

Chun HG, Leyland-Jones B, Cheson BD (1991) Fludarabine phosphate: a synthetic purine antimetabolite with significant activity against lymphoid malignancies. J Clin Oncol 9:175–188

Cohen PA, Jhingran A, Oaknin A, Denny L (2019) Cervical cancer. Lancet (London, England) 393:169–182

Coleman WB (2018) Chapter 4 - Neoplasia. In: Coleman WB, Tsongalis GJ (eds) Molecular pathology, 2nd edn. Academic Press, pp 71–97

Collins CD (2004) Multiple myeloma. Cancer Imaging 4(Spec No A):S47–S53

Colvin ME, Sasaki JC, Tran NL (1999) Chemical factors in the action of phosphoramidic mustard alkylating anticancer drugs: roles for computational chemistry. Curr Pharm Des 5:645–663

Comella P, Casaretti R, Sandomenico C, Avallone A, Franco L (2009) Role of oxaliplatin in the treatment of colorectal cancer. Ther Clin Risk Manag 5:229–238

Corsi A, Calabresi F, Greco C (1978) Comparative effects of cyclophosphamide and isophosphamide on Lewis lung carcinoma. Br J Cancer 38:631–633

Cortes-Funes H, Coronado C (2007) Role of anthracyclines in the era of targeted therapy. Cardiovasc Toxicol 7:56–60

Costa G, Engle RL, Schilling A, Carbone P, Kochwa S, Nachman RL, Glidewell O (1973) Melphalan and prednisone: an effective combination for the treatment of multiple myeloma. Am J Med 54:589–599

Cragg GM, Kingston D, Newman D (2005) Anticancer agents from natural products, pp 1–593

Delaney G, Jacob S, Featherstone C, Barton M (2005) The role of radiotherapy in cancer treatment: estimating optimal utilization from a review of evidence-based clinical guidelines. Cancer 104:1129–1137

Desoize B (2004) Metals and metal compounds in cancer treatment. Anticancer Res 24:1529–1544

Devanandan P, Chowdary PR (2013) A review on the use of bleomycin-cisplatin-vinblastine combinations in therapy of testicular cancer. Indian J Res Pharm Biotechnol 1:793–796

Di Lonardo A, Nasi S, Pulciani S (2015) Cancer: we should not forget the past. J Cancer 6:29–39

Di Micco P, Di Micco B (2005) Up-date on solitary plasmacytoma and its main differences with multiple myeloma. Exp Oncol 27:7–12

Dubinsky MC, Lamothe S, Yang HY, Targan SR, Sinnett D, Theoret Y, Seidman EG (2000) Pharmacogenomics and metabolite measurement for 6-mercaptopurine therapy in inflammatory bowel disease. Gastroenterology 118:705–713

Dunsford ML, Mead GM, Bateman AC, Cook T, Tung K (1999) Severe pulmonary toxicity in patients treated with a combination of docetaxel and gemcitabine for metastatic transitional cell carcinoma. Ann Oncol 10:943–947

Ece B, Eyup B, Aylin Sendemir U (2014) Implementation of nanoparticles in cancer therapy. In: Handbook of research on nanoscience, nanotechnology, and advanced materials. IGI Global, Hershey, pp 447–491

Elson LA (1958) Hematological effects of the alkylating agents. Ann N Y Acad Sci 68:826–833

Emadi A, Jones RJ, Brodsky RA (2009) Cyclophosphamide and cancer: golden anniversary. Nat Rev Clin Oncol 6:638–647

Emami B, Lyman J, Brown A, Coia L, Goitein M, Munzenrider JE, Shank B, Solin LJ, Wesson M (1991) Tolerance of normal tissue to therapeutic irradiation. Int J Radiat Oncol Biol Phys 21:109–122

Fanale D, Bronte G, Passiglia F, Calò V, Castiglia M, Di Piazza F, Barraco N, Cangemi A, Catarella MT, Insalaco L, Listì A, Maragliano R, Massihnia D, Perez A, Toia F, Cicero G, Bazan V (2015) Stabilizing versus destabilizing the microtubules: a double-edge sword for an effective cancer treatment option? Anal Cell Pathol 2015:690916

Farber S, Diamond LK (1948) Temporary remissions in acute leukemia in children produced by folic acid antagonist, 4-aminopteroyl-glutamic acid. N Engl J Med 238:787–793

Feng D, Fidele NB, Agustin MM, Jian G, Bourleyi SI, Augustin L, Olivier NK (2015) Carcinosarcoma of parotid gland (malignant mixed tumor). Ann Maxillofac Surg 5:240–243

Flaherty KT (2006) Chemotherapy and targeted therapy combinations in advanced melanoma. Clin Cancer Res 12:2366s–2370s

Fox EJ (2004) Mechanism of action of mitoxantrone. Neurology 63:S15–S18

Freres P, Jerusalem G, Moonen M (2017) Chapter 2 - Categories of anticancer treatments. In: Lancellotti P, Zamorano Gómez JL, Galderisi M (eds) Anti-cancer treatments and cardiotoxicity. Academic Press, Boston, pp 7–11

Fried W, Kedo A, Barone J (1977) Effects of cyclophosphamide and of busulfan on spleen colony-forming units and on hematopoietic stroma. Cancer Res 37:1205–1209

Friedman OM, Seligman AM (1954) Preparation of N-phosphorylated derivatives of Bis-β-chloroethylamine1a. J Am Chem Soc 76:655–658

Galm U, Hager MH, Van Lanen SG, Ju J, Thorson JS, Shen B (2005) Antitumor antibiotics: bleomycin, enediynes, and mitomycin. Chem Rev 105:739–758

Galton DA (1953) Myleran in chronic myeloid leukaemia; results of treatment. Lancet (London, England) 264:208–213

Galton DA, Till M, Wiltshaw E (1958) Busulfan (1, 4-dimethanesulfonyloxybutane, myleran); summary of clinical results. Ann N Y Acad Sci 68:967–973

Ganai SA (2016a) Histone deacetylase inhibitor givinostat: the small-molecule with promising activity against therapeutically challenging haematological malignancies. J Chemother (Florence, Italy) 28:247–254
Ganai SA (2016b) Histone deacetylase inhibitor sulforaphane: the phytochemical with vibrant activity against prostate cancer. Biomed Pharmacother 81:250–257
Ganai SA (2016c) Panobinostat: the small molecule metalloenzyme inhibitor with marvelous anticancer activity. Curr Top Med Chem 16:427–434
Glynne-Jones R, Chau I (2013) Neoadjuvant therapy before surgical treatment. EJC Suppl 11:45–59
Goa KL, Faulds D (1994) Vinorelbine. A review of its pharmacological properties and clinical use in cancer chemotherapy. Drugs Aging 5:200–234
Goedhals L, Wiyk AL, Smith BL, Fourie S (2006) Pemetrexed (Alimta (R), LY231514) demonstrates clinical activity in chemonaive patients with cervical cancer in a phase II single-agent trial. Int J Gynecol Cancer 16:1172–1178
Gore ME, Fryatt I, Wiltshaw E, Dawson T, Robinson BA, Calvert AH (1989) Cisplatin/carboplatin cross-resistance in ovarian cancer. Br J Cancer 60:767–769
Gradishar WJ, Krasnojon D, Cheporov S, Makhson AN, Manikhas GM, Clawson A, Bhar P, McGuire JR, Iglesias J (2012) Phase II trial of nab-paclitaxel compared with docetaxel as first-line chemotherapy in patients with metastatic breast cancer: final analysis of overall survival. Clin Breast Cancer 12:313–321
Granados-Romero J, Valderrama-Treviño A, Contreras Flores E, Barrera-Mera B, Herrera M, Uriarte-Ruíz K, Ceballos-Villalva JC, Estrada A, Rodríguez C, Arauz-Peña G (2017) Colorectal cancer: a review. Int J Res Med Sci 5:4667
Grant J (2001) Man and medicine: a history.: Farokh Erach Udwadia. (518 pages, £31.50.) Oxford University Press, 2000. ISBN 0-19-565457-9. Fam Pract 18:649–650
Grochow LB, Rowinsky EK, Johnson R, Ludeman S, Kaufmann SH, McCabe FL, Smith BR, Hurowitz L, DeLisa A, Donehower RC et al (1992) Pharmacokinetics and pharmacodynamics of topotecan in patients with advanced cancer. Drug Metab Dispos 20:706–713
Groninger E, Meeuwsen-de Boar T, Koopmans P, Uges D, Sluiter W, Veerman A, Kamps W, de Graaf S (2002) Pharmacokinetics of vincristine monotherapy in childhood acute lymphoblastic leukemia. Pediatr Res 52:113–118
Gueritte F (2001) General and recent aspects of the chemistry and structure-activity relationships of taxoids. Curr Pharm Des 7:1229–1249
Guimaraes I, Guimarães S, Daltoé R, Herlinger A, Klesia P, Madeira T, Ladislau I, Valadão P, Morais L, Junior S, Teixeira G, Amorim D, Zipinotti D, Santos K, Demuth L, Batista A, Rangel (2013) Conventional cancer treatment. https://doi.org/10.5772/55282
Haddow A, Timmis GM (1953) Myleran in chronic myeloid leukæmia chemical constitution and biological action. Lancet 261:207–208
Hagner N, Joerger M (2010) Cancer chemotherapy: targeting folic acid synthesis. Cancer Manag Res 2:293–301
Hajdu SI (2011) A note from history: landmarks in history of cancer, part 1. Cancer 117:1097–1102
Haldar S, Chintapalli J, Croce CM (1996) Taxol induces bcl-2 phosphorylation and death of prostate cancer cells. Cancer Res 56:1253–1255
Hall EJ (2007) Cancer caused by x-rays--a random event? The lancet. Oncology 8:369–370
Hall MD, Hambley TW (2002) Platinum(IV) antitumour compounds: their bioinorganic chemistry. Coord Chem Rev 232:49–67
Hall AG, Tilby MJ (1992) Mechanisms of action of, and modes of resistance to, alkylating agents used in the treatment of haematological malignancies. Blood Rev 6:163–173
Han R, Yang YM, Dietrich J, Luebke A, Mayer-Proschel M, Noble M (2008) Systemic 5-fluorouracil treatment causes a syndrome of delayed myelin destruction in the central nervous system. J Biol 7:12
Hande KR (1998) Etoposide: four decades of development of a topoisomerase II inhibitor. Eur J Cancer (Oxford, England: 1990) 34:1514–1521

Harding M, Kennedy R, Mill L, MacLean A, Duncan I, Kennedy J, Soukop M, Kaye SB (1988) A pilot study of carboplatin (JM8, CBDCA) and chlorambucil in combination for advanced ovarian cancer. Br J Cancer 58:640–643
Hartwell JL, Schrecker AW (1951) Components of podophyllin. V. The constitution of podophyllotoxin1. J Am Chem Soc 73:2909–2916
Hassanpour SH, Dehghani M (2017) Review of cancer from perspective of molecular. J Cancer Res Pract 4:127–129
Hazarika M, White RM, Johnson JR, Pazdur R (2004) FDA drug approval summaries: pemetrexed (Alimta). Oncologist 9:482–488
Horwitz SB, Cohen D, Rao S, Ringel I, Shen HJ, Yang CP (1993) Taxol: mechanisms of action and resistance. J Natl Cancer Inst Monogr:55–61
Hsiao Y-L, Chang P-C, Huang H-J, Kuo C-C, Chen CY-C (2014) Treatment of acute lymphoblastic leukemia from traditional Chinese medicine. Evid Based Complement Alternat Med 2014:601064–601064
Hui YF, Reitz J (1997) Gemcitabine: a cytidine analogue active against solid tumors. Am J Health Syst Pharm 54:162–170. quiz 197-168
Jackson SP, Bartek J (2009) The DNA-damage response in human biology and disease. Nature 461:1071–1078
Janke C, Magiera MM (2020) The tubulin code and its role in controlling microtubule properties and functions. Nat Rev Mol Cell Biol 21:307–326
Johnstone TC, Suntharalingam K, Lippard SJ (2016) The next generation of platinum drugs: targeted Pt(II) agents, nanoparticle delivery, and Pt(IV) prodrugs. Chem Rev 116:3436–3486
Jordan MA, Wilson L (2004) Microtubules as a target for anticancer drugs. Nat Rev Cancer 4:253–265
Jordan MA, Thrower D, Wilson L (1991) Mechanism of inhibition of cell proliferation by Vinca alkaloids. Cancer Res 51:2212–2222
Kacprzak KM (2013) Chemistry and biology of camptothecin and its derivatives. In: Ramawat KG, Mérillon J-M (eds) Natural products: phytochemistry, botany and metabolism of alkaloids, phenolics and terpenes. Springer, Berlin, pp 643–682
Kakde D, Jain D, Shrivastava V, Kakde R, Patil AT (2011) Cancer therapeutics—opportunities, challenges and advances in drug delivery. J Appl Pharm Sci 1:1–10
Kane BJ, Kuhn JG, Roush MK (1992) Pentostatin: an adenosine deaminase inhibitor for the treatment of hairy cell leukemia. Ann Pharmacother 26:939–947
Karahalil B, Yardım-Akaydin S, Baytas S (2019) An overview of microtubule targeting agents for cancer therapy. Arch Ind Hyg Toxicol 70:160–172
Karasawa T, Steyger PS (2015) An integrated view of cisplatin-induced nephrotoxicity and ototoxicity. Toxicol Lett 237:219–227
Kartner N, Riordan JR, Ling V (1983) Cell surface P-glycoprotein associated with multidrug resistance in mammalian cell lines. Science (New York, N.Y.) 221:1285–1288
Kim EH, Andriole GL (2018) Prostate cancer review. Mo Med 115:131–131
Kim KI, Huh IS, Kim IW, Park T, Ahn KS, Yoon SS, Yoon JH, Oh JM (2013) Combined interaction of multi-locus genetic polymorphisms in cytarabine arabinoside metabolic pathway on clinical outcomes in adult acute myeloid leukaemia (AML) patients. Eur J Cancer (Oxford, England: 1990) 49:403–410
Kingston DGI (2009) Tubulin-interactive natural products as anticancer agents. J Nat Prod 72:507–515
Klohs WD, Kraker AJ (1992) Pentostatin: future directions. Pharmacol Rev 44:459–477
Koutinos G, Stathopoulos GP, Dontas I, Perrea-Kotsarelis D, Couris E, Karayannacos PE, Deliconstantinos G (2002) The effect of doxorubicin and its analogue mitoxantrone on cardiac muscle and on serum lipids: an experimental study. Anticancer Res 22:815–820
Kripp M, Hofheinz R-D (2008) Treatment of lymphomatous and leukemic meningitis with liposomal encapsulated cytarabine. Int J Nanomed 3:397–401

Krouwer D, McDermott M, Prados M (1990) Postoperative radiotherapy and radiotherapy combined with CCNU chemotherapy for treatment of brain gliomas. J Neuro-Oncol 8:189–191
Kuppers R, Hansmann ML (2005) The Hodgkin and Reed/Sternberg cell. Int J Biochem Cell Biol 37:511–517
Lansiaux A (2011) Antimetabolites. Bull Cancer 98:1263–1274
Lee J-I (2005) Systemic overview of 5-FU based chemotherapy in stomach cancer. Korean J Clin Oncol 1:82–91
Lefkowitz IB, Packer RJ, Siegel KR, Sutton LN, Schut L, Evans AE (1990) Results of treatment of children with recurrent medulloblastoma/primitive neuroectodermal tumors with lomustine, cisplatin, and vincristine. Cancer 65:412–417
Lemoine A, Lucas C, Ings RMJ (1991) Metabolism of the chloroethylnitrosoureas. Xenobiotica 21:775–791
Li F, Jiang T, Li Q, Ling X (2017) Camptothecin (CPT) and its derivatives are known to target topoisomerase I (Top1) as their mechanism of action: did we miss something in CPT analogue molecular targets for treating human disease such as cancer? Am J Cancer Res 7:2350–2394
Links M, Brown R (1999) Clinical relevance of the molecular mechanisms of resistance to anticancer drugs. Expert Rev Mol Med 1999:1–21
Liu LF, Desai SD, Li TK, Mao Y, Sun M, Sim SP (2000) Mechanism of action of camptothecin. Ann N Y Acad Sci 922:1–10
Loehrer PJ Sr, Lauer R, Roth BJ, Williams SD, Kalasinski LA, Einhorn LH (1988) Salvage therapy in recurrent germ cell cancer: ifosfamide and cisplatin plus either vinblastine or etoposide. Ann Intern Med 109:540–546
Longley DB, Harkin DP, Johnston PG (2003) 5-fluorouracil: mechanisms of action and clinical strategies. Nat Rev Cancer 3:330–338
Lucas DM, Still PC, Pérez LB, Grever MR, Kinghorn AD (2010) Potential of plant-derived natural products in the treatment of leukemia and lymphoma. Curr Drug Targets 11:812–822
Ludwig T, Riethmüller C, Gekle M, Schwerdt G, Oberleithner H (2004) Nephrotoxicity of platinum complexes is related to basolateral organic cation transport. Kidney Int 66:196–202
Lv S, Li A, Wu H, Wang X (2019) Observation of clinical efficacy and toxic and side effects of pirarubicin combined with cytarabine on acute myeloid leukemia. Oncol Lett 17:3411–3417
Malavalli G, Karra S, Muniyappa B (2013) Teratocarcinoma in a non seminomatous, mixed germ cell tumour of the testis-a rare entity. J Clin Diagn Res 7:1439–1440
Marchesi F, Turriziani M, Tortorelli G, Avvisati G, Torino F, Vecchis L (2007) Triazene compounds: mechanism of action and related DNA repair systems. Pharmacol Res 56:275–287
Marinello J, Delcuratolo M, Capranico G (2018) Anthracyclines as topoisomerase II poisons: from early studies to new perspectives. Int J Mol Sci 19:3480
Martin M (2006) Clinical experience with pemetrexed in breast cancer. Semin Oncol 33:S15–S18
Martino E, Casamassima G, Castiglione S, Cellupica E, Pantalone S, Papagni F, Rui M, Siciliano AM, Collina S (2018) Vinca alkaloids and analogues as anti-cancer agents: looking back, peering ahead. Bioorg Med Chem Lett 28:2816–2826
Meegan MJ, O'Boyle NM (2019) Special issue "anticancer drugs". Pharmaceuticals (Basel) 12:134
Mikkelsen TS, Thorn CF, Yang JJ, Ulrich CM, French D, Zaza G, Dunnenberger HM, Marsh S, McLeod HL, Giacomini K, Becker ML, Gaedigk R, Leeder JS, Kager L, Relling MV, Evans W, Klein TE, Altman RB (2011) PharmGKB summary: methotrexate pathway. Pharmacogenet Genomics 21:679–686
Miura K, Kinouchi M, Ishida K, Fujibuchi W, Naitoh T, Ogawa H, Ando T, Yazaki N, Watanabe K, Haneda S, Shibata C, Sasaki I (2010) 5-fu metabolism in cancer and orally-administrable 5-fu drugs. Cancers 2:1717–1730
More GS, Thomas AB, Chitlange SS, Nanda RK, Gajbhiye RL (2019) Nitrogen mustards as alkylating agents: a review on chemistry, mechanism of action and current USFDA status of drugs. Anti Cancer Agents Med Chem 19:1080–1102
Morrison BA (2003) Soft tissue sarcomas of the extremities. Proc (Bayl Univ Med Cent) 16:285–290

Moudi M, Go R, Yien CYS, Nazre M (2013) Vinca alkaloids. Int J Prev Med 4:1231–1235
Mukhtar E, Adhami VM, Mukhtar H (2014) Targeting microtubules by natural agents for cancer therapy. Mol Cancer Ther 13:275–284
Nagai H, Kim YH (2017) Cancer prevention from the perspective of global cancer burden patterns. J Thorac Dis 9:448–451
Nenot JC, Stather JW (1979) Chapter 6 - Pathological effects in animals. In: Nenot JC, Stather JW (eds) The toxicity of plutonium, americium and curium. Pergamon, pp 103–141
Neville BW, Damm DD, Allen CM, Chi AC (2019) 10 - Epithelial pathology. In: Neville BW, Damm DD, Allen CM, Chi AC (eds) Color atlas of oral and maxillofacial diseases. Elsevier, Philadelphia, pp 223–271
Newton HB (2006) Chapter 2 - Clinical pharmacology of brain tumor chemotherapy. In: Newton HB (ed) Handbook of brain tumor chemotherapy. Academic Press, San Diego, pp 21–43
Neyns B, Tosoni A, Hwu WJ, Reardon DA (2010) Dose-dense temozolomide regimens: antitumor activity, toxicity, and immunomodulatory effects. Cancer 116:2868–2877
Nicolle A, Proctor SJ, Summerfield GP (2004) High dose chlorambucil in the treatment of lymphoid malignancies. Leuk Lymphoma 45:271–275
Ogura M (2003) [Cladribine]. Gan to kagaku ryoho. Cancer Chemother 30:309–317
Ojima I, Lichtenthal B, Lee S, Wang C, Wang X (2016) Taxane anticancer agents: a patent perspective. Expert Opin Ther Pat 26:1–20
Ormrod D, Spencer CM (1999) Topotecan: a review of its efficacy in small cell lung cancer. Drugs 58:533–551
Osborn MJ, Freeman M, Huennekens FM (1958) Inhibition of dihydrofolic reductase by aminopterin and amethopterin. Proc Soc Exp Biol Med 97:429–431
Pacheco DY, Stratton NK, Gibson NW (1989) Comparison of the mechanism of action of busulfan with hepsulfam, a new antileukemic agent, in the L1210 cell line. Cancer Res 49:5108–5110
Pai VB, Nahata MC (2000) Cardiotoxicity of chemotherapeutic agents: incidence, treatment and prevention. Drug Saf 22:263–302
Pasquier E, Kavallaris M (2008) Microtubules: a dynamic target in cancer therapy. IUBMB Life 60:165–170
Peters G (2014) Novel developments in the use of antimetabolites. Nucleosides Nucleotides Nucleic Acids 33:358–374
Piller GJ (2001) Leukaemia – a brief historical review from ancient times to 1950. Br J Haematol 112:282–292
Plunkett W, Huang P, Gandhi V (1990) Metabolism and action of fludarabine phosphate. Semin Oncol 17:3–17
Pommier Y (2006) Topoisomerase I inhibitors: camptothecins and beyond. Nat Rev Cancer 6:789–802
Portlock CS, Fischer DS, Cadman E, Lundberg WB, Levy A, Bobrow S, Bertino JR, Farber L (1987) High-dose pulse chlorambucil in advanced, low-grade non-Hodgkin's lymphoma. Cancer Treat Rep 71:1029–1031
Pratt CB, Douglass EC, Etcubanas E, Goren MP, Green AA, Hayes FA, Horowitz ME, Meyer WH, Thompson EI, Wilimas JA (1989) Clinical studies of ifosfamide/mesna at St Jude Children's Research Hospital, 1983–1988. Semin Oncol 16:51–55
Quasthoff S, Hartung HP (2002) Chemotherapy-induced peripheral neuropathy. J Neurol 249:9–17
Ralhan R, Kaur J (2007) Alkylating agents and cancer therapy. Expert Opin Ther Pat 17:1061–1075
Raymond E, Chaney SG, Taamma A, Cvitkovic E (1998) Oxaliplatin: a review of preclinical and clinical studies. Ann Oncol 9:1053–1071
Read WL, Mortimer JE, Picus J (2002) Severe interstitial pneumonitis associated with docetaxel administration. Cancer 94:847–853
Rivkin SE, Green S, Metch B, Glucksberg H, Gad-el-Mawla N, Constanzi JJ, Hoogstraten B, Athens J, Maloney T, Osborne CK et al (1989) Adjuvant CMFVP versus melphalan for operable breast cancer with positive axillary nodes: 10-year results of a Southwest Oncology Group Study. J Clin Oncol 7:1229–1238

Robak T, Korycka A, Lech-Maranda E, Robak P (2009) Current status of older and new purine nucleoside analogues in the treatment of lymphoproliferative diseases. Molecules (Basel, Switzerland) 14:1183–1226

Rowinsky EK, Eisenhauer EA, Chaudhry V, Arbuck SG, Donehower RC (1993) Clinical toxicities encountered with paclitaxel (Taxol). Semin Oncol 20:1–15

Rundles RW, Grizzle J, Bell WN, Corley CC, Frommeyer WB, Greenberg BG, Huguley CM, Watson James G, Jones R, Larsen WE, Loeb V, Leone LA, Palmer JG, Riser WH, Wilson SJ (1959) Comparison of chlorambucil and Myleran® in chronic lymphocytic and granulocytic leukemia. Am J Med 27:424–432

Saeidnia S (2015) Anticancer antibiotics. In: Saeidnia S (ed) New approaches to natural anticancer drugs. Springer, Cham, pp 51–66

Salliot C, van der Heijde D (2009) Long-term safety of methotrexate monotherapy in patients with rheumatoid arthritis: a systematic literature research. Ann Rheum Dis 68:1100–1104

Saris CP, van de Vaart PJ, Rietbroek RC, Blommaert FA (1996) In vitro formation of DNA adducts by cisplatin, lobaplatin and oxaliplatin in calf thymus DNA in solution and in cultured human cells. Carcinogenesis 17:2763–2769

Schabel FM Jr, Johnston TP, Mc CG, Montgomery JA, Laster WR, Skipper HE (1963) Experimental evaluation of potential anticancer agents VIII. Effects of certain nitrosoureas on intracerebral L1210 leukemia. Cancer Res 23:725–733

Schwarzerová K, Bellinvia E, Martinek J, Sikorová L, Dostál V, Libusová L, Bokvaj P, Fischer L, Schmit AC, Nick P (2019) Tubulin is actively exported from the nucleus through the Exportin1/CRM1 pathway. Sci Rep 9:5725

Shaikh AY, Suryadevara S, Tripathi A, Ahmed M, Kane JL, Escobar J, Cerny J, Nath R, McManus DD, Shih J, McGuiness ME, Tighe DA, Meyer TE, Ramanathan M, Aurigemma GP (2016) Mitoxantrone-induced cardiotoxicity in acute myeloid leukemia-a velocity vector imaging analysis. Echocardiography (Mount Kisco, N.Y.) 33:1166–1177

Shewach DS, Kuchta RD (2009) Introduction to cancer chemotherapeutics. Chem Rev 109:2859–2861

Siddik ZH (2005) Mechanisms of action of cancer chemotherapeutic agents: DNA-interactive alkylating agents and antitumour platinum-based drugs. In: Alison MR (ed) The cancer handbook. https://doi.org/10.1002/0470025077.chap84b

Siegel RL, Miller KD, Jemal A (2020) Cancer statistics, 2020. CA Cancer J Clin 70:7–30

Smorenburg CH, Sparreboom A, Bontenbal M, Verweij J (2001) Combination chemotherapy of the taxanes and antimetabolites: its use and limitations. Eur J Cancer (Oxford, England: 1990) 37:2310–2323

Snyder JP, Nettles JH, Cornett B, Downing KH, Nogales E (2001) The binding conformation of taxol in β-tubulin: a model based on electron crystallographic density. Proc Natl Acad Sci 98:5312–5316

Springer CJ, Antoniw P, Bagshawe KD, Searle F, Bisset GM, Jarman M (1990) Novel prodrugs which are activated to cytotoxic alkylating agents by carboxypeptidase G2. J Med Chem 33:677–681

Sreerama L (2014) Alkylating agents. In: Schwab M (ed) Encyclopedia of cancer. Springer, Berlin, pp 1–6

Stanczyk M, Sliwinski T, Trelinska J, Cuchra M, Markiewicz L, Dziki L, Bieniek A, Bielecka-Kowalska A, Kowalski M, Pastorczak A, Szemraj J, Mlynarski W, Majsterek I (2012) Role of base-excision repair in the treatment of childhood acute lymphoblastic leukaemia with 6-mercaptopurine and high doses of methotrexate. Mutat Res 741:13–21

Stentoft J (1990) The toxicity of cytarabine. Drug Saf 5:7–27

Stewart T, Pavlakis N, Ward M (2010) Cardiotoxicity with 5-fluorouracil and capecitabine: more than just vasospastic angina. Intern Med J 40:303–307

Sun H, Huang K, Tang F, Li X, Wang X, Long S, Zhou S, Suolangquzhen, Zhang J, Ning R, Li S, Wang S, Ma D (2018) Adjuvant chemotherapy after surgery can improve clinical outcomes for patients with IB2-IIB cervical cancer with neoadjuvant chemotherapy followed by radical surgery. Sci Rep 8:6443–6443

Szucs Z, Jones RL (2016) Chapter 2 - Introduction to systemic antineoplastic treatments for cardiologists. In: Herrmann J (ed) Clinical cardio-oncology. Elsevier, pp 15–38

Tagliamento M, Genova C, Rossi G, Coco S, Rijavec E, Dal Bello MG, Boccardo S, Grossi F, Alama A (2019) Microtubule-targeting agents in the treatment of non-small cell lung cancer: insights on new combination strategies and investigational compounds. Expert Opin Investig Drugs 28:513–523

Thorn CF, Oshiro C, Marsh S, Hernandez-Boussard T, McLeod H, Klein TE, Altman RB (2011) Doxorubicin pathways: pharmacodynamics and adverse effects. Pharmacogenet Genomics 21:440–446

Toschi L, Finocchiaro G, Bartolini S, Gioia V, Cappuzzo F (2005) Role of gemcitabine in cancer therapy. Future Oncol 1:7–17

Tounekti O, Kenani A, Foray N, Orlowski S, Mir LM (2001) The ratio of single- to double-strand DNA breaks and their absolute values determine cell death pathway. Br J Cancer 84:1272–1279

Valachis A, Mauri D, Polyzos NP, Mavroudis D, Georgoulias V, Casazza G (2010) Fulvestrant in the treatment of advanced breast cancer: a systematic review and meta-analysis of randomized controlled trials. Crit Rev Oncol Hematol 73:220–227

van Maanen JM, Retel J, de Vries J, Pinedo HM (1988) Mechanism of action of antitumor drug etoposide: a review. J Natl Cancer Inst 80:1526–1533

Verheij M (2008) Clinical biomarkers and imaging for radiotherapy-induced cell death. Cancer Metastasis Rev 27:471–480

Vogelzang NJ, Rusthoven JJ, Symanowski J, Denham C, Kaukel E, Ruffie P, Gatzemeier U, Boyer M, Emri S, Manegold C, Niyikiza C, Paoletti P (2003) Phase III study of pemetrexed in combination with cisplatin versus cisplatin alone in patients with malignant pleural mesothelioma. J Clin Oncol 21:2636–2644

Walker MD, Alexander E Jr, Hunt WE, MacCarty CS, Mahaley MS Jr, Mealey J Jr, Norrell HA, Owens G, Ransohoff J, Wilson CB, Gehan EA, Strike TA (1978) Evaluation of BCNU and/or radiotherapy in the treatment of anaplastic gliomas. A cooperative clinical trial. J Neurosurg 49:333–343

Walko CM, Lindley C (2005) Capecitabine: a review. Clin Ther 27:23–44

Wall ME, Wani MC, Cook CE, Palmer KH, McPhail AT, Sim GA (1966) Plant antitumor agents. I. The isolation and structure of camptothecin, a novel alkaloidal leukemia and tumor inhibitor from Camptotheca acuminata. J Am Chem Soc 88:3888–3890

Wani MC, Taylor HL, Wall ME, Coggon P, McPhail AT (1971) Plant antitumor agents. VI. The isolation and structure of taxol, a novel antileukemic and antitumor agent from Taxus brevifolia. J Am Chem Soc 93:2325–2327

Wiltshaw E (1965) Chlorambucil in the treatment of primary adenocarcinoma of the ovary. J Obstet Gynaecol Br Commonw 72:586–594

Wong A, Soo R, Yong W-P, Innocenti F (2009) Clinical pharmacology and pharmacogenetics of gemcitabine. Drug Metab Rev 41:77–88

Wosikowski K, Biedermann E, Rattel B, Breiter N, Jank P, Löser R, Jansen G, Peters GJ (2003) In vitro and in vivo antitumor activity of methotrexate conjugated to human serum albumin in human Cancer cells 5. Clin Cancer Res 9:1917–1926

Yabar CS, Winter JM (2016) Pancreatic cancer: a review. Gastroenterol Clin N Am 45:429–445

Yang SW, Huang P, Plunkett W, Becker FF, Chan JY (1992) Dual mode of inhibition of purified DNA ligase I from human cells by 9-beta-D-arabinofuranosyl-2-fluoroadenine triphosphate. J Biol Chem 267:2345–2349

Yoneda KY, Cross CE (2010) The pulmonary toxicity of anticancer agents. In: McQueen CA (ed) Comprehensive toxicology, 2nd edn. Elsevier, Oxford, pp 477–510

Young RC, Walton LA, Ellenberg SS, Homesley HD, Wilbanks GD, Decker DG, Miller A, Park R, Major F Jr (1990) Adjuvant therapy in stage I and stage II epithelial ovarian cancer. Results of two prospective randomized trials. N Engl J Med 322:1021–1027

Zhang N, Yin Y, Xu S-J, Chen W-S (2008) 5-Fluorouracil: mechanisms of resistance and reversal strategies. Molecules 13:1551–1569

Zhang DS, Jin Y, Luo HY, Wang ZQ, Qiu MZ, Wang FH, Li YH, Xu RH (2015) Pemetrexed for previously treated patients with metastatic gastric cancer: a prospective phase II study. Br J Cancer 112:266–270

Zhang F, Zhang Y, Jia Z, Wu H, Gu K (2019) Oxaliplatin-based regimen is superior to cisplatin-based regimen in tumour remission as first-line chemotherapy for advanced gastric cancer: a meta-analysis. J Cancer 10:1923–1929

7 Modulating Epigenetic Modification Enzymes Through Relevant Epidrugs as a Timely Strategy in Anticancer Therapy

Epigenetic mechanisms play a critical role in governing DNA-templated processes such as transcription. These mechanisms involve modifications on nucleosomal histones and the overlying DNA (Gibney and Nolan 2010). Though histone substrates undergo a variety of post-translational modifications, a single modification occurs to DNA (methylation at C-5 of cytosine). Modifications on histone proteins are primarily directed to unstructured N-terminal tails (Bannister and Kouzarides 2011; Kouzarides 2007). Among the histone modifications acetylation, ubiquitination, ADP-ribosylation, SUMOylation, methylation, phosphorylation and biotinylation are well known. DNA undergoes the only modification that is methylation (Bird 1992; Cao and Yan 2012; Robertson and Jones 2000; Verheugd et al. 2016). Certain modifications such as acetylation unanimously lead to transcriptional activation through chromatin remodelling by passive method (Grunstein 1997). The effect of histone methylation on gene expression depends on the site and degree of this epigenetic mark (Greer and Shi 2012). While histone H3K4me3 favours transcription, H3K9me3 and H3K27me3 have opposite effect (Barski et al. 2007). DNA methylation has silencing effect on gene expression (Kass et al. 1997; Siegfried et al. 1999). Transcriptional silencing due to DNA methylation may be either due to prevention of transcription factor binding or through recruitment of HDAC by way of methyl-CpG-binding protein 2 (MECP2) (Curradi et al. 2002; Jones et al. 1998).

7.1 Epigenetic Modification Enzymes as Guardians of Epigenetic Modifications

All these histone post-translational modifications and DNA methylation are precisely regulated by antagonistic enzyme families or antagonistic enzymes. For instance, histone acetylation is kept in equilibrium by two antagonistic enzyme families: histone acetyltransferases (HATs) and histone deacetylases (HDACs) (Yang and Seto 2007). Lysine methylation is regulated by lysine methyltransferases (KMTs) and lysine demethylases (KDMs) (Bannister et al. 2002). Similarly

S. A. Ganai, *Histone Deacetylase Inhibitors in Combinatorial Anticancer Therapy*,
https://doi.org/10.1007/978-981-15-8179-3_7

phosphorylation is regulated by kinases and protein phosphatase 1 (Rossetto et al. 2012). Moreover ADP-ribosylation is also controlled by ADP-ribosyltransferase and hydrolase (Fontana et al. 2017; Martinez-Zamudio and Ha 2012). DNA methylation being dynamic is regulated by DNMTs and TET/thymine DNA glycosylase (Jin et al. 2011; Kohli and Zhang 2013). Thus DNA methylation together with histone post-translational modifications comes under the umbrella of epigenetic modifications (Handy et al. 2011). Enzymes regulating the post-translational modifications of histone and DNA methylation are known as epigenetic modification enzymes or epigenetic enzymes or epigenetic players (Cheng et al. 2019; Lu 2013).

7.2 Brief Introduction to Epidrugs

Mounting evidences suggest that anomalous epigenetic regulation has strong cross-talk with genesis of cancer. Aberrant activity of epigenetic modification enzymes is responsible for the deregulated epigenetic landscape of cancer (Copeland et al. 2010; Maleszewska and Kaminska 2015). Abnormal activity of these enzymes contributes not only to cancer initiation but also fuels progression (Jackson-Grusby et al. 2001; Muntean and Hess 2009). As discussed in Chap. 5 epigenetic modification enzymes regulate various cellular processes such as apoptosis, transcription of tumour suppressor genes, stability of non-histone proteins, expression of cyclin-dependent kinase inhibitors (Fang and Lu 2002; Seto and Yoshida 2014; Wang et al. 2001). Moreover these enzymes have significant contribution in cell differentiation as they control the level of differentiation-related transcription factors (Cai et al. 2009; Caslini et al. 2006). Further these epigenetic players regulate the expression of cell adhesion molecules such as E-cadherin and extracellular matrix proteins including cystatin (Choi et al. 2016; Whetstine et al. 2005). Anomalous activity of epigenetic enzymes especially HDACs facilitates cancer initiation by hampering apoptosis, silencing tumour suppressor genes, cyclin-dependent kinase inhibitors and by augmenting genomic instability (Bhaskara et al. 2010; Kang et al. 2014; Ma et al. 2015; Wu et al. 2013). By way of enhancing angiogenesis and metastasis, the aberrant activity of these enzymes accentuates cancer progression (Kanno et al. 2012; Kim et al. 2001). Thus it is quite evident that aberrant activity of epigenetic modification enzymes/epigenetic enzymes has crucial implications in cancer onset and advancement. Drugs modulating the activity of epigenetic modifying enzymes or epigenetic enzymes are emerging as promising agents for anticancer therapy. These drugs termed as epigenetic drugs/epidrugs target the epigenetic route for bringing therapeutic effect. Enzymes installing various epigenetic modifications such as HATs, HMTs, DNMTs, kinases, ADP-ribosyl transferases and ubiquitinases are collectively termed as writers. These modifications being dynamic are removed by corresponding antagonistic enzymes of writers, collectively named as writers (Biswas and Rao 2018). Among the erasers HDACs, KDMs, JMJD6, TET proteins, PP1, DUBs are prominent (Chang et al. 2007; Ganai 2019b). From this explanation it is now quite evident that epidrugs target writers and erasers of epigenetic modifications. Bromodomains, the chief readers of acetylated lysine residues are also emerging as epi-targets. They play a key role in directing chromatin modifying enzymes to particular sites. The bromodomain fold possesses acetyl-lysine binding site, the eye catching target for therapeutic intervention (Muller et al. 2011; Zeng and Zhou 2002). Certain bromodomain inhibitors are under preclinical studies and have been tested against inflammation and cancer (Delmore et al. 2011; Magistri et al. 2016). Speaking in few words epidrugs include drugs not only targeting writers and erasers but also drugs targeting reader proteins (bromodomain proteins) (Heerboth et al. 2014).

7.3 Classification and Status of Epidrugs

Epigenetic modifications show plasticity and are reversible in character (Patnaik and Anupriya 2019). Targeting the enzymes regulating these modifications is emerging as an excellent therapeutic strategy against cancer (Andreoli et al. 2013) Epidrugs show strong healing potential by reinstating the epigenetic balance which is dysregulated in cancer (Roberti et al. 2019). Classification of epidrugs is based on the respective target enzyme/enzymes modulated by these drugs. Among known epidrugs only some DNMT and HDAC inhibitors have gained US Food and Drug Administration approval for treating certain cancers (Bohl et al. 2018; Ganesan et al. 2019; Giri and Aittokallio 2019). Other epidrugs are at different stages of clinical trials and some are at preclinical testing stage.

7.3.1 DNA Methyltransferase Inhibitors

The first epidrug approved was the DNA methyltransferase inhibitor with brand name Vidaza and generic name Azacitidine. This antineoplastic agent manufactured by Pharmion Corporation was approved on May 19, 2004 for treating myelodysplastic syndrome patients (Kaminskas et al. 2005). FDA and European Medicines Agency also approved this inhibitor for treating chronic myelomonocytic leukaemia (Gros et al. 2012). Almost 2 years after the approval of Azacitidine, second DNMT inhibitor was approved on May 2, 2006. This inhibitor with brand name Dacogen and generic name Decitabine, manufactured by MGI Pharma Inc. and SuperGen Inc., was also approved for treating the myelodysplastic diseases (Saba 2007). These inhibitors are also undergoing clinical trials for solid tumours (Cowan et al. 2010). Both Azacytidine and Decitabine being structural analogues of cytidine possess nitrogen instead of carbon at position 5 of the pyrimidine ring. These DNMT inhibitors after entering into cells through proper transport are transformed into triphosphate active forms. Being analogues they are able to mimic cytosine and on incorporation into DNA in the synthetic phase of interphase trap DNMTs promoting their proteasomal degradation (Yang et al. 2010). This creates hypomethylation and subsequent activation of silenced tumour suppressor genes. It is noteworthy that Decitabine being deoxyribose analogue gets incorporated only into DNA strands, whereas Azacytidine, a ribonucleoside analogue gets incorporated predominantly into RNA and to a some extent into DNA (Lund et al. 2014; Stresemann and Lyko 2008; Wong et al. 2019).

Zebularine, a cytidine analogue, is another DNA hypomethylating agent. This second generation methyltransferase inhibitor being highly stable and hydrophilic has oral bioavailability suitable for preferential targeting of cancer cells (Andersen et al. 2010; Cheng et al. 2003). Zebularine-substituted DNA and DNMTs form strong covalent complexes resulting in trapping of these enzymes (Hurd et al. 1999). Developed primarily as inhibitor of cytidine-deaminase, Zebularine showed low toxicity even after long term administration (Cheng et al. 2003, 2004). This inhibitor exhibited antitumour effect in hepatocellular carcinoma cells through apoptosis induction (Nakamura et al. 2013). Desired therapeutic efficacy of Azacytidine and Decitabine is strongly hampered by their low plasma half-life. This instability of these inhibitors has been attributed to degradation of these DNMT inhibitors by cytidine-deaminase (Bohl et al. 2018; Estey 2013; Navada et al. 2014). Guadecitabine, another DNMT inhibitor has shown nice stability in plasma as it is resistant to cytidine-deaminase degradation (Roboz et al. 2018; Yoo et al. 2007). Guadecitabine which is dinucleotide of Decitabine and next generation hypomethylating agent has been granted orphan drug status by USFDA for acute myeloid leukaemia (Fig. 7.1) (Griffiths et al. 2013; Roberti et al. 2019).

7.3.2 Histone Methyltransferase Inhibitors

DNA methylation as discussed promotes gene silencing like HDAC overactivity. Histone methylation can either activate or silence the genes depending on the site and degree of this modification (Barski et al. 2007). Histone methyltransferases include lysine methyltransferases (KMTs) and protein arginine methyltransferases (PRMTs). Almost 100 KMTs are known and such enzymes utilize *S*-adenosyl methionine as methyl donor. Sinefungin, a natural product competes with the defined methyl donor for its binding site and as such may non-selectively bind all SAM using methyltransferases (Couture et al. 2006; Wang and Patel 2013). UNC1999, a dual inhibitor of enhancer of zeste homologue 2 (EZH2) and enhancer of zeste homologue 2 (EZH1) showed effective growth inhibition in MLL-rearranged leukaemia, while this was not the case with selective EZH2 inhibitor GSK126 (Xu et al. 2015). In addition to this inhibitor two other inhibitors of EZH2, namely tazemetostat and CPI-1205 have entered into clinical trials (Gulati et al. 2018; Italiano et al. 2018; Taplin et al. 2019).

Disruptor of telomeric silencing 1-like (DOT1L) is unusual methyltransferase as it lacks SET domain but instead possesses an AdoMet binding motif. Using SAM as cofactor, this methyltransferase methylates H3K79. Different inhibitors of DOT1L which are in fact SAM—mimetic molecules have been discovered. EPZ004777, a selective and strong inhibitor of this enzyme resulted in selective killing of MLL-translocated cells (Daigle et al. 2011). Another selective inhibitor of DOT1L, EPZ-5676 (Pinometostat) is under clinical trials for relapsed or refractory leukaemia (Campbell et al. 2017; Stein et al. 2015). Two more inhibitors with high

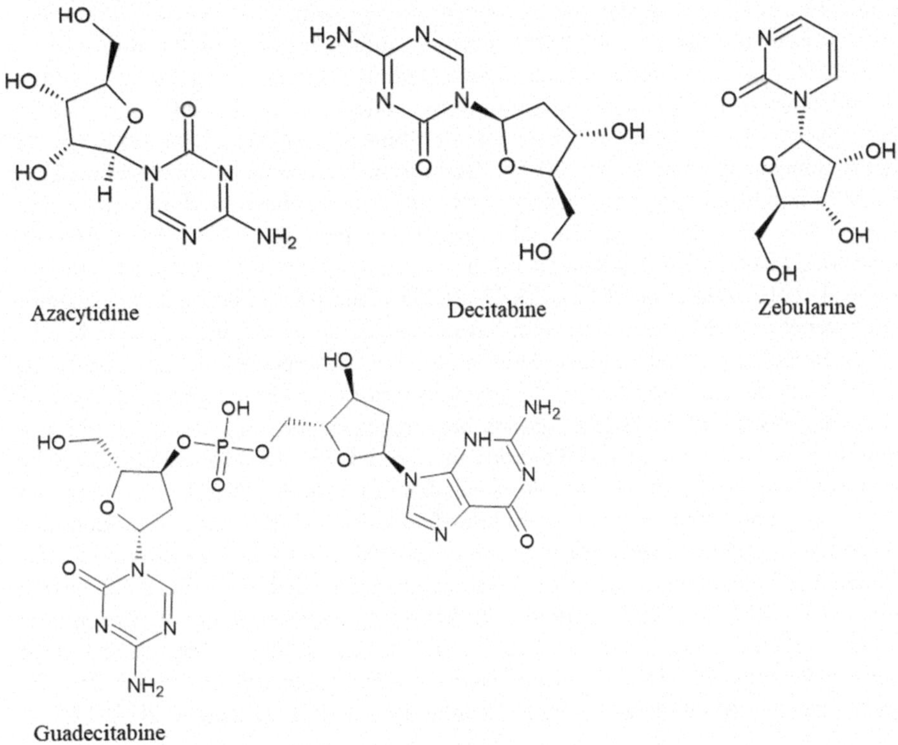

Fig. 7.1 Chemical structures of various DNA methyltransferase (DNMT) inhibitors. Azacytidine (Vidaza) was the first DNMT inhibitor as well as the first epidrug that was approved by USFDA for myelodysplastic syndrome. Later its approval was extended to chronic myelomonocytic leukaemia treatment by FDA and European Medicines Agency. Following this Decitabine (Dacogen) was approved for myelodysplastic diseases. Both of them are cytidine analogues. Zebularine is a second generation DNMT with good stability and oral bioavailability. Like Azacytidine and Decitabine, this inhibitor is also cytidine analogue. Guadecitabine (SGI-110) has comparatively complex structure and is resistant to degradation by cytidine-deaminase. This DNMT inhibitor has been granted orphan drug designation by FDA. All the structures were generated through ACD/ChemSketch (Freeware)

specificity in terms of DOT1L targeting (SGC0946 and SYC-522) have also been certified (Liu et al. 2014). SGC0946, a brominated analogue of EPZ004777 resulted in selective cytotoxicity of mixed lineage leukaemia cells by way of obstructing DOT1L (Wood et al. 2018; Yu et al. 2012). Only recently a novel inhibitor of DOT1L has been identified. This inhibitor, in fact a psammaplin A analogue (PsA-3091), showed encouraging results in breast cancer model (Byun et al. 2019).

Inhibitors of G9a (EHMT2), a lysine methyltransferase methylating H3K9, have been discovered or designed. BIX-01294, a diazepin-quinazolin-amine derivative inhibits this enzyme and has been proved through cellular assays (Kubicek et al. 2007). However, recent studies have proved that BIX-01294 is tenfold selective for methyltransferase GLP (EHMT1) over EHMT2/G9a (Quinn et al. 2010). Chaetocin, a fungal metabolite with specific inhibitory activity against SU (VAR) 3-9 has been characterized (Greiner et al. 2005). Further studies resulted in the discovery of UNC0224, another strong and selective G9a/EHMT2 inhibitor (Liu et al. 2009). Optimization of the 7-dimethylaminopropoxy side chain of this inhibitor ended in the discovery of the most-potent inhibitor UNC0321 for this lysine methyltransferase (Liu et al. 2010). Certain protein arginine methyltransferase (PRMT)-inhibitors have also been discovered. Among these inhibitors three have entered into the journey of clinical trials. These include GSK3326595 (PRMT5 inhibitor), JNJ-64619178 (PRMT5 inhibitor) and GSK3368715 (PRMT1 inhibitor) (Li et al. 2019; Zhu et al. 2019). GSK3326595 is not only potent but also selective and reversible PRMT5 inhibitor that has demonstrated encouraging results in solid and haematological cell models. This inhibitor is undergoing clinical activity and safety evaluation in subjects having myelodysplastic syndrome and acute myeloid leukaemia (Watts et al. 2019). Studies have revealed the synergistic inhibition of tumour growth on combinatorial therapy involving GSK3368715 and GSK3326595 (Fedoriw et al. 2019). Novel inhibitors capable of inhibiting PRMT4 also termed as coactivator-associated arginine methyltransferase 1 (CARM1) have been identified. TP-064, the potent and selective inhibitor of CARM1 showed antiproliferative effect against multiple myeloma models (Nakayama et al. 2018). EZM 2302 another selective CARM1 inhibitor has demonstrated strong antitumour activity in xenograft model of multiple myeloma (Drew et al. 2017). Decamidine, inhibitor of arginine methyltransferase PRMT1, has also been discovered. Molecular docking and in vitro studies have certified its potent PRMT1 inhibitory activity (Zhang et al. 2017).

7.3.3 Histone Demethylase Inhibitors

A variety of histone demethylase inhibitors have been identified. Several inhibitors including IMG-7289, INCB059872, GSK-2879552, TCP, CC-90011, ORY-1001 and ORY-2001 of LSD1 are presently under clinical assessment for their use in anticancer therapy (Fang et al. 2019). Cell-based assays have proved the inhibitory activity of ryuvidine against KDM5A. This inhibitor also obstructed recombinant KDM5B and KDM5C but more pronounced was seen against KDM5B (Mitsui et al. 2019). JIB-04 inhibits various demethylase enzymes and is thus considered as pan-selective inhibitor. Based on its encouraging studies it may prove as a promising drug for tackling colorectal cancer (Kim et al. 2018). Therapeutic intervention of Jumonji demethylases with small molecule JIB-04 makes the cancer cells radiosensitive. Tumour bearing mice models on co-administration of radiation along with this inhibitor markedly prolonged survival (Fig. 7.2) (Bayo et al. 2018). Tranylcypromine, a monoamine oxidase inhibitor has been proved to be LSD1 inactivator. However, the defined small molecule inhibits LSD1 only modestly (Barth et al. 2019; Zheng et al. 2016).

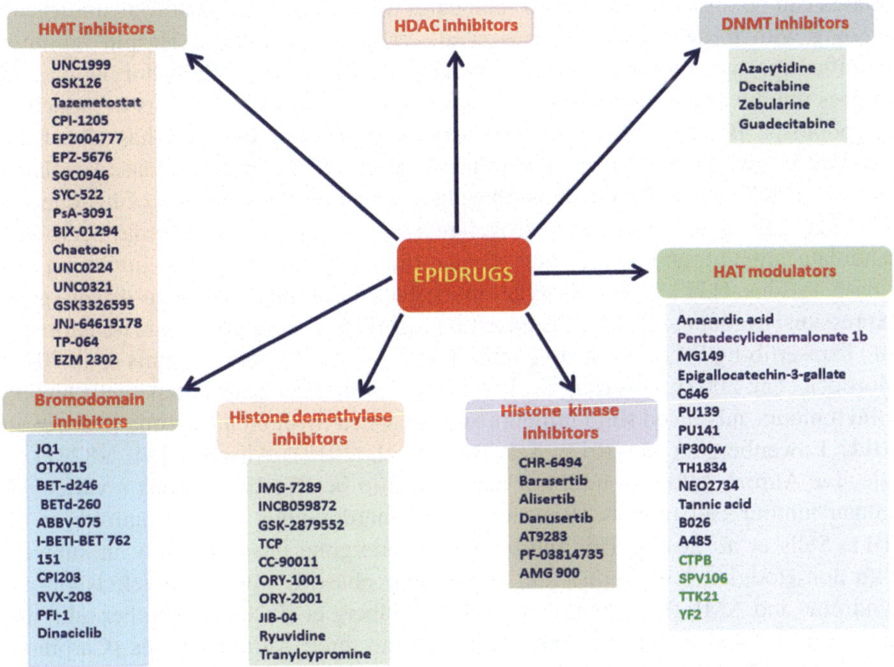

Fig. 7.2 Overview of different types of epidrugs and their respective members. DNMT inhibitors trap DNMTs and result in their degradation. While Azacytidine and Decitabine have USFDA approval, Guadecitabine has FDA orphan designation and is comparatively stable than the aforementioned DNMT inhibitors. Zebularine is orally bioavailable and has high stability. Histone acetyltransferase (HAT) modulators include HAT inhibitors and HAT activators. HAT inhibitors including anacardic acid, MG149 and so on are shown in dark blue colour, while HAT activators are represented by green. Pentadecylidenemalonate 1b shown in purple is inhibitor for p300/CBP while activator for PCAF. Histone kinase inhibitors are also tested and some are undergoing clinical trials. CHR-6494 obstructs Haspin. Barasertib being prodrug after transformation selectively inhibits Aurora B kinase. Alisertib is Aurora A selective, whereas Danusertib inhibits all the three kinases. AT9283 has more inclination towards Aurora B. PF-03814735, orally bioavailable kinase inhibitor targets both Aurora A and Aurora B. AMG 900 interferes with Aurora A, B and C. Histone demethylase inhibitors target histone demethylase enzymes and the member details have been provided in vertical rectangular box. Bromodomain inhibitors target acetyl-lysine readers. BET proteins have two bromodomains and certain inhibitors bind preferentially to one of them. JQ1 the first bromodomain inhibitor targets both bromodomains. Histone methyltransferase inhibitors inhibit the activity of histone methyltransferases (HMTs). The various members of HMT inhibitors have been summarized in the figure. HDAC inhibitors target HDACs and the member details of these inhibitors will be provided in the next chapter dealing solely with such inhibitors

7.3.4 Histone Kinase Inhibitors

Deregulation of histone phosphorylation has implications in cancer. Haspin, a protein kinase phosphorylates histone H3 at threonine 3 (H3T3). For mitotic progression this modification plays a significant role. The activity of these mitotic kinases can be obstructed by small molecules and currently clinical trial studies are ongoing with these agents (Huertas et al. 2012). CHR-6494, a Haspin inhibitor identified through high-throughput screening declines H3T3 phosphorylation and induces mitotic catastrophe. This inhibitor blocks cell cycle and results in induction of apoptosis. Besides the antitumour activity of Haspin inhibitor has also been recorded in xenografted nude mice (Huertas et al. 2012). Protein kinase C related kinase 1 (PRK1 or PKN1) installs phosphate group on threonine 11 of histone H3 (H3T11). This kinase regulates androgen receptor signalling and thus may serve as a candidate target for anticancer drugs. Lestaurtinib, a clinical candidate inhibiting this kinase lessens H3T11 phosphorylation besides inhibiting androgen-driven gene expression (Kohler et al. 2012). Barasertib (AZD1152) being prodrug is transformed into barasertib-hQPA, the selective inhibitor of Aurora B kinase (Dennis et al. 2012; Mortlock et al. 2007). This drug in clinical studies has been tested on patients having solid tumours, advanced solid tumours and advanced myeloid leukaemia (Boss et al. 2011; Lowenberg et al. 2011; Schwartz et al. 2013). Alisertib (MLN8237), a selective Aurora A kinase inhibitor has proved to be effective against a variety of human tumour cell models (Bavetsias and Linardopoulos 2015; Manfredi et al. 2011; Sells et al. 2015). This inhibitor has undergone phase II study in subjects with non-Hodgkin lymphomas and exploratory phase II study in myelodysplastic syndrome and AML (Friedberg et al. 2014; Goldberg et al. 2014). Another inhibitor danusertib proved to be effective against all the three Aurora kinases (Carpinelli et al. 2007; Fancelli et al. 2006). AT9283, a multitargeted kinase inhibitor is comparatively more effective towards Aurora B than Aurora A (Howard et al. 2009). Another orally bioavailable and potent inhibitor of Aurora A and B kinases PF-03814735 has also been studied. Apart from these kinases it impacts many other kinases including FLT3 and MST3 (Jani et al. 2010). Further, AMG 900 although targeting all the three Aurora kinases has more inhibitory activity against Aurora C (Fig. 7.2) (Payton et al. 2010). This inhibitor is under clinical evaluation and is thus tested on patients with advanced cancers (solid and haematological) (Geuns-Meyer et al. 2015).

7.3.5 Bromodomain Inhibitors

Lysine acetylation and its consequent effects are regulated by three protein types including HATs, HDACs and bromodomain (BRD) proteins (Dhalluin et al. 1999; Filippakopoulos and Knapp 2014; Ganai 2014). As mentioned above bromodomain proteins are readers unlike HATs (writers) and HDACs (erasers). Acetyl-lysine binding pocket of bromodomain proteins is an attractive target for pharmacological intervention (Sanchez et al. 2014). BET (bromodomain and extraterminal) proteins belonging to BRD family of proteins have two N-terminal bromodomains (BD1 and BD2) besides having one extra-C terminal domain. BD1 and BD2 have the potential to interact with acetylated lysine residues located on tails of histone proteins. In mammals the BET family has four members within its confines. These include

BRD2 (BRD containing 2), BRD3 in addition to BRD4 and BRDT (Lovén et al. 2013).

JQ1, a small molecule and cell permeable BET inhibitor binds specifically and competitively to BD1 and BD2 bromodomains with strong propensity (Filippakopoulos et al. 2010). An oral BET inhibitor OTX015 is in clinical trial (Phase Ib) against subjects bearing solid tumours. This inhibitor has shown encouraging results against lymphoma models and modulates several cellular processes and molecular players. With other anticancer agents including mTOR inhibitor everolimus, OTX015 works synergistically (Boi et al. 2015; Vázquez et al. 2017). Besides this inhibitor in combination with DNMT inhibitor Azacytidine and HDAC inhibitor Panobinostat synergistically hampers the growth of leukaemia cells (Coudé et al. 2015). BET-d246 causes selective depletion of three BET family members BRD2-BRD4 and has shown high efficacy against triple negative breast cancer. Effective depletion of BET proteins followed by potent antitumour effect was noted in various murine xenograft models (human breast cancer) on BETd-246 and BETd-260 administration (Bai et al. 2017).

ABBV-075, a BET family bromodomain inhibitor targets both bromodomains of BRD2 in addition to BRD4 and BRDT. This inhibitor is currently undergoing clinical trial studies for safety evaluation and pharmacokinetic parameters in patients having advanced tumours solid and haematological (Bui et al. 2017). Another pan-BET inhibitor I-BET151 hampers proliferation of glioblastoma cells (Daniel et al. 2014). Hedgehog activity-driven growth in case of medulloblastoma cells is also attenuated by this inhibitor (Long et al. 2014). In addition I-BET151 strongly suppressed proliferation of myeloma cells both under in vitro and in vivo set-up (Chaidos et al. 2013). I-BET 762 by way of BET inhibition provoked apoptosis in neuroblastoma models. This effect was accompanied by silencing of MYCN (encoding N-myc proto-oncogene protein) and Bcl2 (Wyce et al. 2013). I-BET 762 also affected proliferation of myeloma cells and offered survival benefit in xenograft model of systemic myeloma (Chaidos et al. 2013). CPI203, a JQ1 analogue has enhanced oral as well as intraperitoneal bioavailability. In combination with lenalidomide, this inhibitor has shown synergistic effect in bortezomib resistant cells (Moros et al. 2014). This inhibitor potentiated the antiproliferative effects of lenalidomide and dexamethasone in multiple myeloma cells (Díaz et al. 2017). RVX-208, a resveratrol derivative binding selectively to second bromodomain of BRD2 and BRD3 is currently undergoing phase II clinical studies (Alqahtani et al. 2019; Picaud et al. 2013b). PFI-1, a dihydroquinazolinone binds to BET bromodomain in a chemically distinct manner not usual with proven BET inhibitors (Picaud et al. 2013a). This inhibitor, in leukemic cells, attenuated the expression of Aurora B kinase inducing caspase-driven apoptosis and differentiation (Picaud et al. 2013a). Dinaciclib, a strong inhibitor of CDKs (cyclin-dependent kinases) is undergoing phase III clinical study for leukaemia. Its interaction with one of the bromodomains of BRDT has been revealed by cocrystallization studies (Fig. 7.2) (Martin et al. 2013).

7.3.6 Histone Acetyltransferase Modulators

Through modulation of chromatin topology by transferring acetyl group to the histone lysines, histone acetyltransferases (HATs) promote gene expression. Aberrant activity or expression of these HATs drives various neurological and oncological disorders (Richters and Koehler 2017; Schneider et al. 2013). Several inhibitors of HATs are known and are currently being tested against different disease models by various laboratories globally. Anacardic acid, a small molecule inhibitor of p300/(CREB binding protein) associated factor (PCAF) and p300 obstructs NF-kB dependent gene transcription (Balasubramanyam et al. 2003; Sung et al. 2008). Pentadecylidenemalonate 1b, an analogue of the anacardic acid has proved to be both activator and inhibitor. While this analogue enhances PCAF activity, it inhibits the p300/CBP (Sbardella et al. 2008). Novel derivative of anacardic acid MG149 has binding inclination for Tip60 and MOF (males absent on the first), the MYST family HATs (Ghizzoni et al. 2012). Further studies showed the inhibition of HAT activity of PCAF and p300 by isothiazolones (Stimson et al. 2005). For activation of nuclear factor-kappaB (NF-kappaB), hyperacetylation of RelA is crucial. This hyperacetylation is mediated by p300/CBP and facilitates chronic inflammation. Epigallocatechin-3-gallate (EGCG), a natural HAT inhibitor, inhibiting most of the HATs silences NF-kappaB-triggered inflammatory signalling by way of suppressing RelA acetylation (Choi et al. 2009). Through virtual screening, a novel and active site directed inhibitor of p300, C646, has been identified. This inhibitor showed high potency and selectivity towards the HAT p300 (Bowers et al. 2010).

Another study through structure based virtual screening of one lakh drug like molecules followed by experimental validation identified a novel p300 inhibitor, 4-acetyl-2-methyl-*N*-morpholino-3,4-dihydro-2H-benzo[b][1,4]thiazine-7-sulfonamide (Dekker et al. 2014; Zeng et al. 2013). Two HAT inhibitors, namely PU139 and PU141 through induction of histone hypoacetylation have shown growth inhibitory effects against a variety of neoplastic cell lines including neuroblastoma. While PU139 inhibited Gcn5, PCAF and CBP, PU141 showed selectivity towards CBP and p300 (Gajer et al. 2015). Another inhibitor of p300/CBP, iP300w restrains DUX4 induced cytotoxicity and silences the expression of majority of target genes regulated by this transcription factor (Bosnakovski et al. 2019). TH1834, an inhibitor of MYST HAT, TIP60 attenuated progression of breast cancer in xenograft mice model (Idrissou et al. 2019). SPOP (speckle-type POZ protein) is the tumour suppressor frequently mutated in primary human prostate cancer. NEO2734, a dual inhibitor of bromodomain and extraterminal domain (BET) and CBP-p300 proved effective against prostate cancer bearing SPOP mutation (Yan et al. 2019). Tannic acid, a general HAT inhibitor has the potential to hamper non-alcoholic fatty liver disease. This inhibitor inhibits HATs p300, CBP and PCAF to a varying extent (Chung et al. 2019). Only recently through artificial intelligence based drug discovery, a novel and potent inhibitor of p300 and CBP has been reported. This inhibitor, B026, proved relatively more effective towards p300 (IC_{50} 1.8 nM) versus CBP (IC_{50} 9.5 nM) (Yang et al. 2020). A485, an inhibitor of p300 and selective inhibitor of CBP has shown promising results against a variety of cancer cell lines (Fig. 7.2). This inhibitor showed synergistic effect when used in combination with TRAIL in non-small cell lung cancer cells (Zhang et al. 2020).

N-(4-chloro-3-trifluoromethyl-phenyl)-2-ethoxy-6-pentadecyl-benzamide (CTPB), an amide derivative facilitating the activity of p300 has also been synthesized (Balasubramanyam et al. 2003). CTPB (*N*-(4-chloro-3-trifluoromethyl-phenyl)-2-ethoxy-6-pentadecyl-benzamide), the potent activator of p300 histone acetyltransferase activity has also been identified (Mantelingu et al. 2007). SPV106, an anacardic acid-based molecule proved a peculiar by activating KAT2B and inhibiting KAT3A and 3B (Sbardella et al. 2008). Another activator of CBP/p300 (TTK21) crosses the blood–brain barrier on conjugation to glucose based nanosphere of carbon (Chatterjee et al. 2013). Activator of HAT p300, YF2, has been produced. This activator in combination with romidepsin showed synergistic cytotoxic effects against diffuse large B cell lymphoma (DLBCL) cell lines (Liu et al. 2019).

7.3.7 Histone Deacetylase Inhibitors

Abnormal expression of HDACs is most frequent in several solid and haematological malignancies. By way of distorting acetylation homeostasis and subsequent dysregulation of gene expression HDACs drive a multiplex of bellicose malignancies (Wang et al. 2020). These enzymes function antagonistically to HATs and are essential for executing transcriptional events precisely (Yang and Seto 2007). After DNMT inhibitors, histone deacetylase inhibitors (HDACi) have gained approval to be used against certain haematological malignancies (Ganai 2019a). As the whole theme of this book revolves around HDACi so they will be discussed extensively in the forthcoming chapters.

Taken together, I have explained epidrugs in the context of writers, erasers and readers of epigenetic modifications. Besides, the most predominant types of epidrugs including DNMT inhibitors, histone methyltransferase inhibitors, histone demethylase inhibitors, histone kinase inhibitors, bromodomain inhibitors, histone acetyltransferase modulators have been mentioned. Further the different inhibitors or activators belonging to these classes and their therapeutic effect and status in different cancers have been taken into account. Last but no way least the most highly emerging epidrugs, namely HDAC inhibitors have been lightly explained as their thorough details are coming in the upcoming chapters.

References

Alqahtani A, Choucair K, Ashraf M, Hammouda DM, Alloghbi A, Khan T, Senzer N, Nemunaitis J (2019) Bromodomain and extra-terminal motif inhibitors: a review of preclinical and clinical advances in cancer therapy. Future Sci OA 5:Fso372

Andersen JB, Factor VM, Marquardt JU, Raggi C, Lee Y-H, Seo D, Conner EA, Thorgeirsson SS (2010) An integrated genomic and epigenomic approach predicts therapeutic response to zebularine in human liver cancer. Sci transl med 2:54ra77–54ra77

Andreoli F, Barbosa AJM, Parenti MD, Del Rio A (2013) Modulation of epigenetic targets for anticancer therapy: clinicopathological relevance, structural data and drug discovery perspectives. Curr Pharm Des 19:578–613

Bai L, Zhou B, Yang CY, Ji J, McEachern D, Przybranowski S, Jiang H, Hu J, Xu F, Zhao Y, Liu L, Fernandez-Salas E, Xu J, Dou Y, Wen B, Sun D, Meagher J, Stuckey J, Hayes DF, Li S, Ellis MJ, Wang S (2017) Targeted degradation of BET proteins in triple-negative breast cancer. Cancer Res 77:2476–2487

Balasubramanyam K, Swaminathan V, Ranganathan A, Kundu TK (2003) Small molecule modulators of histone acetyltransferase p300. J Biol Chem 278:19134–19140

Bannister AJ, Kouzarides T (2011) Regulation of chromatin by histone modifications. Cell Res 21:381–395

Bannister AJ, Schneider R, Kouzarides T (2002) Histone methylation: dynamic or static? Cell 109:801–806

Barski A, Cuddapah S, Cui K, Roh TY, Schones DE, Wang Z, Wei G, Chepelev I, Zhao K (2007) High-resolution profiling of histone methylations in the human genome. Cell 129:823–837

Barth J, Abou-El-Ardat K, Dalic D, Kurrle N, Maier A-M, Mohr S, Schütte J, Vassen L, Greve G, Schulz-Fincke J, Schmitt M, Tosic M, Metzger E, Bug G, Khandanpour C, Wagner SA, Lübbert M, Jung M, Serve H, Schüle R, Berg T (2019) LSD1 inhibition by tranylcypromine derivatives interferes with GFI1-mediated repression of PU.1 target genes and induces differentiation in AML. Leukemia 33:1411–1426

Bavetsias V, Linardopoulos S (2015) Aurora kinase inhibitors: current status and outlook. Front Oncol 5:278

Bayo J, Tran TA, Wang L, Peña-Llopis S, Das AK, Martinez ED (2018) Jumonji inhibitors overcome radioresistance in cancer through changes in H3K4 methylation at double-strand breaks. Cell Rep 25:1040–1050.e1045

Bhaskara S, Knutson SK, Jiang G, Chandrasekharan MB, Wilson AJ, Zheng S, Yenamandra A, Locke K, Yuan JL, Bonine-Summers AR, Wells CE, Kaiser JF, Washington MK, Zhao Z, Wagner FF, Sun ZW, Xia F, Holson EB, Khabele D, Hiebert SW (2010) Hdac3 is essential for the maintenance of chromatin structure and genome stability. Cancer Cell 18:436–447

Bird A (1992) The essentials of DNA methylation. Cell 70:5–8

Biswas S, Rao CM (2018) Epigenetic tools (the writers, the readers and the erasers) and their implications in cancer therapy. Eur J Pharmacol 837:8–24

Bohl SR, Bullinger L, Rucker FG (2018) Epigenetic therapy: azacytidine and decitabine in acute myeloid leukemia. Expert Rev Hematol 11:361–371

Boi M, Gaudio E, Bonetti P, Kwee I, Bernasconi E, Tarantelli C, Rinaldi A, Testoni M, Cascione L, Ponzoni M, Mensah AA, Stathis A, Stussi G, Riveiro ME, Herait P, Inghirami G, Cvitkovic E, Zucca E, Bertoni F (2015) The BET bromodomain inhibitor OTX015 affects pathogenetic pathways in preclinical B-cell tumor models and synergizes with targeted drugs. Clin Cancer Res 21:1628–1638

Bosnakovski D, da Silva MT, Sunny ST, Ener ET, Toso EA, Yuan C, Cui Z, Walters MA, Jadhav A, Kyba M (2019) A novel P300 inhibitor reverses DUX4-mediated global histone H3 hyperacetylation, target gene expression, and cell death. Sci Adv 5:eaaw7781

Boss DS, Witteveen PO, van der Sar J, Lolkema MP, Voest EE, Stockman PK, Ataman O, Wilson D, Das S, Schellens JH (2011) Clinical evaluation of AZD1152, an i.v. inhibitor of Aurora B kinase, in patients with solid malignant tumors. Ann Oncol 22:431–437

Bowers EM, Yan G, Mukherjee C, Orry A, Wang L, Holbert MA, Crump NT, Hazzalin CA, Liszczak G, Yuan H, Larocca C, Saldanha SA, Abagyan R, Sun Y, Meyers DJ, Marmorstein R, Mahadevan LC, Alani RM, Cole PA (2010) Virtual ligand screening of the p300/CBP histone acetyltransferase: identification of a selective small molecule inhibitor. Chem Biol 17:471–482

Bui MH, Lin X, Albert DH, Li L, Lam LT, Faivre EJ, Warder SE, Huang X, Wilcox D, Donawho CK, Sheppard GS, Wang L, Fidanze S, Pratt JK, Liu D, Hasvold L, Uziel T, Lu X, Kohlhapp F, Fang G, Elmore SW, Rosenberg SH, McDaniel KF, Kati WM, Shen Y (2017) Preclinical characterization of BET family bromodomain inhibitor ABBV-075 suggests combination therapeutic strategies. Cancer Res 77:2976–2989

Byun WS, Kim WK, Han HJ, Chung H-J, Jang K, Kim HS, Kim S, Kim D, Bae ES, Park S, Lee J, Park H-G, Lee SK (2019) Targeting histone methyltransferase DOT1L by a novel psammaplin a analog inhibits growth and metastasis of triple-negative breast cancer. Mol Ther Oncolytics 15:140–152

Cai KQ, Caslini C, Capo-chichi CD, Slater C, Smith ER, Wu H, Klein-Szanto AJ, Godwin AK, Xu X-X (2009) Loss of GATA4 and GATA6 expression specifies ovarian cancer histological subtypes and precedes neoplastic transformation of ovarian surface epithelia. PLoS One 4: e6454–e6454

Campbell CT, Haladyna JN, Drubin DA, Thomson TM, Maria MJ, Yamauchi T, Waters NJ, Olhava EJ, Pollock RM, Smith JJ, Copeland RA, Blakemore SJ, Bernt KM, Daigle SR (2017) Mechanisms of Pinometostat (EPZ-5676) treatment-emergent resistance in MLL-rearranged leukemia. Mol Cancer Ther 16:1669–1679

Cao J, Yan Q (2012) Histone ubiquitination and deubiquitination in transcription, DNA damage response, and cancer. Front Oncol 2:26

Carpinelli P, Ceruti R, Giorgini ML, Cappella P, Gianellini L, Croci V, Degrassi A, Texido G, Rocchetti M, Vianello P, Rusconi L, Storici P, Zugnoni P, Arrigoni C, Soncini C, Alli C, Patton V, Marsiglio A, Ballinari D, Pesenti E, Fancelli D, Moll J (2007) PHA-739358, a potent inhibitor of Aurora kinases with a selective target inhibition profile relevant to cancer. Mol Cancer Ther 6:3158–3168

Caslini C, Capo-chichi CD, Roland IH, Nicolas E, Yeung AT, Xu XX (2006) Histone modifications silence the GATA transcription factor genes in ovarian cancer. Oncogene 25:5446–5461

Chaidos A, Caputo V, Goudevenou K, Liu B, Marigo I, Chaudhry M, Rotolo A, Tough D, Smithers N, Bassil A, Chapman T, Harker N, Barbash O, Tummino P, Al-Mahdi N, Haynes A, Cutler L, Le B, Rahemtulla A, Karadimitris A (2013) Potent antimyeloma activity of the novel bromodomain inhibitors I-BET151 and I-BET762. Blood 123

Chang B, Chen Y, Zhao Y, Bruick RK (2007) JMJD6 is a histone arginine demethylase. Science (New York, N.Y.) 318:444–447

Chatterjee S, Mizar P, Cassel R, Neidl R, Selvi BR, Mohankrishna DV, Vedamurthy BM, Schneider A, Bousiges O, Mathis C, Cassel JC, Eswaramoorthy M, Kundu TK, Boutillier AL (2013) A novel activator of CBP/p300 acetyltransferases promotes neurogenesis and extends memory duration in adult mice. J Neurosci 33:10698–10712

Cheng JC, Matsen CB, Gonzales FA, Ye W, Greer S, Marquez VE, Jones PA, Selker EU (2003) Inhibition of DNA methylation and reactivation of silenced genes by zebularine. J Natl Cancer Inst 95:399–409

Cheng JC, Weisenberger DJ, Gonzales FA, Liang G, Xu GL, Hu YG, Marquez VE, Jones PA (2004) Continuous zebularine treatment effectively sustains demethylation in human bladder cancer cells. Mol Cell Biol 24:1270–1278

Cheng Y, He C, Wang M, Ma X, Mo F, Yang S, Han J, Wei X (2019) Targeting epigenetic regulators for cancer therapy: mechanisms and advances in clinical trials. Signal Transduct Target Ther 4:62

Choi KC, Jung MG, Lee YH, Yoon JC, Kwon SH, Kang HB, Kim MJ, Cha JH, Kim YJ, Jun WJ, Lee JM, Yoon HG (2009) Epigallocatechin-3-gallate, a histone acetyltransferase inhibitor, inhibits EBV-induced B lymphocyte transformation via suppression of RelA acetylation. Cancer Res 69:583–592

Choi SY, Kee HJ, Kurz T, Hansen FK, Ryu Y, Kim GR, Lin MQ, Jin L, Piao ZH, Jeong MH (2016) Class I HDACs specifically regulate E-cadherin expression in human renal epithelial cells. J Cell Mol Med 20:2289–2298

Chung M-Y, Song J-H, Lee J, Shin EJ, Park JH, Lee S-H, Hwang J-T, Choi H-K (2019) Tannic acid, a novel histone acetyltransferase inhibitor, prevents non-alcoholic fatty liver disease both in vivo and in vitro model. Mol Metab 19:34–48

Copeland RA, Olhava EJ, Scott MP (2010) Targeting epigenetic enzymes for drug discovery. Curr Opin Chem Biol 14:505–510

Coudé M-M, Braun T, Berrou J, Dupont M, Bertrand S, Masse A, Raffoux E, Itzykson R, Delord M, Riveiro ME, Herait P, Baruchel A, Dombret H, Gardin C (2015) BET inhibitor OTX015 targets BRD2 and BRD4 and decreases c-MYC in acute leukemia cells. Oncotarget 6:17698–17712

Couture JF, Hauk G, Thompson MJ, Blackburn GM, Trievel RC (2006) Catalytic roles for carbon-oxygen hydrogen bonding in SET domain lysine methyltransferases. J Biol Chem 281:19280–19287

Cowan LA, Talwar S, Yang AS (2010) Will DNA methylation inhibitors work in solid tumors? A review of the clinical experience with azacitidine and decitabine in solid tumors. Epigenomics 2:71–86

Curradi M, Izzo A, Badaracco G, Landsberger N (2002) Molecular mechanisms of gene silencing mediated by DNA methylation. Mol Cell Biol 22:3157–3173

Daigle SR, Olhava EJ, Therkelsen CA, Majer CR, Sneeringer CJ, Song J, Johnston LD, Scott MP, Smith JJ, Xiao Y, Jin L, Kuntz KW, Chesworth R, Moyer MP, Bernt KM, Tseng JC, Kung AL, Armstrong SA, Copeland RA, Richon VM, Pollock RM (2011) Selective killing of mixed lineage leukemia cells by a potent small-molecule DOT1L inhibitor. Cancer Cell 20:53–65

Dekker FJ, van den Bosch T, Martin NI (2014) Small molecule inhibitors of histone acetyltransferases and deacetylases are potential drugs for inflammatory diseases. Drug Discov Today 19:654–660

Delmore JE, Issa GC, Lemieux ME, Rahl PB, Shi J, Jacobs HM, Kastritis E, Gilpatrick T, Paranal RM, Qi J, Chesi M, Schinzel AC, McKeown MR, Heffernan TP, Vakoc CR, Bergsagel PL, Ghobrial IM, Richardson PG, Young RA, Hahn WC, Anderson KC, Kung AL, Bradner JE, Mitsiades CS (2011) BET bromodomain inhibition as a therapeutic strategy to target c-Myc. Cell 146:904–917

Dennis M, Davies M, Oliver S, D'Souza R, Pike L, Stockman P (2012) Phase I study of the Aurora B kinase inhibitor barasertib (AZD1152) to assess the pharmacokinetics, metabolism and excretion in patients with acute myeloid leukemia. Cancer Chemother Pharmacol 70:461–469

Dhalluin C, Carlson JE, Zeng L, He C, Aggarwal AK, Zhou MM (1999) Structure and ligand of a histone acetyltransferase bromodomain. Nature 399:491–496

Díaz T, Rodríguez V, Lozano E, Mena MP, Calderón M, Rosiñol L, Martínez A, Tovar N, Pérez-Galán P, Bladé J, Roué G, de Larrea CF (2017) The BET bromodomain inhibitor CPI203 improves lenalidomide and dexamethasone activity in in vitro and in vivo models of multiple myeloma by blockade of Ikaros and MYC signaling. Haematologica 102:1776–1784

Drew AE, Moradei O, Jacques SL, Rioux N, Boriack-Sjodin AP, Allain C, Scott MP, Jin L, Raimondi A, Handler JL, Ott HM, Kruger RG, McCabe MT, Sneeringer C, Riera T, Shapiro G, Waters NJ, Mitchell LH, Duncan KW, Moyer MP, Copeland RA, Smith J, Chesworth R, Ribich SA (2017) Identification of a CARM1 inhibitor with potent in vitro and in vivo activity in preclinical models of multiple myeloma. Sci Rep 7:17993

Estey EH (2013) Epigenetics in clinical practice: the examples of azacitidine and decitabine in myelodysplasia and acute myeloid leukemia. Leukemia 27:1803–1812

Fancelli D, Moll J, Varasi M, Bravo R, Artico R, Berta D, Bindi S, Cameron A, Candiani I, Cappella P, Carpinelli P, Croci W, Forte B, Giorgini ML, Klapwijk J, Marsiglio A, Pesenti E, Rocchetti M, Roletto F, Severino D, Soncini C, Storici P, Tonani R, Zugnoni P, Vianello P (2006) 1,4,5,6-tetrahydropyrrolo[3,4-c]pyrazoles: identification of a potent Aurora kinase inhibitor with a favorable antitumor kinase inhibition profile. J Med Chem 49:7247–7251

Fang JY, Lu YY (2002) Effects of histone acetylation and DNA methylation on p21(WAF1) regulation. World J Gastroenterol 8:400–405

Fang Y, Liao G, Yu B (2019) LSD1/KDM1A inhibitors in clinical trials: advances and prospects. J Hematol Oncol 12:129

Fedoriw A, Rajapurkar SR, O'Brien S, Gerhart SV, Mitchell LH, Adams ND, Rioux N, Lingaraj T, Ribich SA, Pappalardi MB, Shah N, Laraio J, Liu Y, Butticello M, Carpenter CL, Creasy C, Korenchuk S, McCabe MT, McHugh CF, Nagarajan R, Wagner C, Zappacosta F, Annan R, Concha NO, Thomas RA, Hart TK, Smith JJ, Copeland RA, Moyer MP, Campbell J, Stickland K, Mills J, Jacques-O'Hagan S, Allain C, Johnston D, Raimondi A, Porter Scott M, Waters N, Swinger K, Boriack-Sjodin A, Riera T, Shapiro G, Chesworth R, Prinjha RK, Kruger RG, Barbash O, Mohammad HP (2019) Anti-tumor activity of the type I PRMT inhibitor, GSK3368715, synergizes with PRMT5 inhibition through MTAP loss. Cancer Cell 36:100–114.e125

Filippakopoulos P, Knapp S (2014) Targeting bromodomains: epigenetic readers of lysine acetylation. Nat Rev Drug Discov 13

Filippakopoulos P, Qi J, Picaud S, Shen Y, Smith WB, Fedorov O, Morse EM, Keates T, Hickman TT, Felletar I, Philpott M, Munro S, McKeown MR, Wang Y, Christie AL, West N, Cameron MJ, Schwartz B, Heightman TD, La Thangue N, French CA, Wiest O, Kung AL, Knapp S, Bradner JE (2010) Selective inhibition of BET bromodomains. Nature 468:1067–1073

Fontana P, Bonfiglio JJ, Palazzo L, Bartlett E, Matic I, Ahel I (2017) Serine ADP-ribosylation reversal by the hydrolase ARH3. elife 6:e28533

Friedberg JW, Mahadevan D, Cebula E, Persky D, Lossos I, Agarwal AB, Jung J, Burack R, Zhou X, Leonard EJ, Fingert H, Danaee H, Bernstein SH (2014) Phase II study of alisertib, a selective Aurora A kinase inhibitor, in relapsed and refractory aggressive B- and T-cell non-Hodgkin lymphomas. J Clin Oncol 32:44–50

Gajer JM, Furdas SD, Gründer A, Gothwal M, Heinicke U, Keller K, Colland F, Fulda S, Pahl HL, Fichtner I, Sippl W, Jung M (2015) Histone acetyltransferase inhibitors block neuroblastoma cell growth in vivo. Oncogenesis 4:e137–e137

Ganai SA (2014) HDAC inhibitors entinostat and suberoylanilide hydroxamic acid (SAHA): the ray of hope for cancer therapy. In: Wells RD, Bond JS, Klinman J, Masters BSS, Bell E (eds) Molecular life sciences: an encyclopedic reference. Springer, New York, pp 1–16

Ganai SA (2019a) Different groups of HDAC inhibitors based on various classifications. In: Ganai SA (ed) Histone deacetylase inhibitors — epidrugs for neurological disorders. Springer, Singapore, pp 33–38

Ganai SA (2019b) Epigenetic enzymes and drawbacks of conventional therapeutic regimens. In: Ganai SA (ed) Histone deacetylase inhibitors — epidrugs for neurological disorders. Springer, Singapore, pp 11–19

Ganesan A, Arimondo PB, Rots MG, Jeronimo C, Berdasco M (2019) The timeline of epigenetic drug discovery: from reality to dreams. Clin Epigenetics 11:174

Geuns-Meyer S, Cee VJ, Deak HL, Du B, Hodous BL, Nguyen HN, Olivieri PR, Schenkel LB, Vaida KR, Andrews P, Bak A, Be X, Beltran PJ, Bush TL, Chaves MK, Chung G, Dai Y, Eden P, Hanestad K, Huang L, Lin MH, Tang J, Ziegler B, Radinsky R, Kendall R, Patel VF, Payton M (2015) Discovery of N-(4-(3-(2-aminopyrimidin-4-yl)pyridin-2-yloxy)phenyl)-4-(4-methylthiophen-2-yl)p hthalazin-1-amine (AMG 900), a highly selective, orally bioavailable inhibitor of aurora kinases with activity against multidrug-resistant cancer cell lines. J Med Chem 58:5189–5207

Ghizzoni M, Wu J, Gao T, Haisma HJ, Dekker FJ, George Zheng Y (2012) 6-alkylsalicylates are selective Tip60 inhibitors and target the acetyl-CoA binding site. Eur J Med Chem 47:337–344

Gibney ER, Nolan CM (2010) Epigenetics and gene expression. Heredity 105:4–13

Giri AK, Aittokallio T (2019) DNMT inhibitors increase methylation in the cancer genome. Front Pharmacol 10

Goldberg SL, Fenaux P, Craig MD, Gyan E, Lister J, Kassis J, Pigneux A, Schiller GJ, Jung J, Jane Leonard E, Fingert H, Westervelt P (2014) An exploratory phase 2 study of investigational Aurora A kinase inhibitor alisertib (MLN8237) in acute myelogenous leukemia and myelodysplastic syndromes. Leuk Res Rep 3:58–61

Greer EL, Shi Y (2012) Histone methylation: a dynamic mark in health, disease and inheritance. Nat Rev Genet 13:343–357

Greiner D, Bonaldi T, Eskeland R, Roemer E, Imhof A (2005) Identification of a specific inhibitor of the histone methyltransferase SU(VAR)3-9. Nat Chem Biol 1:143–145

Griffiths EA, Choy G, Redkar S, Taverna P, Azab M, Karpf AR (2013) SGI-110: DNA methyltransferase inhibitor oncolytic. Drugs Future 38:535–543

Gros C, Fahy J, Halby L, Dufau I, Erdmann A, Gregoire J-M, Ausseil F, Vispé S, Arimondo PB (2012) DNA methylation inhibitors in cancer: recent and future approaches. Biochimie 94:2280–2296

Grunstein M (1997) Histone acetylation in chromatin structure and transcription. Nature 389:349–352

Gulati N, Béguelin W, Giulino-Roth L (2018) Enhancer of zeste homolog 2 (EZH2) inhibitors. Leuk Lymphoma 59:1574–1585

Handy DE, Castro R, Loscalzo J (2011) Epigenetic modifications: basic mechanisms and role in cardiovascular disease. Circulation 123:2145–2156

Heerboth S, Lapinska K, Snyder N, Leary M, Rollinson S, Sarkar S (2014) Use of epigenetic drugs in disease: an overview. Genet Epigenet 6:9–19

Howard S, Berdini V, Boulstridge JA, Carr MG, Cross DM, Curry J, Devine LA, Early TR, Fazal L, Gill AL, Heathcote M, Maman S, Matthews JE, McMenamin RL, Navarro EF, O'Brien MA, O'Reilly M, Rees DC, Reule M, Tisi D, Williams G, Vinkovic M, Wyatt PG (2009) Fragment-based discovery of the pyrazol-4-yl urea (AT9283), a multitargeted kinase inhibitor with potent aurora kinase activity. J Med Chem 52:379–388

Huertas D, Soler M, Moreto J, Villanueva A, Martinez A, Vidal A, Charlton M, Moffat D, Patel S, McDermott J, Owen J, Brotherton D, Krige D, Cuthill S, Esteller M (2012) Antitumor activity of a small-molecule inhibitor of the histone kinase Haspin. Oncogene 31:1408–1418

Hurd PJ, Whitmarsh AJ, Baldwin GS, Kelly SM, Waltho JP, Price NC, Connolly BA, Hornby DP (1999) Mechanism-based inhibition of C5-cytosine DNA methyltransferases by 2-H pyrimidinone. J Mol Biol 286:389–401

Idrissou M, Judes G, Daures M, Sanchez A, Ouardi D, Besse S, Degoul F, Penault-Llorca F, Bignon Y-J, Bernard-Gallon D (2019) TIP60 inhibitor TH1834 reduces breast cancer progression in xenografts in mice. OMICS 23:457–459

Italiano A, Soria JC, Toulmonde M, Michot JM, Lucchesi C, Varga A, Coindre JM, Blakemore SJ, Clawson A, Suttle B, McDonald AA, Woodruff M, Ribich S, Hedrick E, Keilhack H, Thomson B, Owa T, Copeland RA, Ho PTC, Ribrag V (2018) Tazemetostat, an EZH2 inhibitor, in relapsed or refractory B-cell non-Hodgkin lymphoma and advanced solid tumours: a first-in-human, open-label, phase 1 study. Lancet Oncol 19:649–659

Jackson-Grusby L, Beard C, Possemato R, Tudor M, Fambrough D, Csankovszki G, Dausman J, Lee P, Wilson C, Lander E, Jaenisch R (2001) Loss of genomic methylation causes p53-dependent apoptosis and epigenetic deregulation. Nat Genet 27:31–39

Jani JP, Arcari J, Bernardo V, Bhattacharya SK, Briere D, Cohen BD, Coleman K, Christensen JG, Emerson EO, Jakowski A, Hook K, Los G, Moyer JD, Pruimboom-Brees I, Pustilnik L, Rossi AM, Steyn SJ, Su C, Tsaparikos K, Wishka D, Yoon K, Jakubczak JL (2010) PF-03814735, an orally bioavailable small molecule Aurora kinase inhibitor for cancer therapy. Mol Cancer Ther 9:883–894

Jin B, Li Y, Robertson KD (2011) DNA methylation: superior or subordinate in the epigenetic hierarchy? Genes Cancer 2:607–617

Jones PL, Veenstra GJ, Wade PA, Vermaak D, Kass SU, Landsberger N, Strouboulis J, Wolffe AP (1998) Methylated DNA and MeCP2 recruit histone deacetylase to repress transcription. Nat Genet 19:187–191

Kaminskas E, Farrell AT, Wang YC, Sridhara R, Pazdur R (2005) FDA drug approval summary: azacitidine (5-azacytidine, Vidaza) for injectable suspension. Oncologist 10:176–182

Kang ZH, Wang CY, Zhang WL, Zhang JT, Yuan CH, Zhao PW, Lin YY, Hong S, Li CY, Wang L (2014) Histone deacetylase HDAC4 promotes gastric cancer SGC-7901 cells progression via p21 repression. PLoS One 9:e98894

Kanno K, Kanno S, Nitta H, Uesugi N, Sugai T, Masuda T, Wakabayashi G, Maesawa C (2012) Overexpression of histone deacetylase 6 contributes to accelerated migration and invasion activity of hepatocellular carcinoma cells. Oncol Rep 28:867–873

Kass SU, Pruss D, Wolffe AP (1997) How does DNA methylation repress transcription? Trends Genet 13:444–449

Kim MS, Kwon HJ, Lee YM, Baek JH, Jang JE, Lee SW, Moon EJ, Kim HS, Lee SK, Chung HY, Kim CW, Kim KW (2001) Histone deacetylases induce angiogenesis by negative regulation of tumor suppressor genes. Nat Med 7:437–443

Kim MS, Cho HI, Yoon HJ, Ahn Y-H, Park EJ, Jin YH, Jang YK (2018) JIB-04, a small molecule histone demethylase inhibitor, selectively targets colorectal cancer stem cells by inhibiting the Wnt/β-catenin signaling pathway. Sci Rep 8:6611–6611

Kohler J, Erlenkamp G, Eberlin A, Rumpf T, Slynko I, Metzger E, Schule R, Sippl W, Jung M (2012) Lestaurtinib inhibits histone phosphorylation and androgen-dependent gene expression in prostate cancer cells. PLoS One 7:e34973

Kohli RM, Zhang Y (2013) TET enzymes, TDG and the dynamics of DNA demethylation. Nature 502:472–479

Kouzarides T (2007) Chromatin modifications and their function. Cell 128:693–705

Kubicek S, O'Sullivan RJ, August EM, Hickey ER, Zhang Q, Teodoro ML, Rea S, Mechtler K, Kowalski JA, Homon CA, Kelly TA, Jenuwein T (2007) Reversal of H3K9me2 by a small-molecule inhibitor for the G9a histone methyltransferase. Mol Cell 25:473–481

Li X, Wang C, Jiang H, Luo C (2019) A patent review of arginine methyltransferase inhibitors (2010–2018). Expert Opin Ther Pat 29:97–114

Liu F, Chen X, Allali-Hassani A, Quinn AM, Wasney GA, Dong A, Barsyte D, Kozieradzki I, Senisterra G, Chau I, Siarheyeva A, Kireev DB, Jadhav A, Herold JM, Frye SV, Arrowsmith CH, Brown PJ, Simeonov A, Vedadi M, Jin J (2009) Discovery of a 2,4-diamino-7-aminoalkoxyquinazoline as a potent and selective inhibitor of histone lysine methyltransferase G9a. J Med Chem 52:7950–7953

Liu F, Chen X, Allali-Hassani A, Quinn AM, Wigle TJ, Wasney GA, Dong A, Senisterra G, Chau I, Siarheyeva A, Norris JL, Kireev DB, Jadhav A, Herold JM, Janzen WP, Arrowsmith CH, Frye SV, Brown PJ, Simeonov A, Vedadi M, Jin J (2010) Protein lysine methyltransferase G9a inhibitors: design, synthesis, and structure activity relationships of 2,4-diamino-7-aminoalkoxy-quinazolines. J Med Chem 53:5844–5857

Liu W, Deng L, Song Y, Redell M (2014) DOT1L inhibition sensitizes MLL-rearranged AML to chemotherapy. PLoS One 9:e98270

Liu Y, Fiorito J, Gonzale Y, Zuccarello E, Calcagno E, Camarillo JM, Thomas PM, Kelleher N, Deng S, Landry D, O'Connor OA, Wolfe AJ, Moyer B, Arancio O, Amengual J (2019) First-in-class hat activator highly synergistic with pan-hdac inhibitor romidepsin leading to profound histone acetylation cytotoxicity. Hematol Oncol 37:125–126

Long J, Li B, Rodriguez-Blanco J, Pastori C, Volmar CH, Wahlestedt C, Capobianco A, Bai F, Pei XH, Ayad NG, Robbins DJ (2014) The BET bromodomain inhibitor I-BET151 acts downstream of smoothened protein to abrogate the growth of hedgehog protein-driven cancers. J Biol Chem 289:35494–35502

Lovén J, Hoke HA, Lin CY, Lau A, Orlando DA, Vakoc CR, Bradner JE, Lee TI, Young RA (2013) Selective inhibition of tumor oncogenes by disruption of super-enhancers. Cell 153:320–334

Lowenberg B, Muus P, Ossenkoppele G, Rousselot P, Cahn JY, Ifrah N, Martinelli G, Amadori S, Berman E, Sonneveld P, Jongen-Lavrencic M, Rigaudeau S, Stockman P, Goudie A, Faderl S, Jabbour E, Kantarjian H (2011) Phase 1/2 study to assess the safety, efficacy, and pharmacokinetics of barasertib (AZD1152) in patients with advanced acute myeloid leukemia. Blood 118:6030–6036

Lu D (2013) Epigenetic modification enzymes: catalytic mechanisms and inhibitors. Acta Pharm Sin B 3:141–149

Lund K, Cole JJ, VanderKraats ND, McBryan T, Pchelintsev NA, Clark W, Copland M, Edwards JR, Adams PD (2014) DNMT inhibitors reverse a specific signature of aberrant promoter DNA methylation and associated gene silencing in AML. Genome Biol 15:406–406

Ma Y, Yue Y, Pan M, Sun J, Chu J, Lin X, Xu W, Feng L, Chen Y, Chen D, Shin VY, Wang X, Jin H (2015) Histone deacetylase 3 inhibits new tumor suppressor gene DTWD1 in gastric cancer. Am J Cancer Res 5:663–673

Magistri M, Velmeshev D, Makhmutova M, Patel P, Sartor GC, Volmar CH, Wahlestedt C, Faghihi MA (2016) The BET-Bromodomain inhibitor JQ1 reduces inflammation and tau phosphorylation at Ser396 in the brain of the 3xTg model of Alzheimer's disease. Curr Alzheimer Res 13:985–995

Maleszewska M, Kaminska B (2015) Deregulation of histone-modifying enzymes and chromatin structure modifiers contributes to glioma development. Future Oncol 11:2587–2601

Manfredi MG, Ecsedy JA, Chakravarty A, Silverman L, Zhang M, Hoar KM, Stroud SG, Chen W, Shinde V, Huck JJ, Wysong DR, Janowick DA, Hyer ML, Leroy PJ, Gershman RE, Silva MD, Germanos MS, Bolen JB, Claiborne CF, Sells TB (2011) Characterization of Alisertib (MLN8237), an investigational small-molecule inhibitor of Aurora A kinase using novel in vivo pharmacodynamic assays. Clin Cancer Res 17:7614–7624

Mantelingu K, Kishore AH, Balasubramanyam K, Kumar GV, Altaf M, Swamy SN, Selvi R, Das C, Narayana C, Rangappa KS, Kundu TK (2007) Activation of p300 histone acetyltransferase by small molecules altering enzyme structure: probed by surface-enhanced Raman spectroscopy. J Phys Chem B 111:4527–4534

Martin MP, Olesen SH, Georg GI, Schönbrunn E (2013) Cyclin-dependent kinase inhibitor Dinaciclib interacts with the acetyl-lysine recognition site of Bromodomains. ACS Chem Biol 8:2360–2365

Martinez-Zamudio R, Ha HC (2012) Histone ADP-ribosylation facilitates gene transcription by directly remodeling nucleosomes. Mol Cell Biol 32:2490–2502

Mitsui E, Yoshida S, Shinoda Y, Matsumori Y, Tsujii H, Tsuchida M, Wada S, Hasegawa M, Ito A, Mino K, Onuki T, Yoshida M, Sasaki R, Mizukami T (2019) Identification of ryuvidine as a KDM5A inhibitor. Sci Rep 9:9952

Moros A, Rodríguez V, Saborit-Villarroya I, Montraveta A, Balsas P, Sandy P, Martínez A, Wiestner A, Normant E, Campo E, Pérez-Galán P, Colomer D, Roué G (2014) Synergistic antitumor activity of lenalidomide with the BET bromodomain inhibitor CPI203 in bortezomib-resistant mantle cell lymphoma. Leukemia 28:2049–2059

Mortlock AA, Foote KM, Heron NM, Jung FH, Pasquet G, Lohmann JJ, Warin N, Renaud F, De Savi C, Roberts NJ, Johnson T, Dousson CB, Hill GB, Perkins D, Hatter G, Wilkinson RW, Wedge SR, Heaton SP, Odedra R, Keen NJ, Crafter C, Brown E, Thompson K, Brightwell S, Khatri L, Brady MC, Kearney S, McKillop D, Rhead S, Parry T, Green S (2007) Discovery, synthesis, and in vivo activity of a new class of pyrazoloquinazolines as selective inhibitors of aurora B kinase. J Med Chem 50:2213–2224

Muller S, Filippakopoulos P, Knapp S (2011) Bromodomains as therapeutic targets. Expert Rev Mol Med 13:e29–e29

Muntean A, Hess J (2009) Epigenetic dysregulation in cancer. Am J Pathol 175:1353–1361

Nakamura K, Aizawa K, Nakabayashi K, Kato N, Yamauchi J, Hata K, Tanoue A (2013) DNA methyltransferase inhibitor zebularine inhibits human hepatic carcinoma cells proliferation and induces apoptosis. PLoS One 8:e54036

Nakayama K, Szewczyk MM, Dela Sena C, Wu H, Dong A, Zeng H, Li F, de Freitas RF, Eram MS, Schapira M, Baba Y, Kunitomo M, Cary DR, Tawada M, Ohashi A, Imaeda Y, Saikatendu KS, Grimshaw CE, Vedadi M, Arrowsmith CH, Barsyte-Lovejoy D, Kiba A, Tomita D, Brown PJ (2018) TP-064, a potent and selective small molecule inhibitor of PRMT4 for multiple myeloma. Oncotarget 9:18480–18493

Navada SC, Steinmann J, Lubbert M, Silverman LR (2014) Clinical development of demethylating agents in hematology. J Clin Invest 124:40–46

Pastori C, Daniel M, Penas C, Volmar CH, Johnstone AL, Brothers SP, Graham RM, Allen B, Sarkaria JN, Komotar RJ, Wahlestedt C, Ayad NG (2014) BET bromodomain proteins are required for glioblastoma cell proliferation. Epigenetics 9:611–620

Patnaik S, Anupriya (2019) Drugs targeting epigenetic modifications and plausible therapeutic strategies against colorectal cancer. Front Pharmacol 10:588

Payton M, Bush TL, Chung G, Ziegler B, Eden P, McElroy P, Ross S, Cee VJ, Deak HL, Hodous BL, Nguyen HN, Olivieri PR, Romero K, Schenkel LB, Bak A, Stanton M, Dussault I, Patel VF, Geuns-Meyer S, Radinsky R, Kendall RL (2010) Preclinical evaluation of AMG 900, a novel potent and highly selective pan-aurora kinase inhibitor with activity in taxane-resistant tumor cell lines. Cancer Res 70:9846–9854

Picaud S, Da Costa D, Thanasopoulou A, Filippakopoulos P, Fish PV, Philpott M, Fedorov O, Brennan P, Bunnage ME, Owen DR, Bradner JE, Taniere P, O'Sullivan B, Müller S, Schwaller J, Stankovic T, Knapp S (2013a) PFI-1, a highly selective protein interaction inhibitor, targeting BET Bromodomains. Cancer Res 73:3336–3346

Picaud S, Wells C, Felletar I, Brotherton D, Martin S, Savitsky P, Diez-Dacal B, Philpott M, Bountra C, Lingard H, Fedorov O, Müller S, Brennan PE, Knapp S, Filippakopoulos P (2013b) RVX-208, an inhibitor of BET transcriptional regulators with selectivity for the second bromodomain. Proc Natl Acad Sci 110:19754–19759

Quinn AM, Allali-Hassani A, Vedadi M, Simeonov A (2010) A chemiluminescence-based method for identification of histone lysine methyltransferase inhibitors. Mol BioSyst 6:782–788

Richters A, Koehler A (2017) Epigenetic modulation using small molecules - targeting histone acetyltransferases in disease. Curr Med Chem 24

Roberti A, Valdes AF, Torrecillas R, Fraga MF, Fernandez AF (2019) Epigenetics in cancer therapy and nanomedicine. Clin Epigenetics 11:81

Robertson KD, Jones PA (2000) DNA methylation: past, present and future directions. Carcinogenesis 21:461–467

Roboz GJ, Kantarjian HM, Yee KWL, Kropf PL, O'Connell CL, Griffiths EA, Stock W, Daver NG, Jabbour E, Ritchie EK, Walsh KJ, Rizzieri D, Lunin SD, Curio T, Chung W, Hao Y, Lowder JN, Azab M, Issa JJ (2018) Dose, schedule, safety, and efficacy of guadecitabine in relapsed or refractory acute myeloid leukemia. Cancer 124:325–334

Rossetto D, Avvakumov N, Côté J (2012) Histone phosphorylation. Epigenetics 7:1098–1108

Saba HI (2007) Decitabine in the treatment of myelodysplastic syndromes. Ther Clin Risk Manag 3:807–817

Sanchez R, Meslamani J, Zhou MM (2014) The bromodomain: from epigenome reader to druggable target. Biochim Biophys Acta 1839:676–685

Sbardella G, Castellano S, Vicidomini C, Rotili D, Nebbioso A, Miceli M, Altucci L, Mai A (2008) Identification of long chain alkylidenemalonates as novel small molecule modulators of histone acetyltransferases. Bioorg Med Chem Lett 18:2788–2792

Schneider A, Chatterjee S, Bousiges O, Selvi BR, Swaminathan A, Cassel R, Blanc F, Kundu TK, Boutillier AL (2013) Acetyltransferases (HATs) as targets for neurological therapeutics. Neurotherapeutics 10:568–588

Schwartz GK, Carvajal RD, Midgley R, Rodig SJ, Stockman PK, Ataman O, Wilson D, Das S, Shapiro GI (2013) Phase I study of barasertib (AZD1152), a selective inhibitor of Aurora B kinase, in patients with advanced solid tumors. Investig New Drugs 31:370–380

Sells TB, Chau R, Ecsedy JA, Gershman RE, Hoar K, Huck J, Janowick DA, Kadambi VJ, LeRoy PJ, Stirling M, Stroud SG, Vos TJ, Weatherhead GS, Wysong DR, Zhang M, Balani SK, Bolen JB, Manfredi MG, Claiborne CF (2015) MLN8054 and Alisertib (MLN8237): discovery of selective oral Aurora A inhibitors. ACS Med Chem Lett 6:630–634

Seto E, Yoshida M (2014) Erasers of histone acetylation: the histone deacetylase enzymes. Cold Spring Harb Perspect Biol 6:a018713

Siegfried Z, Eden S, Mendelsohn M, Feng X, Tsuberi B-Z, Cedar H (1999) DNA methylation represses transcription in vivo. Nat Genet 22:203–206

Stein EM, Garcia-Manero G, Rizzieri DA, Tibes R, Berdeja JG, Jongen-Lavrencic M, Altman JK, Dohner H, Thomson B, Blakemore SJ, Daigle S, Fine G, Waters NJ, Krivstov AV, Koche R, Armstrong SA, Ho PT, Lowenberg B, Tallman MS (2015) A phase 1 study of the DOT1L inhibitor, Pinometostat (EPZ-5676), in adults with relapsed or refractory leukemia: safety, clinical activity, exposure and target inhibition. Blood 126:2547–2547

Stimson L, Rowlands MG, Newbatt YM, Smith NF, Raynaud FI, Rogers P, Bavetsias V, Gorsuch S, Jarman M, Bannister A, Kouzarides T, McDonald E, Workman P, Aherne GW (2005) Isothiazolones as inhibitors of PCAF and p300 histone acetyltransferase activity. Mol Cancer Ther 4:1521–1532

Stresemann C, Lyko F (2008) Modes of action of the DNA methyltransferase inhibitors azacytidine and decitabine. Int J Cancer 123:8–13

Sung B, Pandey MK, Ahn KS, Yi T, Chaturvedi MM, Liu M, Aggarwal BB (2008) Anacardic acid (6-nonadecyl salicylic acid), an inhibitor of histone acetyltransferase, suppresses expression of nuclear factor-kappaB-regulated gene products involved in cell survival, proliferation, invasion, and inflammation through inhibition of the inhibitory subunit of nuclear factor-kappaBalpha kinase, leading to potentiation of apoptosis. Blood 111:4880–4891

Taplin M-E, Hussain A, Shah S, Shore ND, Agrawal M, Clark W, Edenfield WJ, Nordquist LT, Sartor OA, Butrynski JE, Chatta GS, Fleming MT, Oh WK, Bradley B, Piel J, Nash D, Colak G, Li J, Lebedinsky C, Antonarakis ES (2019) ProSTAR: a phase Ib/II study of CPI-1205, a small molecule inhibitor of EZH2, combined with enzalutamide (E) or abiraterone/prednisone (A/P) in patients with metastatic castration-resistant prostate cancer (mCRPC). J Clin Oncol 37: TPS335–TPS335

Vázquez R, Riveiro ME, Astorgues-Xerri L, Odore E, Rezai K, Erba E, Panini N, Rinaldi A, Kwee I, Beltrame L, Bekradda M, Cvitkovic E, Bertoni F, Frapolli R, D'Incalci M (2017) The bromodomain inhibitor OTX015 (MK-8628) exerts anti-tumor activity in triple-negative breast cancer models as single agent and in combination with everolimus. Oncotarget 8:7598–7613

Verheugd P, Bütepage M, Eckei L, Lüscher B (2016) Players in ADP-ribosylation: readers and erasers. Curr Protein Pept Sci 17:654–667

Wang Z, Patel DJ (2013) Small molecule epigenetic inhibitors targeted to histone lysine methyltransferases and demethylases. Q Rev Biophys 46:349–373

Wang C, Fu M, Mani S, Wadler S, Senderowicz AM, Pestell RG (2001) Histone acetylation and the cell-cycle in cancer. Front Biosci 6:D610–D629

Wang P, Wang Z, Liu J (2020) Role of HDACs in normal and malignant hematopoiesis. Mol Cancer 19:5

Watts JM, Bradley TJ, Thomassen A, Brunner AM, Minden MD, Papadantonakis N, Abedin S, Baines AJ, Barbash O, Gorman S, Kremer BE, Borthakur GM (2019) A phase I/II study to investigate the safety and clinical activity of the protein arginine methyltransferase 5 inhibitor GSK3326595 in subjects with myelodysplastic syndrome and acute myeloid leukemia. Blood 134:2656–2656

Whetstine JR, Ceron J, Ladd B, Dufourcq P, Reinke V, Shi Y (2005) Regulation of tissue-specific and extracellular matrix-related genes by a class I histone deacetylase. Mol Cell 18:483–490

Wong KK, Lawrie CH, Green TM (2019) Oncogenic roles and inhibitors of DNMT1, DNMT3A, and DNMT3B in acute myeloid Leukaemia. Biomark Insights 14:1177271919846454

Wood K, Tellier M, Murphy S (2018) DOT1L and H3K79 methylation in transcription and genomic stability. Biomol Ther 8:11

Wu J, Du C, Lv Z, Ding C, Cheng J, Xie H, Zhou L, Zheng S (2013) The up-regulation of histone deacetylase 8 promotes proliferation and inhibits apoptosis in hepatocellular carcinoma. Dig Dis Sci 58:3545–3553

Wyce A, Ganji G, Smitheman KN, Chung CW, Korenchuk S, Bai Y, Barbash O, Le B, Craggs PD, McCabe MT, Kennedy-Wilson KM, Sanchez LV, Gosmini RL, Parr N, McHugh CF, Dhanak D, Prinjha RK, Auger KR, Tummino PJ (2013) BET inhibition silences expression of

MYCN and BCL2 and induces cytotoxicity in neuroblastoma tumor models. PLoS One 8: e72967

Xu B, On DM, Ma A, Parton T, Konze KD, Pattenden SG, Allison DF, Cai L, Rockowitz S, Liu S, Liu Y, Li F, Vedadi M, Frye SV, Garcia BA, Zheng D, Jin J, Wang GG (2015) Selective inhibition of EZH2 and EZH1 enzymatic activity by a small molecule suppresses MLL-rearranged leukemia. Blood 125:346–357

Yang XJ, Seto E (2007) HATs and HDACs: from structure, function and regulation to novel strategies for therapy and prevention. Oncogene 26:5310–5318

Yang X, Lay F, Han H, Jones PA (2010) Targeting DNA methylation for epigenetic therapy. Trends Pharmacol Sci 31:536–546

Yang Y, Zhang R, Li Z, Mei L, Wan S, Ding H, Chen Z, Xing J, Feng H, Han J, Jiang H, Zheng M, Luo C, Zhou B (2020) Discovery of highly potent, selective, and orally efficacious p300/CBP histone acetyltransferases inhibitors. J Med Chem 63:1337–1360

Yoo CB, Jeong S, Egger G, Liang G, Phiasivongsa P, Tang C, Redkar S, Jones PA (2007) Delivery of 5-aza-2′-deoxycytidine to cells using oligodeoxynucleotides. Cancer Res 67:6400–6408

Yu W, Chory EJ, Wernimont AK, Tempel W, Scopton A, Federation A, Marineau JJ, Qi J, Barsyte-Lovejoy D, Yi J, Marcellus R, Iacob RE, Engen JR, Griffin C, Aman A, Wienholds E, Li F, Pineda J, Estiu G, Shatseva T, Hajian T, Al-Awar R, Dick JE, Vedadi M, Brown PJ, Arrowsmith CH, Bradner JE, Schapira M (2012) Catalytic site remodelling of the DOT1L methyltransferase by selective inhibitors. Nat Commun 3:1288

Zeng L, Zhou M-M (2002) Bromodomain: an acetyl-lysine binding domain. FEBS Lett 513:124–128

Zeng F-Q, Peng S, Li L, Mu L-B, Zhang Z-H, Zhang Z, Huang N (2013) Structure-based identification of drug-like inhibitors of p300 histone acetyltransferase. Yao xue xue bao 48:700–708

Zhang J, Qian K, Yan C, He M, Jassim BA, Ivanov I, Zheng YG (2017) Discovery of decamidine as a new and potent PRMT1 inhibitor. Med Chem Commun 8:440–444

Zhang B, Chen D, Liu B, Dekker FJ, Quax WJ (2020) A novel histone acetyltransferase inhibitor A485 improves sensitivity of non-small-cell lung carcinoma cells to TRAIL. Biochem Pharmacol 175:113914

Zheng YC, Yu B, Jiang GZ, Feng XJ, He PX, Chu XY, Zhao W, Liu HM (2016) Irreversible LSD1 inhibitors: application of tranylcypromine and its derivatives in cancer treatment. Curr Top Med Chem 16:2179–2188

Zhu K, Shao J, Tao H, Yan X, Luo C, Zhang H, Duan W (2019) Rational design, synthesis and biological evaluation of novel triazole derivatives as potent and selective PRMT5 inhibitors with antitumor activity. J Comput Aided Mol Des 33:775–785

Conspectus of Structurally Distinct Groups of Histone Deacetylase Inhibitors of Classical Histone Deacetylases and Sirtuins

8

Interplay of genetic and epigenetic dysregulation powers the onset and progression of cancer (Lund and van Lohuizen 2004). Proper execution of gene expression programs is potentially reliant on acetylation homeostasis (Fraga et al. 2005; Li and Seto 2016). Regulated by histone acetyltransferases (HATs) and their functional antagonists histone deacetylases (HDACs), this homeostasis drives cellular processes in a normal manner (Yang and Seto 2007). Anomalous expression/activity of HDACs contributes significantly to tumour onset and progression through epigenetic mechanism (Li and Seto 2016). Moreover, HDAC overactivity, by way of altering stability and functioning of non-histone proteins also facilitates cancer signalling (Ganai 2018; Singh et al. 2010). Although both HAT inactivity and HDAC overactivity have been noted in cancers, the latter is preferred for intervention (Marks et al. 2004). This is because from pharmacological standpoint it is highly straightforward to obstruct an enzyme instead of inducing one (Shabason et al. 2010). Due to this fact HDAC inhibition has gained a massive clinical interest as a potential strategy for subduing cancer.

8.1 Histone Deacetylase Inhibitors and Their Various Classifications

Re-establishing cellular acetylation homeostasis through targeting of HDACs reverses the dysregulation caused by overexpression of these enzymes. This has been achieved by using active site directed small-molecule therapeutics of HDACs (Suraweera et al. 2018). These molecules known as histone deacetylase inhibitors (HDACi) have gained strong importance nowadays in treating several malignancies (Ganai 2014; Zhao et al. 2020). HDACi by way of interfering HDACs modulate cellular events and reinstate conditions feasible for normal cell growth (Fig. 8.1). They trigger apoptosis, autophagy and DNA damage in cancer cells for bringing therapeutic effect (Kiweler et al. 2020; Liao et al. 2020; Richa et al. 2020). Most of the HDACi inhibit HDACs in a reversible manner although few including depudecin

S. A. Ganai, *Histone Deacetylase Inhibitors in Combinatorial Anticancer Therapy*,
https://doi.org/10.1007/978-981-15-8179-3_8

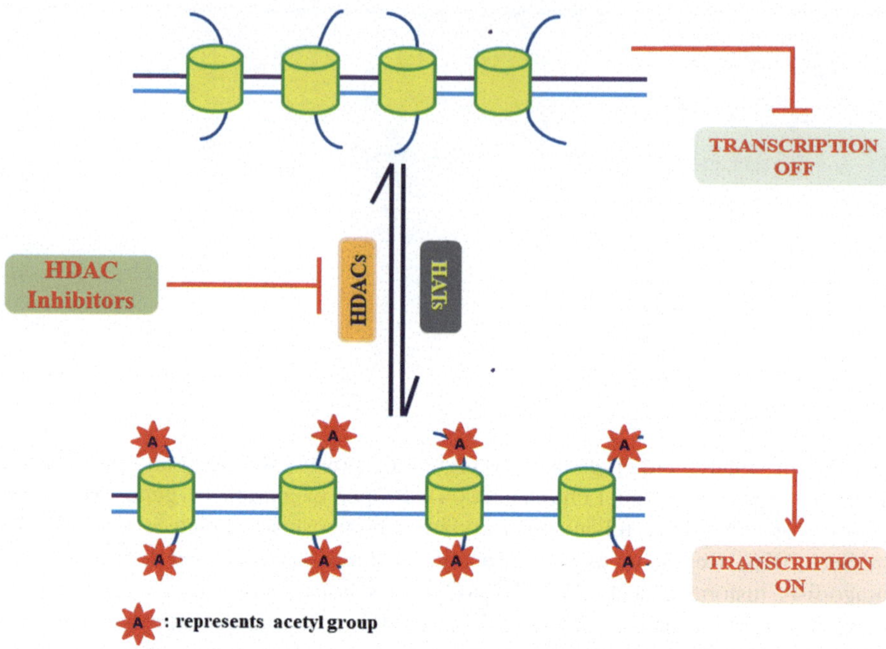

Fig. 8.1 Mechanism of action of histone acetyltransferases (HATs), histone deacetylases (HDACs) and histone deacetylase inhibitors (HDACi). HATs promote transcription by installing acetyl group on ε-lysine residues of unstructured histone tails. This powers electrostatic repulsion between underlying polycationic histones and overlying polyanionic DNA culminating in transcription. HDACs reverse this process by erasing these acetyl tags from underlying nucleosomal histones. This augments electrostatic attraction between the DNA and histone scaffolds resulting in chromatin condensation and subsequent silencing of transcription. Overactivity of HDACs by way of disrupting acetylation/deacetylation equilibrium promotes transcriptional dysfunction, thereby making the platform fertile for tumour onset. Histone deacetylase inhibitors (HDACi) are active site directed inhibitors of HDACs. These inhibitors fine tune the acetylation homeostasis altered by HDAC overactivity

and chlamydocin have been proved to irreversibly bind HDACs (Bhuiyan et al. 2006; Kijima et al. 1993). Based on differences in chemical structure, specificity and origin, HDACi have been classified into various types (Damaskos et al. 2017; Ganai 2019). For the ease of understanding each classification will be discussed under separate headings.

8.1.1 HDACi Groups Based on Chemical Structure Difference

Various HDACi approved or undergoing preclinical or clinical evaluation have been classified into several groups based on variation in their chemical structure. They may be short chain fatty acids such as sodium butyrate, valproic acid, phenylbutyrate; benzamide derivatives like mocetinostat (MGCD-0103), entinostat

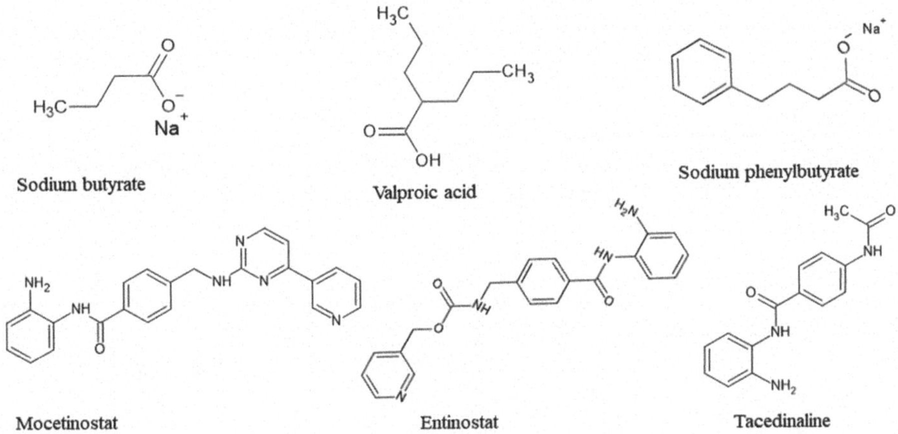

Fig. 8.2 Chemical structures of various inhibitors belonging to short chain fatty acid and benzamide derivative group of histone deacetylase inhibitors (HDACi). While short chain fatty acid HDACi are shown on top row, benzamide derivative group HDACi along with their corresponding names are represented by bottom row. Structures generated by ACD/ChemSketch (Freeware)

(MS 275), tacedinaline (CI-994) and chidamide (Fig. 8.2) (Ganai 2014; Sun et al. 2019); macrocyclic HDACi including cyclic tetrapeptides (apicidin, HC-toxin and trapoxin A) and bicyclic depsipeptides (largazole, romidepsin (FK228), FR901375, Spiruchostatin A and B) (Fig. 8.3) (Mwakwari et al. 2010); hydroxamates like panobinostat, trichostatin A, vorinostat (SAHA), pracinostat, belinostat, givinostat, abexinostat, CUDC-101 and scriptaid (Mottamal et al. 2015; Sanaei and Kavoosi 2019); electrophilic ketones such as trifluoromethyl ketone (Fig. 8.4) (Frey et al. 2002; Madsen and Olsen 2016). Among these groups hydroxamates are potent and have been studied extensively (Eckschlager et al. 2017).

8.1.2 Classification Based on Specificity Towards HDAC Subtypes

Most of the HDACi target almost all classical HDACs non-specifically and are thus known as pan-HDACi. Among the hydroxamates, vorinostat and trichostatin A being authorized pan-inhibitors target HDACs (1–9) with nearly similar potency (Khan et al. 2008). Further studies have proved that vorinostat inhibits Class I, Class IIb and Class IV HDACs but has low effect on Class IIa HDACs (Bradner et al. 2010; Marks and Xu 2009). Taken together broad-spectrum HDACi and therefore obviously non-selective come under the confines of pan-HDACi (Li and Seto 2016; Zhang and Xu 2015). HDACi affecting various HDACs of a particular class or a single HDAC come under selective inhibitors. Selective inhibitors obstructing several HDACs belonging to a single class are known as class selective whereas those inhibiting a single HDAC are isoform selective (Bieliauskas and Pflum 2008). HDACi selective for Class I HDACs (HDAC1–3) possess 2-aminoanilide as their

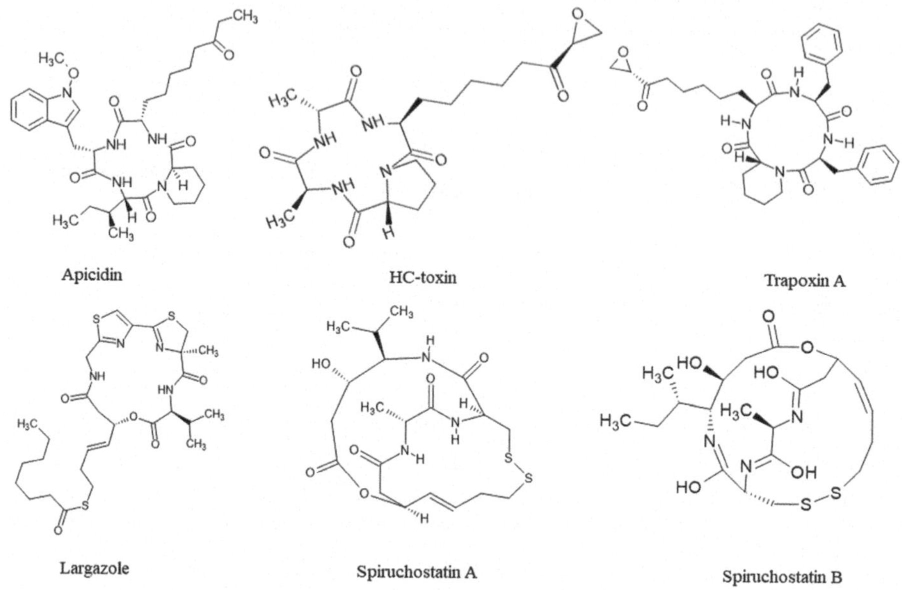

Fig. 8.3 Structures of macrocyclic histone deacetylase inhibitors (HDACi). Macrocyclic HDACi include cyclic tetrapeptides and bicyclic depsipeptides. The top row shows examples of cyclic tetrapeptide HDACi along with their name, while the bottom row designates the members of bicyclic depsipeptides along with their corresponding names

zinc binding group. Such inhibitors include mocetinostat and entinostat of benzamide derivative group (Thaler and Mercurio 2014). Ricolinostat (ACY-1215), targeting HDAC6 is an example of isoform-selective HDACi (Cao et al. 2018; Yee et al. 2015). Among selective HDAC6 inhibitors, this was the first to enter into the journey of clinical trial studies. Following this, citarinostat (ACY-241), another HDAC6 selective, orally available inhibitor has been selected for phase I clinical study (Huang et al. 2017). As isoform selective inhibitors target only a single HDAC; thus, they will be perfect chemical probes for clear understanding of the role played by each HDAC. Further the isoform-selective inhibitors will help in exploring the underlying molecular mechanism by way of which a particular HDAC triggers cancer signalling (Bieliauskas and Pflum 2008). Further due to narrow spectrum of selective HDACi in general and isoform selective inhibitors in particular they may offer improved therapeutic benefit (Yang et al. 2019).

8.1.3 Classification Based on Origin of HDACi

HDACi may be synthetic or natural depending on source. Synthetic HDACi are synthesized in chemical laboratories. HDACi such as vorinostat, entinostat, mocetinostat and others are synthetic. HDACi derived from natural sources like bacteria, fungi, plants, marine organisms come under natural HDACi. For instance,

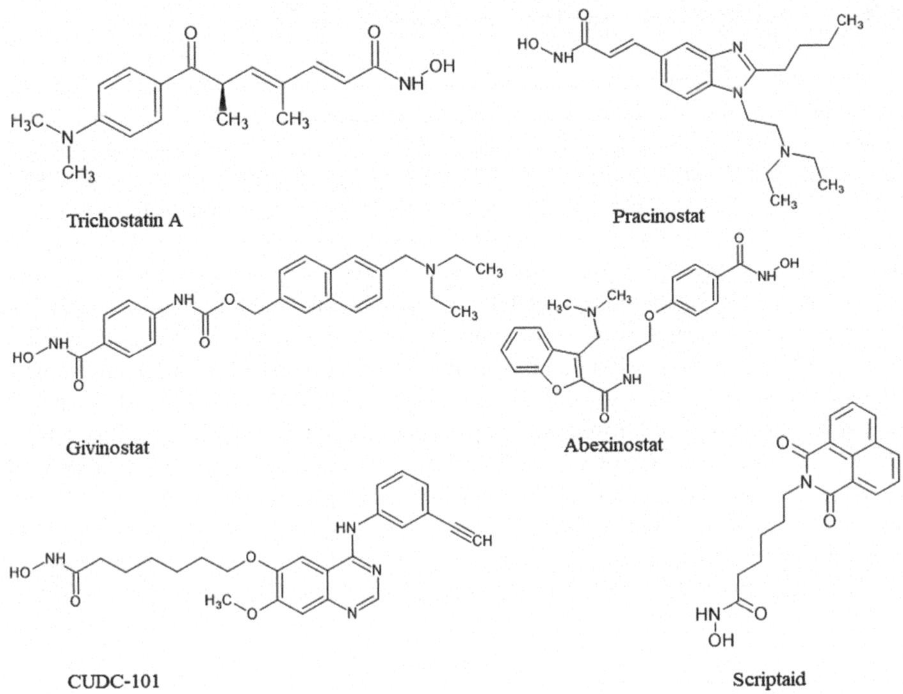

Fig. 8.4 Certain inhibitors of hydroxamate group of histone deacetylase inhibitors along with their chemical structure and name. Majority of hydroxamates inhibit wide range of HDACs and are thus considered as broad-spectrum HDACi or pan-HDACi

TSA derived from the *Streptomyces hygroscopicus* has demonstrated potent inhibitory activity against a broad range of HDACs (Tsuji et al. 1976). Depudecin, another HDACi was first isolated from *Alternaria brassicicola* and afterwards from *Nimbya scirpicola*, a weed pathogen (Matsumoto et al. 1992; Tanaka et al. 2000). Another natural HDAC inhibitor Psammaplin A derived from marine sponge inhibits Class I HDACs (Kim et al. 2007). Apicidin belonging to a cyclic tetrapeptide group of HDACi has been isolated from *Fusarium pallidoroseum* (Singh et al. 1996). Another potent HDAC inhibitor romidepsin (FK228) has been obtained from *Chromobacterium violaceum* (Cheng et al. 2007; Potharla et al. 2011). Isolated from Symploca sp (marine cyanobacterium), largazole, a bicyclic depsipeptide has shown strong antiproliferative activity against cancer cells (Taori et al. 2008). Pomiferin, a natural HDAC inhibitor with strong growth inhibitory activity against colon cancer cells has *Maclura pomifera* fruits as its source (Son et al. 2007). Apigenin, luteolin and sulforaphane are other examples of natural HDAC inhibitors. All these inhibitors have shown promising results in various cancer models (Ganai 2016a; Imran et al. 2019; Yan et al. 2017).

8.2 Highlights on Current Status of HDACi

As of now four HDACi have cleared the tough journey of clinical trials by gaining approval from the US Food and Drug Administration (USFDA). Among these inhibitors vorinostat/SAHA (brand name Zolinza) was the first in achieving approval for cutaneous T-cell lymphoma (CTCL) treatment (Marks and Breslow 2007; Ververis et al. 2013). This approval was conferred on vorinostat in the month of October 2006. Following this a bicyclic depsipeptide romidepsin (Istodax) became first among macrocyclic HDACi and second overall in gaining approval for treating cutaneous T-cell lymphoma (CTCL) and peripheral T-cell lymphoma (PTCL). These approvals were bestowed on romidepsin in the month of November 2009 and May 2011, respectively (Campas-Moya 2009; Ververis et al. 2013). Belinostat (Beleodaq) occupied third place in the list of approved HDACi on July 2014 when it was granted permission for treating relapsed/refractory PTCL patients (Chun 2015). On 23 February 2015, panobinostat (Farydak) was included in the list of approved HDACi. This inhibitor received approval for treating multiple myeloma, a cancer related to plasma cells (Ganai 2016b). China Food and Drug Administration (CFDA) has approved a benzamide HDAC inhibitor chidamide (Epidaza) for treating recurrent or refractory PTCL (Fig. 8.5) (Ganai 2019; Lu et al. 2016). Pracinostat, a

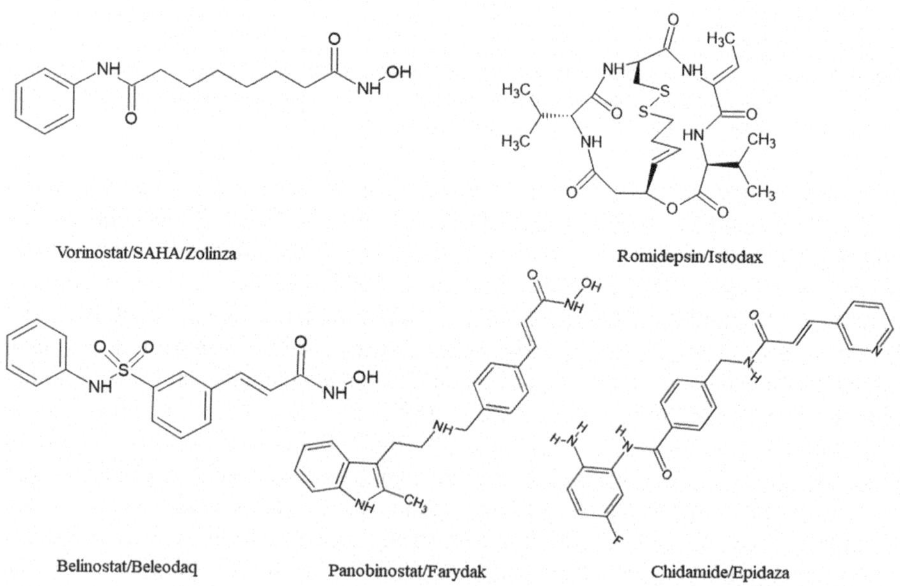

Fig. 8.5 Structural details of various inhibitors approved by United States Food and Drug Administration (USFDA) and China FDA (CFDA). The first four inhibitors starting from top are USFDA approved, whereas the last inhibitor chidamide has been approved by CFDA. Vorinostat, belinostat and panobinostat belong to hydroxamate group of histone deacetylase inhibitors (HDACi), romidepsin is from bicyclic depsipeptide group and chidamide is benzamide derivative group. Thus majority of approved inhibitors are hydroxamates

hydroxamate group HDAC inhibitor has been granted orphan designation by USFDA for acute myeloid leukaemia treatment (Bose and Grant 2014). Mocetinostat, a benzamide group HDAC inhibitor has been conferred the orphan drug designation for diffuse large B-cell lymphoma and myelodysplastic syndrome treatment (Avendaño and Menéndez 2015).

8.3 Brief Overview of Sirtuin Inhibitors

Sirtuins being NAD^+-dependent are mechanistically different from zinc-dependent or classical HDACs. Mounting evidences suggest the crucial role of these HDACs in various biological processes. Small molecule inhibitors developed against sirtuins are nicotinamide and its analogues and thioacyllysine-containing compounds (Jackson et al. 2003; Smith and Denu 2007). While nicotinamide analogues are not most likely mechanism based inhibitors, thioacyllysine compounds are mechanism based inhibitors (Hu et al. 2014; Smith and Denu 2007). Other sirtuin inhibitors may work by competitive inhibition and thus may block substrate binding. Nicotinamide, the sirtuin inhibitor obstructs various sirtuins ranging from SIRT1–SIRT3 and SIRT5–SIRT6 with varying potency (Hu et al. 2013; Tervo et al. 2004; Yuan et al. 2012). Analogues of this inhibitor and benzamide have also shown sirtuin inhibitory activity. A couple of 3′-phenethyloxy-2-anilino benzamide analogues have proved potent and selective inhibitors of SIRT2 (Suzuki et al. 2012). Further certain thioacyllysine possessing compounds have proved as mechanism based sirtuin inhibitors.

Certain β-naphthol compounds have also proved to be inhibitors of sirtuins. Sirtinol, as an example of such inhibitors proved to be effective against human SIRT1 (Grozinger et al. 1999). JGB1741, a sirtinol based compound showed selective inhibitory activity against SIRT1 and induced antiproliferative in three distinct cancer cell lines (Kalle et al. 2010). Salermide, another inhibitor based on sirtinol structure has proved to be more potent than latter in terms of inhibiting SIRT1 and SIRT2. By way of restraining SIRT1, salermide induced apoptosis in a broad range of cancer cell lines of human origin. Cambinol, inhibiting SIRT1 and SIRT2 is also a β-naphthol compound (Heltweg et al. 2006). Besides, analogues of this inhibitor have also shown activity against sirtuins.

Moreover, certain indole derivatives have proved to be sirtuin inhibitors. These include EX527 (Selisistat), inauhzin, AC-93253, GW5074 and Ro31-8220 (Huber et al. 2010; Napper et al. 2005; Trapp et al. 2006; Zhang et al. 2012, 2009). Derivatives of splitomycin have also proved to be inhibitors activity against sirtuins. Some of them inhibit SIRT1, some SIRT2, while others affect both (Hu et al. 2014). Suramin, a well-known drug for treating sleeping sickness and onchocerciasis and its analogues also inhibit sirtuins. Suramin itself inhibits the SIRT5 activity (Schuetz et al. 2007; Trapp et al. 2007). Tenovin-6, another inhibitor with comparatively strong inclination towards SIRT2 than SIRT1 has also been reported (McCarthy et al. 2012). Other worth commenting sirtuin inhibitors are AGK2, a selective

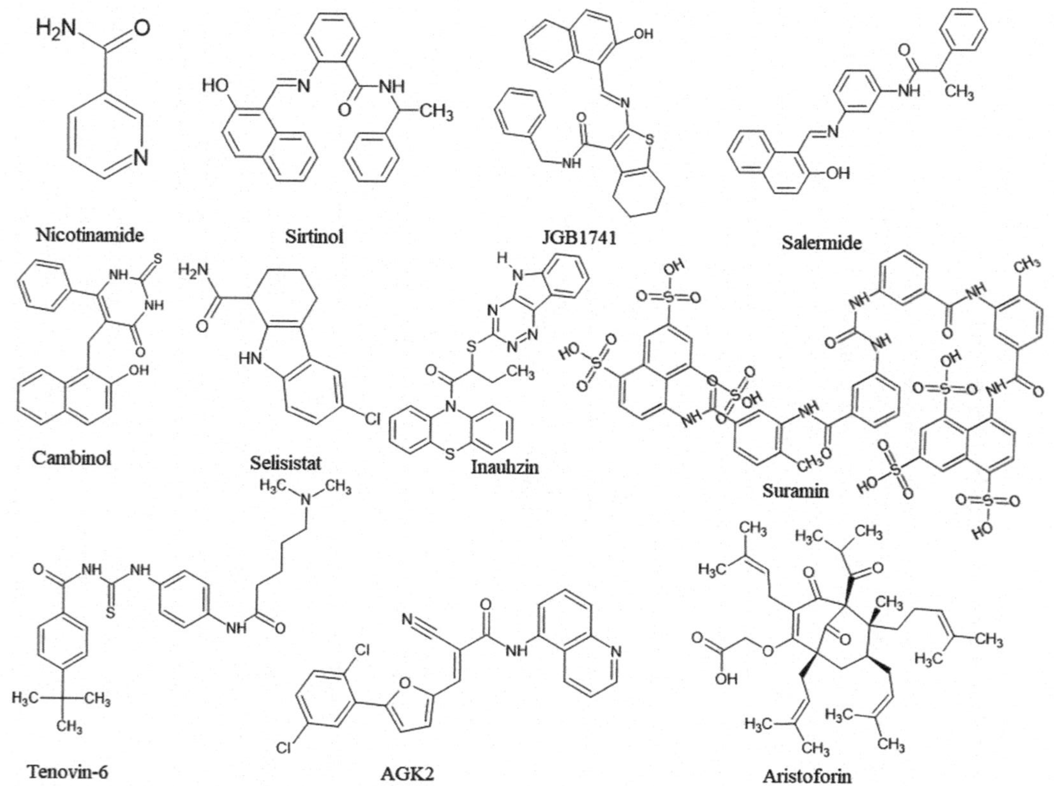

Fig. 8.6 Various sirtuin inhibitor structures belonging to different groups. Nicotinamide and its derivatives serve as sirtuin inhibitors. Among them only the structure of nicotinamide has been shown. Inhibitors such as sirtinol, JGB1741, salermide and cambinol are β-naphthol compounds. Certain sirtuin inhibitors like selisistat, inauhzin are indole derivatives. Suramin, the sleeping sickness has sirtuin inhibitory activity. Tenovin and its analogues also obstruct sirtuins. Here the structure of tenovin-6, a tenovin analogue has been shown. Chemical structure of AGK2, a SIRT2 selective inhibitor and a phloroglucinol derivative aristoforin lies at the bottom row

inhibitor of SIRT2, and aristoforin, a phloroglucinol derivative which inhibits both SIRT1 and SIRT2 (Fig. 8.6) (Gey et al. 2007; Hu et al. 2014; Outeiro et al. 2007).

8.4 Different Components of HDAC Inhibitor Structure

Most of the HDACi possess three components in their structure. These components include cap group and zinc binding group (ZBG) connected by a linker (Miller et al. 2003). Cap group present in majority of HDACi serves as surface-binding group. This group interacts with the rim amino acid residues of HDACs. Macrocyclic HDACi like romidepsin have bulky cap region (Mwakwari et al. 2010). Through modifications in cap group, a selective HDAC6 inhibitor tubacin has been developed (Haggarty et al. 2003). Tubacin has very large capping group compared to pan-HDAC inhibitor vorinostat (SAHA) (Bieliauskas and Pflum 2008). Linker region lies in between cap group and ZBG. Thus linker region connects cap region of inhibitor with its ZBG. The linker region of typical HDAC inhibitor trichostatin A is five-carbon in length. The linker region spans the active site tunnel and makes interactions with the tunnel residues (Miller et al. 2003). Zinc binding group is involved in chelating zinc ion situated deep in the active site of HDACs (Miller et al. 2003; Yang et al. 2019). The classic ZBGs include hydroxamic acid, benzamide, carboxylic acid and thiols (Zhang et al. 2018). Novel ZBGs such as imidazole thione, chelidamic group, benzoylhydrazide, trifluoromethyloxadiazolyl (TFMO) moiety and 2-(oxazol-2-yl) phenol moiety have been reported and HDACi with these groups have also been synthesized and tested for selectivity through activity assay (Kleinschek et al. 2016; Li and Woster 2015; Lobera et al. 2013; Valente et al. 2012; Wang et al. 2015).

Taken together, I have discussed different groups of HDACi of classical HDACs based on chemical structure differences. Inhibitors of these HDACs have been classified on the basis of specificity towards HDAC subtypes. Further the synthetic and natural HDACi have also been taken into account. This was followed by overview of inhibitors approved by USFDA and CFDA in addition to inhibitors having orphan designation. Further I have given an outline of sirtuin inhibitors by discussing the compounds and their analogues capable of inhibiting these mechanistically contrasting HDACs. Moreover, a brief idea about the typical structural features present in majority of HDACi and their respective interactions with different regions of HDAC active site has been provided.

References

Avendaño C, Menéndez JC (2015) Chapter 8 - Epigenetic therapy of cancer. In: Avendaño C, Menéndez JC (eds) Medicinal chemistry of anticancer drugs, 2nd edn. Elsevier, Boston, pp 325–358

Bhuiyan MP, Kato T, Okauchi T, Nishino N, Maeda S, Nishino TG, Yoshida M (2006) Chlamydocin analogs bearing carbonyl group as possible ligand toward zinc atom in histone deacetylases. Bioorg Med Chem 14:3438–3446

Bieliauskas AV, Pflum MKH (2008) Isoform-selective histone deacetylase inhibitors. Chem Soc Rev 37:1402–1413

Bose P, Grant S (2014) Orphan drug designation for pracinostat, volasertib and alvocidib in AML. Leuk Res 38

Bradner JE, West N, Grachan ML, Greenberg EF, Haggarty SJ, Warnow T, Mazitschek R (2010) Chemical phylogenetics of histone deacetylases. Nat Chem Biol 6:238–243

Campas-Moya C (2009) Romidepsin for the treatment of cutaneous T-cell lymphoma. Drugs Today (Barcelona, Spain: 1998) 45:787–795

Cao J, Lv W, Wang L, Xu J, Yuan P, Huang S, He Z, Hu J (2018) Ricolinostat (ACY-1215) suppresses proliferation and promotes apoptosis in esophageal squamous cell carcinoma via miR-30d/PI3K/AKT/mTOR and ERK pathways. Cell Death Dis 9:817

Cheng YQ, Yang M, Matter AM (2007) Characterization of a gene cluster responsible for the biosynthesis of anticancer agent FK228 in Chromobacterium violaceum no. 968. Appl Environ Microbiol 73:3460–3469

Chun P (2015) Histone deacetylase inhibitors in hematological malignancies and solid tumors. Arch Pharm Res 38:933–949

Damaskos C, Garmpis N, Valsami S, Kontos M, Spartalis E, Kalampokas T, Kalampokas E, Athanasiou A, Moris D, Daskalopoulou A, Davakis S, Tsourouflis G, Kontzoglou K, Perrea D, Nikiteas N, Dimitroulis D (2017) Histone deacetylase inhibitors: an attractive therapeutic strategy against breast cancer. Anticancer Res 37:35–46

Eckschlager T, Plch J, Stiborova M, Hrabeta J (2017) Histone deacetylase inhibitors as anticancer drugs. Int J Mol Sci 18

Fraga MF, Ballestar E, Villar-Garea A, Boix-Chornet M, Espada J, Schotta G, Bonaldi T, Haydon C, Ropero S, Petrie K, Iyer NG, Pérez-Rosado A, Calvo E, Lopez JA, Cano A, Calasanz MJ, Colomer D, Piris MA, Ahn N, Imhof A, Caldas C, Jenuwein T, Esteller M (2005) Loss of acetylation at Lys16 and trimethylation at Lys20 of histone H4 is a common hallmark of human cancer. Nat Genet 37:391–400

Frey RR, Wada CK, Garland RB, Curtin ML, Michaelides MR, Li J, Pease LJ, Glaser KB, Marcotte PA, Bouska JJ, Murphy SS, Davidsen SK (2002) Trifluoromethyl ketones as inhibitors of histone deacetylase. Bioorg Med Chem Lett 12:3443–3447

Ganai SA (2014) HDAC inhibitors entinostat and suberoylanilide hydroxamic acid (SAHA): the ray of hope for cancer therapy. In: Wells RD, Bond JS, Klinman J, Masters BSS, Bell E (eds) Molecular life sciences: an encyclopedic reference. Springer, New York, pp 1–16

Ganai SA (2016a) Histone deacetylase inhibitor sulforaphane: the phytochemical with vibrant activity against prostate cancer. Biomed Pharmacother 81:250–257

Ganai SA (2016b) Panobinostat: the small molecule metalloenzyme inhibitor with marvelous anticancer activity. Curr Top Med Chem 16:427–434

Ganai SA (2018) Histone deacetylase inhibitors modulating non-epigenetic players: the novel mechanism for small molecule based therapeutic intervention. Curr Drug Targets 19:593–601

Ganai SA (2019) Different groups of HDAC inhibitors based on various classifications. In: Ganai SA (ed) Histone deacetylase inhibitors — epidrugs for neurological disorders. Springer, Singapore, pp 33–38

Gey C, Kyrylenko S, Hennig L, Nguyen LH, Buttner A, Pham HD, Giannis A (2007) Phloroglucinol derivatives guttiferone G, aristoforin, and hyperforin: inhibitors of human sirtuins SIRT1 and SIRT2. Angew Chem Int Ed Engl 46:5219–5222

Grozinger CM, Hassig CA, Schreiber SL (1999) Three proteins define a class of human histone deacetylases related to yeast Hda1p. Proc Natl Acad Sci U S A 96:4868–4873

Haggarty SJ, Koeller KM, Wong JC, Grozinger CM, Schreiber SL (2003) Domain-selective small-molecule inhibitor of histone deacetylase 6 (HDAC6)-mediated tubulin deacetylation. Proc Natl Acad Sci U S A 100:4389–4394

Heltweg B, Gatbonton T, Schuler AD, Posakony J, Li H, Goehle S, Kollipara R, DePinho RA, Gu Y, Simon JA, Bedalov A (2006) Antitumor activity of a small-molecule inhibitor of human silent information regulator 2 enzymes. Cancer Res 66:4368–4377

Hu J, He B, Bhargava S, Lin H (2013) A fluorogenic assay for screening Sirt6 modulators. Org Biomol Chem 11:5213–5216

Hu J, Jing H, Lin H (2014) Sirtuin inhibitors as anticancer agents. Future Med Chem 6:945–966

Huang P, Almeciga-Pinto I, Jarpe M, van Duzer JH, Mazitschek R, Yang M, Jones SS, Quayle SN (2017) Selective HDAC inhibition by ACY-241 enhances the activity of paclitaxel in solid tumor models. Oncotarget 8:2694–2707

Huber K, Schemies J, Uciechowska U, Wagner JM, Rumpf T, Lewrick F, Süss R, Sippl W, Jung M, Bracher F (2010) Novel 3-arylideneindolin-2-ones as inhibitors of NAD + -dependent histone deacetylases (sirtuins). J Med Chem 53:1383–1386

Imran M, Rauf A, Abu-Izneid T, Nadeem M, Shariati MA, Khan IA, Imran A, Orhan IE, Rizwan M, Atif M, Gondal TA, Mubarak MS (2019) Luteolin, a flavonoid, as an anticancer agent: a review. Biomed Pharmacother 112:108612

Jackson MD, Schmidt MT, Oppenheimer NJ, Denu JM (2003) Mechanism of nicotinamide inhibition and transglycosidation by Sir2 histone/protein deacetylases. J Biol Chem 278:50985–50998

Kalle AM, Mallika A, Badiger J, Alinakhi, Talukdar P, Sachchidanand (2010) Inhibition of SIRT1 by a small molecule induces apoptosis in breast cancer cells. Biochem Biophys Res Commun 401:13–19

Khan N, Jeffers M, Kumar S, Hackett C, Boldog F, Khramtsov N, Qian X, Mills E, Berghs SC, Carey N, Finn PW, Collins LS, Tumber A, Ritchie JW, Jensen PB, Lichenstein HS, Sehested M (2008) Determination of the class and isoform selectivity of small-molecule histone deacetylase inhibitors. Biochem J 409:581–589

Kijima M, Yoshida M, Sugita K, Horinouchi S, Beppu T (1993) Trapoxin, an antitumor cyclic tetrapeptide, is an irreversible inhibitor of mammalian histone deacetylase. J Biol Chem 268:22429–22435

Kim DH, Shin J, Kwon HJ (2007) Psammaplin a is a natural prodrug that inhibits class I histone deacetylase. Exp Mol Med 39:47–55

Kiweler N, Wünsch D, Wirth M, Mahendrarajah N, Schneider G, Stauber RH, Brenner W, Butter F, Krämer OH (2020) Histone deacetylase inhibitors dysregulate DNA repair proteins and antagonize metastasis-associated processes. J Cancer Res Clin Oncol 146:343–356

Kleinschek A, Meyners C, Digiorgio E, Brancolini C, Meyer-Almes FJ (2016) Potent and selective non-hydroxamate histone deacetylase 8 inhibitors. ChemMedChem 11:2598–2606

Li Y, Woster PM (2015) Discovery of a new class of histone deacetylase inhibitors with a novel zinc binding group. Medchemcomm 6:613–618

Li Y, Seto E (2016) HDACs and HDAC inhibitors in cancer development and therapy. Cold Spring Harb Perspect Med 6:a026831

Liao B, Sun Q, Yuan Y, Yin Y, Qiao J, Jiang P (2020) Histone deacetylase inhibitor MGCD0103 causes cell cycle arrest, apoptosis, and autophagy in liver cancer cells. J Cancer 11:1915–1926

Lobera M, Madauss KP, Pohlhaus DT, Wright QG, Trocha M, Schmidt DR, Baloglu E, Trump RP, Head MS, Hofmann GA, Murray-Thompson M, Schwartz B, Chakravorty S, Wu Z, Mander PK, Kruidenier L, Reid RA, Burkhart W, Turunen BJ, Rong JX, Wagner C, Moyer MB, Wells C, Hong X, Moore JT, Williams JD, Soler D, Ghosh S, Nolan MA (2013) Selective class IIa histone deacetylase inhibition via a nonchelating zinc-binding group. Nat Chem Biol 9:319–325

Lu X, Ning Z, Li Z, Cao H, Wang X (2016) Development of chidamide for peripheral T-cell lymphoma, the first orphan drug approved in China. Intractable Rare Dis Res 5:185–191

Lund AH, van Lohuizen M (2004) Epigenetics and cancer. Genes Dev 18:2315–2335

Madsen AS, Olsen CA (2016) A potent trifluoromethyl ketone histone deacetylase inhibitor exhibits class-dependent mechanism of action. Med Chem Commun 7:464–470

Marks PA, Breslow R (2007) Dimethyl sulfoxide to vorinostat: development of this histone deacetylase inhibitor as an anticancer drug. Nat Biotechnol 25:84–90

Marks PA, Xu WS (2009) Histone deacetylase inhibitors: potential in cancer therapy. J Cell Biochem 107:600–608

Marks PA, Richon VM, Miller T, Kelly WK (2004) Histone deacetylase inhibitors. Adv Cancer Res 91:137–168

Matsumoto M, Matsutani S, Sugita K, Yoshida H, Hayashi F, Terui Y, Nakai H, Uotani N, Kawamura Y, Matsumoto K et al (1992) Depudecin: a novel compound inducing the flat phenotype of NIH3T3 cells doubly transformed by ras- and src-oncogene, produced by Alternaria brassicicola. J Antibiot 45:879–885

McCarthy AR, Pirrie L, Hollick JJ, Ronseaux S, Campbell J, Higgins M, Staples OD, Tran F, Slawin AM, Lain S, Westwood NJ (2012) Synthesis and biological characterisation of sirtuin inhibitors based on the tenovins. Bioorg Med Chem 20:1779–1793

Miller TA, Witter DJ, Belvedere S (2003) Histone deacetylase inhibitors. J Med Chem 46:5097–5116

Mottamal M, Zheng S, Huang TL, Wang G (2015) Histone deacetylase inhibitors in clinical studies as templates for new anticancer agents. Molecules (Basel, Switzerland) 20:3898–3941

Mwakwari SC, Patil V, Guerrant W, Oyelere AK (2010) Macrocyclic histone deacetylase inhibitors. Curr Top Med Chem 10:1423–1440

Napper AD, Hixon J, McDonagh T, Keavey K, Pons JF, Barker J, Yau WT, Amouzegh P, Flegg A, Hamelin E, Thomas RJ, Kates M, Jones S, Navia MA, Saunders JO, DiStefano PS, Curtis R (2005) Discovery of indoles as potent and selective inhibitors of the deacetylase SIRT1. J Med Chem 48:8045–8054

Outeiro TF, Kontopoulos E, Altmann SM, Kufareva I, Strathearn KE, Amore AM, Volk CB, Maxwell MM, Rochet J-C, McLean PJ, Young AB, Abagyan R, Feany MB, Hyman BT, Kazantsev AG (2007) Sirtuin 2 inhibitors rescue α-synuclein-mediated toxicity in models of Parkinson's disease. Science (New York, N.Y.) 317:516–519

Potharla VY, Wesener SR, Cheng Y-Q (2011) New insights into the genetic organization of the FK228 biosynthetic gene cluster in Chromobacterium violaceum no. 968. Appl Environ Microbiol 77:1508–1511

Richa S, Dey P, Park C, Yang J, Son JY, Park JH, Lee SH, Ahn MY, Kim IS, Moon HR, Kim HS (2020) A new histone deacetylase inhibitor, MHY4381, induces apoptosis via generation of reactive oxygen species in human prostate cancer cells. Biomol Ther 28:184–194

Sanaei M, Kavoosi F (2019) Histone deacetylases and histone deacetylase inhibitors: molecular mechanisms of action in various cancers. Adv Biomed Res 8:63–63

Schuetz A, Min J, Antoshenko T, Wang CL, Allali-Hassani A, Dong A, Loppnau P, Vedadi M, Bochkarev A, Sternglanz R, Plotnikov AN (2007) Structural basis of inhibition of the human NAD + -dependent deacetylase SIRT5 by suramin. Structure 15:377–389

Shabason JE, Tofilon PJ, Camphausen K (2010) HDAC inhibitors in cancer care. Oncology (Williston Park, N.Y.) 24:180–185

Singh S, Zink D, Polishook J, Dombrowski A, Rattray S, Schmatz D, Goetz M (1996) Apicidins: novel cyclic tetrapeptides as coccidiostats and antimalarial agents from Fusarium pallidoroseum. Tetrahedron Lett 37:8077–8080

Singh BN, Zhang G, Hwa YL, Li J, Dowdy SC, Jiang S-W (2010) Nonhistone protein acetylation as cancer therapy targets. Expert Rev Anticancer Ther 10:935–954

Smith BC, Denu JM (2007) Mechanism-based inhibition of Sir2 deacetylases by thioacetyl-lysine peptide. Biochemistry 46:14478–14486

Son IH, Chung IM, Lee SI, Yang HD, Moon HI (2007) Pomiferin, histone deacetylase inhibitor isolated from the fruits of Maclura pomifera. Bioorg Med Chem Lett 17:4753–4755

Sun Y, Li J, Xu Z, Xu J, Shi M, Liu P (2019) Chidamide, a novel histone deacetylase inhibitor, inhibits multiple myeloma cells proliferation through succinate dehydrogenase subunit a. Am J Cancer Res 9:574–584

Suraweera A, O'Byrne KJ, Richard DJ (2018) Combination therapy with histone deacetylase inhibitors (HDACi) for the treatment of cancer: achieving the full therapeutic potential of HDACi. Front Oncol 8

Suzuki T, Khan MN, Sawada H, Imai E, Itoh Y, Yamatsuta K, Tokuda N, Takeuchi J, Seko T, Nakagawa H, Miyata N (2012) Design, synthesis, and biological activity of a novel series of human sirtuin-2-selective inhibitors. J Med Chem 55:5760–5773

Tanaka M, Fujimori T, Nabeta K (2000) Biosynthesis of depudecin, a metabolite of Nimbya scirpicola. Biosci Biotechnol Biochem 64:244–247

Taori K, Paul V, Luesch H (2008) Structure and activity of largazole, a potent antiproliferative agent from the Floridian marine Cyanobacterium Symploca sp. J Am Chem Soc 130:1806–1807

Tervo AJ, Kyrylenko S, Niskanen P, Salminen A, Leppanen J, Nyronen TH, Jarvinen T, Poso A (2004) An in silico approach to discovering novel inhibitors of human sirtuin type 2. J Med Chem 47:6292–6298

Thaler F, Mercurio C (2014) Towards selective inhibition of histone deacetylase isoforms: what has been achieved, where we are and what will be next. ChemMedChem 9:523–526

Trapp J, Jochum A, Meier R, Saunders L, Marshall B, Kunick C, Verdin E, Goekjian P, Sippl W, Jung M (2006) Adenosine mimetics as inhibitors of NAD + -dependent histone deacetylases, from kinase to sirtuin inhibition. J Med Chem 49:7307–7316

Trapp J, Meier R, Hongwiset D, Kassack MU, Sippl W, Jung M (2007) Structure-activity studies on suramin analogues as inhibitors of NAD + -dependent histone deacetylases (sirtuins). ChemMedChem 2:1419–1431

Tsuji N, Kobayashi M, Nagashima K, Wakisaka Y, Koizumi K (1976) A new antifungal antibiotic, trichostatin. J Antibiot 29:1–6

Valente S, Conte M, Tardugno M, Nebbioso A, Tinari G, Altucci L, Mai A (2012) Developing novel non-hydroxamate histone deacetylase inhibitors: the chelidamic warhead. MedChemComm 3:298–304

Ververis K, Hiong A, Karagiannis TC, Licciardi PV (2013) Histone deacetylase inhibitors (HDACIs): multitargeted anticancer agents. Biologics 7:47–60

Wang Y, Stowe RL, Pinello CE, Tian G, Madoux F, Li D, Zhao LY, Li JL, Wang Y, Wang Y, Ma H, Hodder P, Roush WR, Liao D (2015) Identification of histone deacetylase inhibitors with benzoylhydrazide scaffold that selectively inhibit class I histone deacetylases. Chem Biol 22:273–284

Yan X, Qi M, Li P, Zhan Y, Shao H (2017) Apigenin in cancer therapy: anti-cancer effects and mechanisms of action. Cell Biosci 7:50–50

Yang XJ, Seto E (2007) HATs and HDACs: from structure, function and regulation to novel strategies for therapy and prevention. Oncogene 26:5310–5318

Yang F, Zhao N, Ge D, Chen Y (2019) Next-generation of selective histone deacetylase inhibitors. RSC Adv 9:19571–19583

Yee AJ, Bensinger W, Voorhees PM, Berdeja JG, Richardson PG, Supko J, Tamang D, Jones SS, Wheeler C, Markelewicz RJ Jr, Raje NS (2015) Ricolinostat (ACY-1215), the first selective HDAC6 inhibitor, in combination with lenalidomide and dexamethasone in patients with relapsed and relapsed-and-refractory multiple myeloma: phase 1b results (ACE-MM-101 study). Blood 126:3055–3055

Yuan H, Wang Z, Li L, Zhang H, Modi H, Horne D, Bhatia R, Chen W (2012) Activation of stress response gene SIRT1 by BCR-ABL promotes leukemogenesis. Blood 119:1904–1914

Zhang Y, Xu W (2015) Isoform-selective histone deacetylase inhibitors: the trend and promise of disease treatment. Epigenomics 7:5–7

Zhang Y, Au Q, Zhang M, Barber J, Ng S, Zhang B (2009) Identification of a small molecule SIRT2 inhibitor with selective tumor cytotoxicity. Biochem Biophys Res Commun 386:729–733

Zhang Q, Zeng SX, Zhang Y, Zhang Y, Ding D, Ye Q, Meroueh SO, Lu H (2012) A small molecule Inauhzin inhibits SIRT1 activity and suppresses tumour growth through activation of p53. EMBO Mol Med 4:298–312

Zhang L, Zhang J, Jiang Q, Zhang L, Song W (2018) Zinc binding groups for histone deacetylase inhibitors. J Enzyme Inhib Med Chem 33:714–721

Zhao C, Dong H, Xu Q, Zhang Y (2020) Histone deacetylase (HDAC) inhibitors in cancer: a patent review (2017-present). Expert Opin Ther Pat 30:263–274

9 Singlet Anticancer Therapy Through Epi-Weapons Histone Deacetylase Inhibitors and Its Shortcomings

Appalling toxicities and chemoresistance associated with conventional anticancer therapies raise applicability related concerns (Howard et al. 2016). This emphasizes the dire need of novel therapies which are well tolerated and concurrently effective (Ganai et al. 2017; Guimaraes et al. 2013). Experimental evidences certify the implications of epigenetic mechanisms in tumour onset, progression and drug resistance (Hrabeta et al. 2014; Wilting and Dannenberg 2012). Histone deacetylases (HDACs), the metalloenzymes altering the chromatin topology through deacetylation of nucleosomal histones epigenetically silence genes involved in inhibition of tumour development (Ma et al. 2015). Anomalous expression of HDACs through alteration of cellular acetylation homeostasis causes transcriptional deregulation thereby making the conditions convenient for genesis and advancement of cancer (Li and Seto 2016; Saha and Pahan 2006). Thus fine tuning the gene expression by restoration of acetylation homeostasis through intervention of these HDACs has emerged as a promising strategy in epigenetic-based anticancer therapy. These small molecules, by name histone deacetylase inhibitors (HDACi), block the enzymatic function of HDACs through binding at their active site (Lakshmaiah et al. 2014; Lombardi et al. 2011).

9.1 HDACi in Anticancer Monotherapy

From molecular level studies it is quite evident that HDACs are often overexpressed in most of the cancers (Sanaei and Kavoosi 2019). Thus therapeutic intervention of HDACs with HDACi has gained importance in treating cancer from over a decade. Four HDACi including three hydroxamates (vorinostat/SAHA, belinostat and panobinostat) and one macrocyclic HDAC inhibitor romidepsin have already

S. A. Ganai, *Histone Deacetylase Inhibitors in Combinatorial Anticancer Therapy*,
https://doi.org/10.1007/978-981-15-8179-3_9

acquired USFDA approval for treating various malignancies and many are undergoing clinical and preclinical trials. Chidamide, a Class I selective benzamide derivative group HDAC inhibitor was approved as orphan drug by China FDA for treatment (oral) of recurrent or refractory peripheral T-cell lymphoma (PTCL) (Lu et al. 2016). In monotherapy HDACi are used as single agents for treating cancer or for studying their individual anticancer effect against cell, preclinical and clinical models. This will become more understandable by focussing on single cancer at a given time.

9.1.1 HDACi Against Pancreatic Cancer

Pancreatic cancer comes within the confines of most bellicose human cancers and it is expected to become third premium cancer-related death cause by outperforming breast cancer. Only marginal improvement has been noted in relative 5 year survival rate on employing conventional therapeutic approaches (Ganai et al. 2017). CG200745, a novel hydroxamate group HDAC inhibitor has been found to reduce pancreatic cancer cell viability dose dependently (Lee et al. 2017). On treatment with SAHA/vorinostat (5 μM) 50% growth decrease reportedly occurred in human pancreatic cancer cell line (BxPC-3). This effect of SAHA has been attributed mainly to inhibition of HDAC1 and HDAC3. Class I selective inhibitor entinostat (MS-275) also reduced the cell growth by 50% at lower concentration (1 μM) whereas complete growth abolition was reported at relatively higher concentration (5 μM) of entinostat (Peulen et al. 2013). Therapeutic intervention with panobinostat induced apoptosis in various pancreatic cancer cell lines, the effect being more pronounced against BxPC-3. This inhibitor also caused a marked reduction in pancreatic tumour growth in xenograft mouse model (Mehdi et al. 2012).

Cytotoxic effect of two HDAC inhibitors namely NVP-LAQ824 (dacinostat) and NVP-LBH589 (another name of panobinostat) has been studied against a series of pancreatic cancer cell models. Both inhibitors resulted in substantial growth suppression in all the models under in vitro condition. This growth inhibition was found to be associated with hyperacetylation of histone H4 (nucleosomal), induction of $p21^{(WAF-1)}$, cell cycle blockade and enhanced programmed cell death. However, under in vivo set-up only panobinostat showed marked effect in reducing tumour mass (Haefner et al. 2008). Another study involving three pancreatic cancer cell models tested the antiproliferative effect of pan-HDAC inhibitor (panobinostat), Class IIa specific inhibitor (MC1568), Class I-selective inhibitor (mocetinostat) and a single HDAC specific inhibitor tubastatin A (HDAC6 specific) on them. Among these inhibitors most promising antiproliferative and apoptotic effect was seen on intervention with panobinostat. Mocetinostat also showed similar effect but to a different (relatively lesser) extent compared to panobinostat. On the other hand only modest antiproliferative and cytotoxic effect was noted with tubastatin A and MC1568 (Wang et al. 2012). Reduction in pancreatic cell growth and induction of apoptosis has been seen on treatment with HDACi sulforaphane. Moreover, in xenograft mouse model (pancreatic cancer) this inhibitor markedly inhibited tumour growth without imparting any noticeable toxicity. Sulforaphane in pancreatic cancer cells triggered degradation of client proteins of heat shock protein 90 and prevented its interaction with $p50^{Cdc37}$ (cochaperone) (Li et al. 2012).

Another HDAC inhibitor AR-42 showed antiproliferative effect against pancreatic cancer cells by modulating the expression of genes and proteins having cross-talk with cell cycle. Besides this inhibitor triggered apoptosis in pancreatic cancer cells by way of inducing reactive oxygen species (ROS) production and DNA damage. Expression levels of various molecular players including miR33, miR-30D and miR-125b, negatively regulating p53, were elevated on intervention with defined inhibitor. Significant reduction in growth of BxPC-3 xenograft tumour also occurred on administration of AR-42 (Chen et al. 2017b). Antiproliferative effect of 4-phenylbutyrate, a short chain fatty acid group HDAC inhibitor has been studied on pancreatic cancer models. Marked growth inhibitory effects were recorded on pharmacological intervention with this inhibitor. This inhibitor caused reduction of xenograft tumour volume through obstruction of histone deacetylation and cell proliferation. Substantial enhancement in the expression of connexin 43 (Cx43) on intervention with 4-phenylbutyrate was also observed in human pancreatic cancer cell line *T3M4* (Dovzhanskiy et al. 2012). *By enhancing the expression of* major histocompatibility complex class I-related chain A (MICA) and MICB by way of PI3K/Akt signalling pathway in pancreatic cancer cells HDAC inhibitor valproic acid sensitized these cells to natural killer cell driven cytotoxicity (Shi et al. 2014).

9.1.2 Role of HDACi in Overcoming Prostate Cancer

After lung cancer, prostate cancer in American men is the second chief cause of cancer-related deaths. This malignancy has more inclination towards older men (65 or older) and is uncommon in men below 40 years of age (Ganai 2016b). After it became evident that abnormal expression/activity of HDACs has close association with prostate cancer development, studies were performed involving HDAC inhibitors (Abbas and Gupta 2008; Ganai 2016b; Sanaei and Kavoosi 2019). It has been revealed that attenuation of prostate cancer signalling occurs on sulforaphane treatment. This inhibitor destabilizes androgen receptor by promoting the dissociation of its chaperone HSP90 through inhibition of HDAC6 deacetylase activity. Hyperacetylation of HSP90 on HDAC6 inhibition by this inhibitor is critical for androgen receptor-HSP90 segregation (Gibbs et al. 2009).

While growth arrest has been reported in PC-3 and LNCaP cell lines, cell death induction was noticed in DU-145 cells on entinostat exposure. Protein levels of prostate specific antigen in LAPC4 cell line were lowered on entinostat treatment. Growth reduction of subcutaneous xenografts of these pancreatic cell lines (excluding LAPC4) evident after the administration of this inhibitor has been attributed to histone hyperacetylation and escalated p21 expression (Qian et al. 2007). Pro-apoptotic miRNA by name miR31 is supressed in prostate cancer through epigenetic mechanism. This miRNA has E2F6, an antiapoptotic protein as its downstream target. Benzamide derivative mocetinostat induced apoptosis in prostate cancer cells through upregulation of miR31 and subsequent downregulation of the E2F6 (Zhang et al. 2016b). Vorinostat induced apoptosis in a couple of prostate cancer cell lines through Akt/FOXO3a signalling pathway. By way of restraining Akt activation this inhibitor facilitated the FOXO3a activation (Shi et al. 2017). Another short chain fatty acid group HDAC inhibitor sodium phenylbutyrate when applied on prostate cancer cells markedly reduced their viability. Migration and colony formation abilities of both prostate cancer cell lines (DU145 and PC3) were hampered by the above-mentioned inhibitor. While survivin expression was markedly alleviated, the phosphorylation of ERK and p-38 was significantly elevated on sodium phenylbutyrate intervention. Apart from this, sodium phenylbutyrate administration supressed the in vivo tumour development of the above prostate cancer cell lines (Xu et al. 2016).

Preferential anticancer effect has been noted in androgen dependent prostate cancer cells on belinostat treatment suggesting the importance of androgen receptor in mediating the effect of this HDAC inhibitor (Gravina et al. 2012). Studies have shown that chromic valproic acid treatment significantly reduces prostate cancer cell growth through enhanced activation of caspase-2 and caspase-3. This effect was recorded not only in androgen receptor positive but also in androgen receptor negative prostate cancer cells. In addition, reduction in xenograft tumour growth on chronic valproic acid administration has also been certified (Xia et al. 2006). N-Myc downstream regulated 1 (NRDG1) encodes NRDG1 protein whose function is suppression of metastasis. By upregulating NRDG1 in metastatic prostate cancer (PC3) cells valproic acid hampered their invasion. However, no NRDG1 escalation was seen in non-metastatic prostate cancer cells on the defined intervention suggesting the differential effects of this valproic acid in metastatic and non-metastatic prostate cancer cell lines (Lee and Kim 2015). Chromopeptide A, a depsipeptide HDAC inhibitor isolated from Chromobacterium sp. HS-13-94 (bacteria derived from marine sediment) has shown antiproliferative effect against pancreatic cancer cells. This inhibitor selectively obstructed the activities of Class I HDACs but in a substrate non-competitive fashion. Chromopeptide A triggered apoptosis in these cells and in a dose-escalation manner impeded the migration of PC3 cells. Substantial reduction of tumour growth in prostate cancer (PC3) xenograft mice has been observed on intravenous injection of this inhibitor (Sun et al. 2017). Growth inhibitory responses in prostate cancer cells on apigenin exposure have been ascribed to inhibition of two Class I HDACs namely HDAC1 and HDAC3 (Table 9.1) (Pandey et al. 2012).

9.1.3 HDACi in Anti-Lung Cancer Therapy

Lung cancer includes both non-small cell lung cancer (NSCLC) and small cell lung cancer (SCLC). While 84% cases of lung cancer fall under NSCLC only 13% of all cases are SCLC. Lung cancer occurs at the age of 65 or above and the average age has been put forward as 70 years. Twenty five percent (25%) of all cancer deaths

Table 9.1 Different molecular mechanisms modulated by various HDAC inhibitors as single agent therapeutics for eliciting cytotoxic effect in pancreatic and prostate cancer models

HDAC inhibitor/ inhibitors used singly	Cancer name	Cell line/animal model used for research	Molecular targets or players activated or upregulated for cytotoxic effect	Enzymes and other molecular players impeded or downregulated for engendering cytotoxicity	Literature evidence
Vorinostat, entinostat	Pancreatic cancer	BxPC-3		HDAC1 and HDAC3	Peulen et al. (2013)
Panobinostat		BxPC-3, xenograft model (mouse)			Mehdi et al. (2012)
NVP-LAQ824, LBH589			$p21^{(WAF\text{-}1)}$		Haefner et al. (2008)
Sulforaphane		Xenograft model mouse		Heat shock protein 90	Li et al. (2012)
AR-42		BxPC-3 xenograft	ROS, miR-30D and miR33, miR-125b		Chen et al. (2017b)
4-Phenylbutyrate		*T3M4*	Cx43		Dovzhanskiy et al. (2012)
Valproic acid		MIA PaCa-2, PANC-1, BxPC-3	MICB, MICA		Shi et al. (2014)
Sulforaphane	Prostate cancer	LNCaP and VCaP	Androgen receptor	HDAC6	Gibbs et al. (2009)
Entinostat		PC-3, LNCaP, DU-145	p21		Qian et al. (2007)
Mocetinostat		DU-145	miR31	E2F6	Zhang et al. (2016b)
Vorinostat		PC-3, DU145	FOXO3a activation	Akt activation	Shi et al. (2017)
Sodium phenylbutyrate		PC3 and DU145	ERK and p-38 phosphorylation	Survivin	Xu et al. (2016)

(continued)

Table 9.1 (continued)

HDAC inhibitor/ inhibitors used singly	Cancer name	Cell line/animal model used for research	Molecular targets or players activated or upregulated for cytotoxic effect	Enzymes and other molecular players impeded or downregulated for engendering cytotoxicity	Literature evidence
Valproic acid		C4-2, LNCaP, PC3, DU145, xenograft model	Caspase-2 and caspase-3		Xia et al. (2006)
Valproic acid		PC3	NRDG1		Lee and Kim (2015)
Chromopeptide A		PC3, Xenograft mice		Class I HDACs	Sun et al. (2017)
Apigenin		PC-3 and 22Rv1		HDAC3, HDAC1	Pandey et al. (2012)

have been attributed to lung cancer (Zappa and Mousa 2016). Two structurally similar histone deacetylase inhibitors were tested against a series (16) of NSCLC cell lines and in 50% of the cases both inhibitors exhibited potent antitumour activity (Miyanaga et al. 2008). Panobinostat showed modest activity as single agent in small cell lung cancer but was well tolerated although no partial or complete responses were recorded (de Marinis et al. 2013). In NSCLC cells sulforaphane was found to impede EGFR signalling pointing towards its anti-metastatic ability (Chen et al. 2015). Therapeutic intervention with sulforaphane not only showed antiproliferative effect against NSCLC cell lines but also supressed migration and invasion of 95D and H1299 cells in a dose-dependent manner. It has been observed that metastasis in these cells is highly reliant on miR-616-5p levels. Sulforaphane epigenetically lessened miR-616-5p levels in the above-mentioned cell lines and as such inhibited their migration and invasion (Wang et al. 2017).

OSU-HDAC-44, a novel HDAC inhibitor attenuated tumour growth in NSCLC preclinical models through disruption of F-actin and induction of intrinsic programmed cell death (Tang et al. 2010). HDAC inhibitor CG200745 showed antiproliferative effect in NSCLC cell models through epigenetic mechanism. Induction of cell apoptosis on CG200745 exposure was observed in Calu6 cells (Lee et al. 2017). Impairing Notch1 signalling in SCLC impairs their viability. SCLC (DMS53) cells on valproic acid treatment showed significant morphological changes and induction of Notch1 protein active forms. Most importantly exposure to valproic acid inhibited proliferation of these cells in a dose-escalation fashion (Platta et al. 2008). Studies have shown the critical involvement of Notch1 receptor in tumour development in NSCLC. Regulation of this receptor by HDAC6 through ubiquitin proteasome system has been studied in three NSCLC cell lines. Inhibition of HDAC6 by ACY1215 reduced Notch1 receptor levels dose dependently. In all the three cell lines (H1299, LL2 and A549), ACY1215 enhanced apoptosis and PARP1 cleavage. Under conditions of in vivo, HDAC6 inhibition with the defined inhibitor lessened the growth rate of LL2 tumour. Thus it is highly evident that HDAC6 facilitates Notch1 signalling in NSCLC cells which in turn augments their proliferation (Deskin et al. 2020).

Growth inhibitory effects of HDAC inhibitor sodium butyrate were tested on two NSCLC cell lines one with canonical p53 (NCI460) and other with its mutated form (NCI-H23). This inhibitor attenuated the growth of both the models by way of modulating cell cycle associated proteins. While $p21^{waf1}$ was induced in NCI-H23, p16ink4 and p27kip1 induction was noted in NCI-H460 following sodium butyrate exposure (Pellizzaro et al. 2001). Nm23-H1 and CD44v6 genes have critical role in the metastasis of a variety of malignancies (Carbognani et al. 1998; Tee et al. 2006). The impact of histone deacetylase inhibitor valproic acid on the expression profile of these genes has been studied in NSCLC cell line namely A549. While the expression of CD44v6, MMP-9 and MMP-2 was found to be escalated the expression of Nm23-H1 showed opposite trend (enhanced expression). These findings favour the use of valproic acid in restraining lung cancer metastasis and its treatment (Niknamian 2019). Effect of tacedinaline/CI-994, belonging to benzamide derivative group of HDACi has been studied in a couple of lung cancer cell lines. These NSCLC cell lines were LX-1 and A-549 representing squamous cell carcinoma and adenocarcinoma, respectively. Survival inhibition was noted which was found to be concentration dependent and as proved by recovery experiments the effect of this inhibitor was cytostatic (Loprevite et al. 2005). While testing its efficacy in a broad range of lung cancers and mesotheliomas, panobinostat, the hydroxamate proved to be cytotoxic in all the model cell lines. Substantial reduction in tumour growth 62% was noted on administration of panobinostat as compared to vehicle treated control and specifically this inhibitor proved to be more effective in xenografts of SCLC (Crisanti et al. 2009). Pan-inhibitor belinostat induced cytotoxic effect in a series of lung squamous cell carcinoma cells (SK-MES-1, H2170, H520, SW900 and Calu-1) through downregulation of MAPK pathway. Activation of this pathway has crucial role in induction of cisplatin resistance (Kong et al. 2017).

9.1.4 Tackling Breast Cancer with HDACi

Excluding lung cancer, breast cancer is the first chief death cause among American women and it has been estimated that 81% of breast cancers are invasive (DeSantis et al. 2019). Above 75% of invasive breast cancers are "no special type" previously known as ductal carcinomas. Invasive lobular carcinoma, the most frequent "special type" of breast cancer forms 15% of the invasive breast cancer cases (Thomas et al. 2019). Although genetic and epigenetic dysregulation together facilitate the onset and advancement of breast cancer, the latter being reversible is preferred for pharmacological intervention (Kohler et al. 2012). Breast cancer genesis and development is triggered by imbalance in the activity of histone acetyl transferases (HATs) and their antagonistic players HDACs. As supported by experimental evidences HDAC overactivity distorts cellular acetylation homeostasis thereby supporting the breast cancer signalling (Ediriweera et al. 2019).

HDAC inhibitor pracinostat hampered the growth and metastasis of breast cancer cells by impairing their chemotactic motility and through alteration of epithelial to mesenchymal transition. While this inhibitor lowered the expression of vimentin, N-cadherin, HDAC4 and HDAC5, elevated E-cadherin expression was noted. Compared to FDA approved HDAC inhibitor SAHA, pracinostat exhibited superlative antitumour properties in a pair of breast cancer in vivo models (Chen et al. 2020). Panobinostat impeded the survival and proliferation of four triple negative breast cancer cell lines (BT-549, MDA-MB-231, MDA-MB-157 and MDA-MB-468) and excluding MDA-MB-468 induced apoptosis in all of them. This inhibitor markedly attenuated tumour formation (BT-549 and MDA-MB-231) in mice models. Under both conditions panobinostat escalated the CDH1 protein besides inducing the morphological alterations in MDA-MB-231 cells (Tate et al. 2012). Panobinostat enhanced the E-cadherin expression on MDA-MB-231 plasma membranes without alteration of oestrogen pathway and enhanced the expression of Snail and Slug (Fortunati et al. 2014). Oestrogen receptor 1 (ESR1) gene encodes the oestrogen receptor alpha (ERα) protein. Strong anti-breast cancer activity of TSA has been observed under both conditions and it has been demonstrated that this inhibitor promotes repression of ESR1 and through proteasome mediated pathway lessens ERα expression (Noh et al. 2016). TSA exerted the potent anticancer effect in breast cancer cells. Another study has proved that TSA induced cytotoxicity in breast cancer cells is due to enhancement of mitochondrial reactive oxygen species production (Sun et al. 2014).

Sulforaphane under in vitro and in vivo set-up hampered the growth of stem-like cells of triple negative breast cancer (Castro et al. 2019). In MDA-MB-231, metastatic breast cancer model, sulforaphane supressed the expression of tubulin deacetylase HDAC6. While this inhibition elevated the acetylation status of HSP90, the expression of c-myc underwent downregulation. Suppression of c-myc in turn downregulated hTERT at mRNA levels besides inducing p21. VEGF and MMPs having central role in metastasis were also downregulated due to sulforaphane. Thus the antimetastasis effect of sulforaphane in breast cancer cells has been ascribed to modulation of the above-mentioned molecular players (Sarkar and Chakraborty Mukherjee 2015). While studying the effects of sodium butyrate on two breast cancer cell lines (MDA-MB-468 and MCF-7) and a normal breast cell line it was found that this inhibitor selectively exerts cytotoxic effect in the cancer lines with no significant effect on normal line (MCF-10A). Further, apoptosis triggered by sodium butyrate was found to be associated with increased ROS production and caspase activity in addition to a drop in mitochondrial membrane potential (Salimi et al. 2017). Sodium butyrate suppressed the growth of both hormone independent and hormone dependent cell lines in a dose and time-dependent manner. This inhibitor showed antiproliferative effect by cell cycle blockade resulting in apoptosis only in oestrogen receptor possessing breast cancer line only (Coradini et al. 1997). Intensive studies dealing with the effect of valproic acid on the HER-2 overexpressing (SKBR3) breast cancer model have proved that this inhibitor exerts antiproliferative effect by inciting cell cycle blockade and apoptosis by way of significantly escalating HSP70 acetylation time dependently (Table 9.2) (Mawatari et al. 2015). Other evidence based study has revealed that valproic acid is nicely tolerated and hampers burgeoning of breast tumours (Cohen et al. 2017).

9.1.5 Subduing Colorectal/Colon Cancer with HDACi

This cancer initiates in the colon or rectum of the large intestines. While colon cancer develops from colon the rectal cancer originates from rectum. As these cancers share multiple features in common thus they are usually named together as colorectal cancer. These cancers start as polyps on the colon or rectum (inner lining) and with the passage of time some polyps though not all transform into cancer depending on the polyp type (Delavari et al. 2014). While adenomatous polyps are precancerous the hyperplastic and inflammatory polyps although more prevalent are not generally precancerous (Conteduca et al. 2013). Within a decade almost 15% of the adenomas expectedly get advanced to carcinoma state (Mármol et al. 2017). Nowadays this cancer occupies the fourth rank in being the world's most deadliest cancer taking

Table 9.2 Monotherapy with HDAC inhibitors modulates multiple molecular targets in experimental models of lung and breast cancer for invoking cytotoxic signalling

Name of HDAC inhibitor	Type of cancer	Disease models used	Underlying molecular targets upregulated or activated for triggering cell death	Cellular molecules obstructed or lowered for induction of cancer cell cytotoxicity	Reference
Sulforaphane	Lung cancer	H1299 and 95D		miR-616-5p	Wang et al. (2017)
OSU-HDAC-44		A549, CL1-1, H1299		F-actin	Tang et al. (2010)
Valproic acid		DMS53	Notch1		Platta et al. (2008)
ACY1215		A549, LL2, H1299		Notch1 receptor, HDAC6	Deskin et al. (2020)
Sodium butyrate		NCI460, NCI-H23	$p21^{waf}$, p27kip1, p16ink4		Pellizzaro et al. (2001)
Valproic acid		A549	MMP-9, MMP-2, CD44v6	Nm23-H1	Niknamian (2019)
Tacedinaline		A-549, LX-1	–	–	Loprevite et al. (2005)
Belinostat		H520, SW900 Calu-1, SK-MES-1, H2170		MAPK	Kong et al. (2017)
Pracinostat	Breast cancer	MDA-MB-231, nude mice tumor model	E-cadherin	N-cadherin, vimentin, HDAC4, HDAC5	Chen et al. (2020)
Panobinostat		BT-549, MDA-MB-157, MDA-MB-231, mice model	E-cadherin, snail, slug	CDH1	Tate et al. (2012)
TSA		MCF-7		ESR1, ERα	Noh et al. (2016)
Sulforaphane			HSP90 acetylation, p21	HDAC6, c-myc, hTERT, VEGF, MMPs	Sarkar and Chakraborty Mukherjee (2015)
Sodium butyrate		MCF-7, MDA-MB-468	ROS, Caspase		Salimi et al. (2017)
Valproic acid		SKBR3	HSP70 acetylation		Mawatari et al. (2015)

about 9 lakh lives annually (Dekker et al. 2019). Currently certain therapeutics targeting epigenetic route are tested in a variety of colorectal cancer models (Parizadeh et al. 2018).

HDACi have shown encouraging anti-colorectal cancer effects (Mariadason 2008). Anticancer activity of vorinostat/SAHA has been studied in colon cancer cells and tumours using nude mice. Intervention with this inhibitor at concentration of 3 μM reduced the expression of different HDAC proteins. While SAHA elevated the expression of Rb and p53 proteins, suppression of c-myc (oncogenic protein) was recorded. Intraperitoneal administration of SAHA (100 mg/kg) in murine models suppressed tumour growth through induction of tumour necrosis. Immunohistochemistry based studies revealed that this inhibitor lowers the expression levels of different HDAC subtypes, cyclin D1, Ki67 (cell proliferation marker) and survivin (Jin et al. 2012). Exposure of 320 HSR (colon cancer cells) to this inhibitor attenuated proliferation of these cells time and concentration dependently. It has been observed that 50% growth reduction occurs following 72 h of SAHA (5 μM) treatment. This treatment was associated with substantial inhibition of Bcl-xL and survivin, the antiapoptosis proteins (Sun et al. 2010).

Panobinostat in nanomolar concentrations has shown antiproliferative effect against colon cancer cell lines. This inhibitor activated death-associated protein kinase (tumour suppressor), the critical molecular player well known for triggering apoptosis and autophagy (Gandesiri et al. 2012). This inhibitor altered the expression profile of 5–7% genes having regulatory roles in apoptosis, angiogenesis, replication of DNA and mitosis (LaBonte et al. 2009). Growth inhibitory effects of panobinostat were observed in colorectal cancer lines with half maximum inhibitory concentration values ranging from 5.5 to 25.9 μmol/L (LaBonte et al. 2011). During phase II clinical studies no objective tumour responses were observed after singlet use of panobinostat and in 4 among 29 patients grade 4 thrombocytopenia was observed (Gold et al. 2012). Pracinostat another hydroxamate group HDAC inhibitor offered enhanced therapeutic benefit as compared to vorinostat in colorectal cancer (xenograft) models. While at highest tolerated dose of SAHA 48% tumour grown suppression was recorded, 94% inhibition was accompanied with pracinostat administration (Novotny-Diermayr et al. 2010). Post exposure of pracinostat in HCT-116 cells was associated with enhanced alpha-tubulin and core histone H3 acetylation. Cyclin-dependent kinase inhibitor $p21^{CIP/WAF-1}$was elevated whereas retinoblastoma protein phosphorylation was lowered at escalated concentration of this therapeutic agent. Apart from this pracinostat triggered PARP cleavage and cell cycle obstruction in the above-mentioned model (Ganai 2016a; Novotny-Diermayr et al. 2010).

Romidepsin as single agent in phase II trial exerted no objective responses in 25 patients. Among these patients 4 showed neither decrease nor increase of their colorectal cancer. In 56% of the patients, grade 3 toxicities especially anorexia or fatigue were observed after romidepsin treatment (Whitehead et al. 2009). Suppression of colorectal cancer cell (HCT116 and LOVO) migration on sodium butyrate treatment was accompanied with downregulation of Bmi-1 (oncogene). This inhibitor enhanced the expression of miR-200c which in turn promoted the oncogene downregulation (Xu et al. 2018). Butyrate-induced suppression of colorectal cancer cell (HCT8, HT29, HCT116, LOVO) motility has been attributed to AKT/ERK deactivation and HDAC3 inhibition (Li et al. 2017). In HT-29 and HCT-116 sodium butyrate triggered endoplasmic stress-induced autophagy. It was also observed that suppression of autophagy by genetic and pharmacology intervention promoted sodium butyrate-induced apoptosis and this was accompanied by enhanced PARP cleavage (Zhang et al. 2016a). Phenylbutyrate belonging to same group of HDACi as that of butyrate exerted cytotoxicity in HT-29 and HCT-116 cell lines in a concentration following trend (Al-Keilani and Darweesh 2017).

9.1.6 HDACi as Liver Cancer Therapeutics

It has been estimated that 42,810 novel liver cancer cases will be confirmed in the year 2020 in the USA. Hepatocellular carcinoma accounts for about 75% of liver cancers and the incidence of liver cancer is more inclined towards men (three times higher) as compared to women. Its risk factors being easy to bridle can strongly prevent its occurrence (Siegel et al. 2020). Various levels of evidence link the transcriptional dysfunction with multiple forms of cancer. Due to certified implications of HDAC overexpression in liver cancer, HDACi have been and are being tested against a variety of liver cancer models. The ability of these epigenetic-route targeting inhibitors to reset transcriptional events makes them appropriate candidates for liver cancer therapy (Coradini and Speranza 2005).

While exploring the effects of vorinostat/SAHA on hepatoma cells (Huh6 and HepG2), it was found that this inhibitor induces apoptosis in these cells but proved to be abortive in primary hepatocytes of human origin. This inhibitor evoked caspase-8, caspase-3 besides PARP degradation (Emanuele et al. 2007). Experiments of vorinostat on hepatocellular carcinoma cell lines clearly indicated the growth inhibitory effects of this inhibitor through stimulation of apoptosis and cell cycle blockage. Depletion of certain proteins like pAKT, Notch and pERK1/2 was also induced by vorinostat in these cells (Kunnimalaiyaan et al. 2017). Involvement of long non-coding RNAs (lncRNAs) has been identified in the liver carcinogenesis. TSA invoked apoptosis in hepatocellular carcinoma cells by way of inducing lncRNA-uc002mbe.2 (Yang et al. 2013). Further studies on Huh7 cells revealed the substantially enhanced levels of this lncRNA after treatment with TSA. In fact TSA-driven p21 induction, cell cycle obstruction and apoptosis and reduction in pAKT in these cells were mediated by uc002mbe.2. Silencing studies involving the knockdown of this lncRNA restrained TSA-triggered effects suggesting the critical role of uc002mbe.2 in TSA-provoked apoptosis (Chen et al. 2017a). ZINC24469384, a novel lead (benzamide derivative) having HDAC inhibitory and thus anticancer activity has been identified through virtual screening and subsequent in vitro assays. This inhibitor not only induced cell cycle arrest but also induced apoptosis in HepG2 cells. The antiproliferative effect of this inhibitor was found to be mediated by NRIH4 upregulation which in turn negatively regulates proliferation (Song et al. 2019).

Resminostat, a selective Class I inhibitor inhibited the proliferation of hepatocellular carcinoma cells. This inhibitor dose dependently impeded proliferation of Hep3B cells. Following 24 h exposure of resminostat (80 nM) in three liver cancer cell lines (HepG2, Hep3B, Huh7) suppression of proliferation was observed. Resminostat in Hep3B induced decline in global HDAC activities and lowered the expression of all Class I HDACs excluding HDAC8 at transcription level (Zhao et al. 2018). Cytotoxic effects of resminostat along with the underlying mechanism being triggered have been studied in two mesenchymal (HLF, HLE) and a single epithelial (Hep3B) hepatocellular carcinoma cell lines. This inhibitor elicited cytotoxicity in all the above-mentioned lines but to a varying degree. While highest sensitivity was seen in HLF (IC_{50} 2.0 μM) least sensitivity was noted in Hep3B cell line (IC_{50} 5.9 μM). Transition of mesenchymal to epithelial phenotype was seen on resminostat treatment in the aforementioned mesenchymal cell lines. Resminostat intervention culminated in upregulation of *CDH1 with collateral downregulation of SNAI2 and TWIST1 (transcription factors for EMT-induction)* (Soukupova et al. 2017).

The therapeutic effect of sulforaphane has been studied in two liver cancer cell lines (Huh-7, HepG2) and xenograft mice model. Sulforaphane in dose-dependent fashion reduced the proliferation of liver cancer cells by way of downregulating the molecular players CCNB1, CDK2, CCND1 and CDK1. Immunohistochemical studies exposed that the marked reduction in tumour burdens is strongly linked to proliferation inhibition of tumour cells. Sulforaphane-treated tumours also showed reduction in the transcription levels of above-mentioned genes (Moriya et al. 2018). Impact of sulforaphane has been studied on the angiogenesis of hepatocellular carcinoma tumour. This inhibitor substantially bridled the HepG2-triggered HUVEC angiogenesis steps ranging from migration to tube formation. SFN-treated HepG2 cells showed depleted levels of VEGF, HIF-1α and STAT3. Moreover it was found in modified chick embryo chorioallantoic membrane model that sulforaphane not only quells tumour growth but also angiogenesis (Liu et al. 2017).

Cytotoxicity inducing potential of mocetinostat has been tested on a pair of liver cancer cell lines (Huh7 and HepG2). Through way of G2/M phase arrest in addition to mitochondria mediated apoptosis, this inhibitor suppressed the proliferation of liver cancer cells. Mocetinostat-induced apoptosis was found to be caspase and ROS dependent as pan-inhibition of caspase and ROS mitigation attenuated the defined cell death process. Importantly, this inhibitor collapsed tumour growth without imparting any marked systemic toxicity to the experimental animal mode (Liao et al. 2020). HDAC1/2 inhibition by romidepsin (macrocyclic HDAC inhibitor) caused apoptosis in liver cancer cells models namely Huh7 and HepG2. Intratumoural injections of romidepsin significantly reduced the growth of implanted tumours in nude mice which were developed through subcutaneous injections of MHCC97H cells. This effect was proved to be mediated by upregulation of $p19^{INK4d}$ in addition to $p21^{Waf1/Cip1}$ which have the critical implications in induction of apoptosis (Zhou et al. 2018). The underlying mechanism by which HDAC inhibitor butyrate induces growth arrest and programmed cell death/apoptosis in liver cancer cells has been explored. After the completion of study in hepatocarcinoma cell line

Huh7, it was found that sodium butyrate downregulates SIRT1 expression by escalating the levels of miR-22. Inhibition of this sirtuin in turn promotes ROS generation thereby triggering cytochrome c release thereby resulting in caspase-3 activation and subsequent apoptosis (Pant et al. 2017). Inhibitory effects on growth/tumourigenicity of hepatocellular carcinoma cells by sodium butyrate have been observed in a series of hepatocellular carcinoma cells (HuH-7, huH-2, Hep3B, huH-1, HLE and PLC/PRF/5). It has been suggested that this effect possibly may be ascribed to p21 induction. Sodium butyrate-induced growth inhibition was accompanied with DNA fragmentation and depletion of alpha-fetoprotein. Most importantly the apoptotic effect of sodium butyrate was independent of the p53 status of the tested liver cancer cells (Yamamoto et al. 1998). Besides sodium butyrate exerts anticancer effect in a pair of hepatic cancer cell lines HepG2 and SMMC-7721. While downregulation of HDAC4 at protein level was noted, acetylation status of histone H3 underwent escalation. Sodium butyrate markedly suppressed the EMT-transition possibly by altering HDAC4 and MMP7 (Wang et al. 2013). In a couple of hepatocellular carcinoma cells (SMMC-7721 and SK-Hep1) sodium butyrate exerted differential effects at varying concentrations. Only high sodium butyrate concentrations were able to generate apoptotic signal (Jiang et al. 2012). Valproic acid also elicited proliferation inhibition in HuH7 cells under both in vitro and in vivo conditions. Antiproliferative effect on tumour cells on oral administration of this inhibitor has been connected to DNA fragmentation, caspase-3 activation and most importantly to downregulation of mRNA levels of Notch-1 (Table 9.3) (Machado et al. 2011).

9.1.7 Using HDACi for Bladder Cancer Therapy

Urothelial carcinoma is the most frequently occurring form of bladder cancer and accounts for over 90% of the urinary tract cancers. Earlier this cancer was designated as transitional cell carcinoma (Chalasani et al. 2009; Yaxley 2016). This cancer initiates from those urothelial cells which line the inner side of the bladder (Chow et al. 2012). While 1–2% of bladder cancers in the US population are squamous cell carcinomas, 1% and less than 1% of the bladder cancers are adenocarcinomas and small cell carcinomas, respectively (Dadhania et al. 2015; Jacobo et al. 1977; Mostofi 1968; Thomas et al. 1971). In the pathogenesis of bladder cancer epigenetic aberrations play a considerable role (Martinez-Zamudio and Ha 2012). Only recently it has been proved that Class IIa HDACs namely HDAC4 and HDAC9 are overexpressed in bladder cancer models suggesting the possible use of HDACi in circumventing the defined cancer (Buckwalter et al. 2019).

Romidepsin, vorinostat and TSA have been tested in a bladder cancer cell line 5637 and it was found that these inhibitors not only restrain cell growth but induce apoptosis in the defined line. These inhibitors modulated the acetylation status of not only histone but also non-histone proteins (Li et al. 2016). Experiments have validated the significant clonogenic growth suppression effects of givinostat and romidepsin in urothelial carcinoma cell lines (UM-UC-3, VM-CUB1, 639-V,

Table 9.3 Summary of the various molecular mechanisms through which HDAC inhibitors induce antiproliferative effect or death in various models of colorectal and liver cancer

Small molecule HDAC inhibitor	Cancer subtype	Experimental models used	Protein/gene or mRNA enhanced, activated for cell death signalling	mRNA or protein or gene suppressed, inhibited or inactivated	Proof from literature
Vorinostat	Colorectal cancer	Colon tumor and colon cancer cell model	Rb, p53	c-myc, cyclin D1, survivin, Ki67	Jin et al. (2012)
		320 HSR		Of Bcl-xL, survivin	Sun et al. (2010)
Panobinostat		HCT116	Death-associated protein kinase		Gandesiri et al. (2012)
		Xenograft model			Novotny-Diermayr et al. (2010)
		HCT-116	Alpha-tubulin acetylation, H3 acetylation, $p21^{CIP/WAF-1}$	Retinoblastoma protein	Novotny-Diermayr et al. (2010)
Sodium butyrate		LOVO, HCT116	miR-200c	Bmi-1	Xu et al. (2018)
Butyrate		LOVO, HCT8, HCT116, HT29		AKT/ERK, HDAC3	Li et al. (2017)
Sodium butyrate		HT-29, HCT-116	–	–	Zhang et al. (2016a)
Phenylbutyrate		HCT-116, HT-29	–	–	Al-Keilani and Darweesh (2017)
Vorinostat	Liver cancer	Huh6, HepG2	Caspase-8, caspase-3		Emanuele et al. (2007)
		Colorectal cancer cells		pERK1/2, pAKT, notch	Kunnimalaiyaan et al. (2017)

(continued)

Table 9.3 (continued)

Small molecule HDAC inhibitor	Cancer subtype	Experimental models used	Protein/gene or mRNA enhanced, activated for cell death signalling	mRNA or protein or gene suppressed, inhibited or inactivated	Proof from literature
TSA		Hepatocellular carcinoma cells	lncRNA-uc002mbe.2		Yang et al. (2013)
		Huh7	lncRNA-uc002mbe.2, p21 induction	pAKT	Chen et al. (2017a)
ZINC24469384		HepG2	NRIH4		Song et al. (2019)
Resminostat		Hep3B, HepG2, Huh7		HDAC1, 2,3	Zhao et al. (2018)
		HLF, HLE, Hep3B	*CDH1*	*TWIST1, SNAI2*	Soukupova et al. (2017)
Sulforaphane		HepG2, Huh-7		CCNB1, CCND1, CDK2, CDK1	Moriya et al. (2018)
		HepG2, HUVEC		STAT3, HIF-1α, VEGF	Liu et al. (2017)
Mocetinostat		Huh7, HepG2	ROS		Liao et al. (2020)
Romidepsin			$p19^{INK4d}$, $p21^{Waf1/Cip1}$	*HDAC1, HDAC2*	Zhou et al. (2018)
Sodium butyrate		Huh7	miR-22, caspase-3, ROS	SIRT1	Pant et al. (2017)
		huH-1, PLC/PRF/5, HLE, Hep3B	p21	Alpha-fetoprotein	Yamamoto et al. (1998)
		SMMC-7721, HepG2	*H3 acetylation*	HDAC4	Wang et al. (2013)
Valproic acid		HuH7	Caspase-3	Notch-1	Machado et al. (2011)

RT-112 and SW-1710). Moreover, these inhibitors markedly declined the proliferation of these cell models. This promising anticancer effect of above-mentioned HDACi has been related to combinatorial inhibition of HDAC1/HDAC2 (Pinkerneil et al. 2016). In a panel of bladder cancer cell lines (HT1376, T24, RT4 and TCCSUP) valproic acid in a dose increasing trend enhanced the acetylation status of histone H3 in addition to $p21^{WAF1}$ expression. This inhibitor impeded the rate of invasion in all these models excluding RT4 although no impact was observed in their migration tendency. Further chronic administration of this inhibitor for a period of 34 days exerted substantial reduction in tumour growth in xenograft models (T24) (Chen et al. 2006). Overexpression of HDAC1 has been quantified from bladder cancer specimens and urinary bladders of BBN (urothelial carcinogen) treated mice. Dose-dependent decline in survival was noted in 5637 and HT-1376 (urinary bladder cancer cells) on valproic acid exposure. This effect has been correlated to enhanced protein expression of $p21^{WAF1}$ post valproic acid treatment (Ozawa et al. 2010). In a series of bladder cancer cell lines valproic acid intervention induced growth-obstruction effects. The intensity of effect was found to be directly proportional to duration of the valproic acid application (Vallo et al. 2011). Antiproliferative effect of vorinostat and valproic acid has been explored on two bladder cancer cell line T24 and UMUC3. Both these inhibitors reduced the proliferation of these cells. Valproic acid triggered antiproliferative effect was found to be mediated by enhanced expression of thrombospondin-1 (Byler et al. 2012). In vivo studies have proved the assembling of valproic acid in bladder, its high uptake by liver and brain (Kim et al. 2013).

Inhibition of proliferation of bladder cancer cells (T24) on vorinostat administration was associated with induction of $p21^{WAF1}$ through epigenetic mechanism (Richon et al. 2000). Vorinostat-induced reduction in urothelial cancer cell viability has been allocated to noticeable arrest of cell cycle, downregulation of thymidylate synthase, upregulation of $p21^{WAF1}$ (Niegisch et al. 2013). Higher expression of Notch3 has been identified in bladder tumour tissues of urothelial cancer patients. Pharmacological intervention with hydroxamate vorinostat declined proliferation of bladder cancer cells (T24) through enhanced levels of acetylated Notch3, reduced Notch3 expression and cell cycle arrest induction (Zhang et al. 2017). Both sodium butyrate and TSA inhibited the growth of bladder cancer cells under in vitro conditions. While the butyrate effect was seen at millimolar concentrations, TSA induced effects were prominent in micromolar concentrations. TSA administration resulted in above 70% decrease in tumour volume in two (UM-UC-3 and EJ) xenograft nude mice models of bladder cancer with no discernible toxicity. Although no E-cadherin expression alteration was observed in HDAC inhibitor treated cells, plakoglobin upregulation was quite evident (Canes et al. 2005). Growth inhibitory effect was observed in response to TSA treatment in BIU-87 cells which on molecular studies proved possibly to be due to enhanced message/mRNA levels of $p21^{WAF1}$ and G1 arrest (Li et al. 2006). This inhibitor invoked cytotoxic effect (apoptosis) in cell model (T24) of bladder cancer by upregulating the transient receptor potential cation channel subfamily M member 2 (TRPM2) epigenetically (enhancing H3K9ac) (Cao et al. 2015). Belinostat also induced cell growth inhibition in four model cell lines of bladder cancer (RT4, T24, 5637 and J82) dose dependently and among these lines 5637 displayed the highest sensitivity. Moreover, reduction in the growth of bladder tumour in Ha-ras transgenic mice that has been ascribed to induction of *p21*^{WAF1} was observed on intraperitoneal administration of this inhibitor with no overt toxicity (Buckley et al. 2007). Belinostat loaded nanoparticles offered long lasting HDAC inhibition and reduced tumour growth by 70%. Besides nanoparticles loaded with the defined HDAC inhibitor escalated intratumoural histone H4 acetylation by 2.5-fold compared to nanoparticles containing no belinostat (Martin et al. 2013).

Tubacin, a selective inhibitor of HDAC6 arrested proliferation and provoked apoptosis through downregulation of MYC, FGFR3 (fibroblast growth factor receptor 3), cyclin D1 and by triggering DNA damage in RT112 cells. In addition to this tubacin hampered tumour growth strikingly in xenoplant assays (Ota et al. 2018). Four HDACi including entinostat, apicidin, TSA and valproic acid were tested on several bladder cancer cell lines. All these inhibitors restrained cell growth in a dose-dependent manner. While in cells lines with high invasive potential these inhibitors were accompanied with γ-catenin upregulation, modest and poorly invasive lines demonstrated the downregulation of desmoglein (Giannopoulou et al. 2019; Gould et al. 2010). Autophagy and miRNAs are critically involved in the development of cancer. Anticancer effects of sodium butyrate were evaluated using T24 and 5637 cell lines. This inhibitor suppressed migration, triggered autophagy and excessive ROS production. For executing these effects sodium butyrate upregulated the expression of miR-139-5p, reduced Bmi-1 and stimulated AMPK-mTOR pathway (Wang et al. 2020). Phase I clinical study has been performed with panobinostat and belinostat. While with belinostat one of the patient having advanced urothelial carcinoma showed complete response, panobinostat treatment induced partial response in another patient having advanced bladder cancer (Agarwal et al. 2016; Sharma et al. 2015). A novel HDAC inhibitor, 19i when used on urothelial carcinoma cells diminished their clonogenic tendency. This HDAC inhibitor was found to exert the anticancer effect by way of inhibiting Class I HDACs. Even at lower concentrations than approved SAHA, this novel inhibitor was found to be active (Kaletsch et al. 2018). Panobinostat, the approved drug for multiple myeloma has been studied on bladder cancer cell lines having canonical/intact TP53 (HT1197) and mutated Tp53 (T24 and UMUC3). In all the mentioned cell lines this inhibitor aroused cell cycle arrest and elevated p21 levels dose dependently (Gupta et al. 2019). Romidepsin decreased cell survival in three bladder cancer cell lines notably HT1376, MBT2 and RT112. Although all these lines were sensitive, the most sensitivity was shown by RT112 and in this model romidepsin enhanced the acetylation of histone H3 at lysine 18 (H3K18ac) (Table 9.4) (Paillas et al. 2020).

Table 9.4 Compendium of different mechanisms facilitated or hampered by HDAC inhibitors for triggering cytotoxicity in bladder cancer cells

HDAC inhibitor tested	Name of the cancer	Models used for experimental work	Downregulated genes, mRNAs or proteins	Upregulated mRNAs, proteins or genes	Evidence from research
Givinostat/ romidepsin	Bladder cancer	SW-1710, UM-UC-3, 639-V, VM-CUB1		HDAC1/ HDAC2	Pinkerneil et al. (2016)
Valproic acid		T24, RT4, TCCSUP, HT1376, xenograft model	$p21^{WAF1}$, H3 acetylation		Chen et al. (2006)
		5637, HT-1376	$p21^{WAF1}$		Ozawa et al. (2010)
		T24, UMUC3	Thrombospondin-1		Byler et al. (2012)
Vorinostat		T24	$p21^{WAF1}$		Richon et al. (2000)
		18 UCCs	$p21^{WAF1}$	Thymidylate synthase	Niegisch et al. (2013)
		T24	Acetylated Notch3	Notch3	Zhang et al. (2017)
Sodium butyrate/ TSA		UM-UC-3, EJ, xenograft nude mice	Plakoglobin		Canes et al. (2005)
TSA		BIU-87	$p21^{WAF1}$		Li et al. (2006)
		T24	TRPM2, H3K9ac		Cao et al. (2015)
Belinostat		5637, J82, T24, RT4,	*$p21^{WAF1}$*		Buckley et al. (2007)
		Xenograft tumors	H4 acetylation		Martin et al. (2013)
Tubacin		RT112		FGFR3, MYC, cyclin D1	Ota et al. (2018)
Entinostat/ Valproic acid/TSA		Multiple bladder cancer cells	γ-Catenin	Desmoglein	Gould et al. (2010)

(continued)

Table 9.4 (continued)

HDAC inhibitor tested	Name of the cancer	Models used for experimental work	Downregulated genes, mRNAs or proteins	Upregulated mRNAs, proteins or genes	Evidence from research
Sodium butyrate		5637	ROS, miR-139-5p, AMPK-mTOR	Reduced Bmi-1	Wang et al. (2020)
19i/ LMK235		VM-CUB1, 639-V		Class I HDACs	Kaletsch et al. (2018)
Panobinostat		HT1197, UMUC3	p21		Gupta et al. (2019)
Romidepsin		MBT2, RT112	H3K18ac		Paillas et al. (2020)

9.2 Limitations of HDACi as Single Agent Therapeutics Against Cancer

Although histone deacetylase inhibitors exert encouraging anticancer effects against various cancer models the desired effects have not been achieved by using these small molecules as single agents (monotherapy/singlet therapy). This is because in response to single agent treatment, cancer cells generate resistance mechanisms sufficient enough for circumventing cytotoxic effects. Majority of HDACi as single agents are not effective against solid tumours due to their low and ineffective concentrations in these tumours (Gryder et al. 2012). In certain cases like vorinostat and panobinostat this inefficiency against solid tumours has been related to pharmacokinetic profile. For instance vorinostat has low solubility in water, weak cell permeability and its oral bioavailability is also low (McClure et al. 2018). Bioavailability of another HDAC inhibitor panobinostat ranges from poor to moderate whereas the poor solubility of romidepsin has not promoted its oral administration (Konsoula et al. 2009). Evidence based study involving vorinostat and TSA has indicated that these inhibitors may enhance cancer severity by invoking epithelial to mesenchymal transition/EMT phenotype (Kong et al. 2012). HDACs are critically involved in multiple pathways and pan-inhibitors like vorinostat, panobinostat and TSA target a wide range of classical HDACs and thus may promote rather than restraining cell growth (Rana et al. 2020). Several HDACs of Class I and Class II have role in inhibition of angiogenesis (Kim and Bae 2011). VEGF activity regulation in breast cancer cells by Kruppel-like factor 4/KLF4 is mediated through two Class I members (HDAC2, HDAC3) (Ray et al. 2013). TP Binding Cassette Subfamily B Member 1(ABCB1) gene encodes P-Glycoprotein (P-gp), a drug effluxing pump on plasma membrane of cells relying on ATP (Zhou 2008). This protein with wide substrate inclination effluxes drugs out of the cancer cells. Multiple HDACi such as TSA, sodium butyrate and vorinostat enhanced the expression of this glycoprotein in various cancer cell lines (Mickley et al. 1989; Wang et al. 2016). Moreover, several dose-limiting toxicities have been reported with HDACi. These include QTcF prolongation, fatigue, atrial fibrillation, febrile neutropenia, epilepticus, renal failure, nausea, diarrhoea and vomiting (Ganai 2015; Garcia-Manero et al. 2008). Neuroconstipation and other toxicities like somnolence and neurocognitive impairment are associated with valproic acid (Subramanian et al. 2010). The active sites of Classical HDACs share similarity to a greater degree and this increases the probability of off-targeting during therapeutic intervention especially in case of pan-HDACi (Ganai 2018).

Up to now I have extensively elaborated the promising anti-neoplastic effects of HDACi as individual agents in the deadliest cancers such as lung cancer, colorectal cancer and so on. The molecular players and signalling mechanisms invoked or alleviated by various structurally similar or distinct HDACi in inducing cytotoxicity effects in cancer models have been discussed. The therapeutic effect of these inhibitors has been considered not in cell lines but also in preclinical and clinical models. Though the effects of these inhibitors in these models are promising but the clinical trial results of these inhibitors against the above discussed cancers are not so encouraging. The toxicity profile of these inhibitors when used in monotherapy has been thoroughly considered. Further some possible mechanisms which hamper the therapeutic efficacy of these inhibitors have also been put forward. Collectively, HDACi offer limited therapeutic benefit when used in singlet therapy and higher concentrations of these inhibitors are required for acquiring therapeutic effect thereby sensitizing typical cells as well. However, the novel therapeutic strategies and HDAC inhibitor designing methods have been employed to improve the efficacy of these inhibitors which will be discussed in the imminent chapters of the book.

References

Abbas A, Gupta S (2008) The role of histone deacetylases in prostate cancer. Epigenetics 3:300–309

Agarwal N, McPherson JP, Bailey H, Gupta S, Werner TL, Reddy G, Bhat G, Bailey EB, Sharma S (2016) A phase I clinical trial of the effect of belinostat on the pharmacokinetics and pharmacodynamics of warfarin. Cancer Chemother Pharmacol 77:299–308

Al-Keilani M, Darweesh R (2017) Abstract 1188: sodium phenylbutyrate has an antineoplastic effect and enhances the cytotoxicity of 5-fluorouracil and irinotecan in colorectal cancer cell lines. Cancer Res 77:1188–1188

Buckley MT, Yoon J, Yee H, Chiriboga L, Liebes L, Ara G, Qian X, Bajorin DF, Sun TT, Wu XR, Osman I (2007) The histone deacetylase inhibitor belinostat (PXD101) suppresses bladder cancer cell growth in vitro and in vivo. J Transl Med 5:49

Buckwalter JM, Chan W, Shuman L, Wildermuth T, Ellis-Mohl J, Walter V, Warrick JI, Wu XR, Kaag M, Raman JD, DeGraff DJ (2019) Characterization of histone deacetylase expression within in vitro and in vivo bladder cancer model systems. Int J Mol Sci 20

Byler TK, Leocadio D, Shapiro O, Bratslavsky G, Stodgell CJ, Wood RW, Messing EM, Reeder JE (2012) Valproic acid decreases urothelial cancer cell proliferation and induces thrombospondin-1 expression. BMC Urol 12:21

Canes D, Chiang GJ, Billmeyer BR, Austin CA, Kosakowski M, Rieger-Christ KM, Libertino JA, Summerhayes IC (2005) Histone deacetylase inhibitors upregulate plakoglobin expression in bladder carcinoma cells and display antineoplastic activity in vitro and in vivo. Int J Cancer 113:841–848

Cao QF, Qian SB, Wang N, Zhang L, Wang WM, Shen HB (2015) TRPM2 mediates histone deacetylase inhibition-induced apoptosis in bladder cancer cells. Cancer Biother Radiopharm 30:87–93

Carbognani P, Spaggiari L, Romani A, Solli P, Corradi A, Cantoni AM, Petronini PG, Borghetti AF, Rusca M, Bobbio P (1998) Expression of human CD44v6 in non-small-cell lung cancer. Eur Surg Res 30:403–408

Castro NP, Rangel MC, Merchant AS, MacKinnon G, Cuttitta F, Salomon DS, Kim YS (2019) Sulforaphane suppresses the growth of triple-negative breast cancer stem-like cells in vitro and in vivo. Cancer Prev Res (Phila) 12:147–158

Chalasani V, Chin JL, Izawa JI (2009) Histologic variants of urothelial bladder cancer and nonurothelial histology in bladder cancer. Can Urol Assoc J 3:S193–S198

Chen C-L, Sung J, Cohen M, Chowdhury WH, Sachs MD, Li Y, Lakshmanan Y, Yung BYM, Lupold SE, Rodriguez R (2006) Valproic acid inhibits invasiveness in bladder cancer but not in prostate cancer cells. J Pharmacol Exp Ther 319:533–542

Chen C-Y, Yu Z-Y, Chuang Y-S, Huang R-M, Wang T-CV (2015) Sulforaphane attenuates EGFR signaling in NSCLC cells. J Biomed Sci 22:38

Chen T, Gu C, Xue C, Yang T, Zhong Y, Liu S, Nie Y, Yang H (2017a) LncRNA-uc002mbe.2 interacting with hnRNPA2B1 mediates AKT deactivation and p21 up-regulation induced by trichostatin in liver cancer cells. Front Pharmacol 8:669–669

Chen Y-J, Wang W-H, Wu W-Y, Hsu C-C, Wei L-R, Wang S-F, Hsu Y-W, Liaw C-C, Tsai W-C (2017b) Novel histone deacetylase inhibitor AR-42 exhibits antitumor activity in pancreatic cancer cells by affecting multiple biochemical pathways. PLoS One 12:e0183368

Chen J, Li N, Liu B, Ling J, Yang W, Pang X, Li T (2020) Pracinostat (SB939), a histone deacetylase inhibitor, suppresses breast cancer metastasis and growth by inactivating the IL-6/STAT3 signalling pathways. Life Sci 248:117469

Chow N-H, Knowles M, Bivalacqua TJ (2012) Urothelial carcinoma. Adv Urol 2012:461370

Cohen AL, Neumayer L, Boucher K, Factor RE, Shrestha G, Wade M, Lamb JG, Arbogast K, Piccolo SR, Riegert J, Schabel M, Bild AH, Werner TL (2017) Window-of-opportunity study of valproic acid in breast cancer testing a gene expression biomarker. JCO Precis Oncol:1–11

Conteduca V, Sansonno D, Russi S, Dammacco F (2013) Precancerous colorectal lesions (review). Int J Oncol 43:973–984

Coradini D, Speranza A (2005) Histone deacetylase inhibitors for treatment of hepatocellular carcinoma. Acta Pharmacol Sin 26:1025–1033

Coradini D, Biffi A, Costa A, Pellizzaro C, Pirronello E, Di Fronzo G (1997) Effect of sodium butyrate on human breast cancer cell lines. Cell Prolif 30:149–159

Crisanti MC, Wallace AF, Kapoor V, Vandermeers F, Dowling ML, Pereira LP, Coleman K, Campling BG, Fridlender ZG, Kao GD, Albelda SM (2009) The HDAC inhibitor panobinostat (LBH589) inhibits mesothelioma and lung cancer cells in vitro and in vivo with particular efficacy for small cell lung cancer. Mol Cancer Ther 8:2221–2231

Dadhania V, Czerniak B, Guo CC (2015) Adenocarcinoma of the urinary bladder. Am J Clin Exp Urol 3:51–63

de Marinis F, Atmaca A, Tiseo M, Giuffreda L, Rossi A, Gebbia V, Antonio CD, Zotto LD, Al-Batran S-E, Marsoni S, Wolf M (2013) A phase II study of the histone deacetylase inhibitor panobinostat (LBH589) in pretreated patients with small-cell lung cancer. J Thorac Oncol 8:1091–1094

Dekker E, Tanis PJ, Vleugels JLA, Kasi PM, Wallace MB (2019) Colorectal cancer. Lancet 394:1467–1480
Delavari A, Mardan F, Salimzadeh H, Bishehsari F, Khosravi P, Khanehzad M, Nasseri-Moghaddam S, Merat S, Ansari R, Vahedi H, Shahbazkhani B, Saberifiroozi M, Sotoudeh M, Malekzadeh R (2014) Characteristics of colorectal polyps and cancer; a retrospective review of colonoscopy data in Iran. Middle East J Dig Dis 6:144–150
DeSantis CE, Ma J, Gaudet MM, Newman LA, Miller KD, Goding Sauer A, Jemal A, Siegel RL (2019) Breast cancer statistics, 2019. CA Cancer J Clin 69:438–451
Deskin B, Yin Q, Zhuang Y, Saito S, Shan B, Lasky JA (2020) Inhibition of HDAC6 attenuates tumor growth of non-small cell lung cancer. Transl Oncol 13:135–145
Dovzhanskiy DI, Hartwig W, Lazar NG, Schmidt A, Felix K, Straub BK, Hackert T, Krysko DV, Werner J (2012) Growth inhibition of pancreatic cancer by experimental treatment with 4-phenylbutyrate is associated with increased expression of Connexin 43. Oncol Res 20:103–111
Ediriweera MK, Tennekoon KH, Samarakoon SR (2019) Emerging role of histone deacetylase inhibitors as anti-breast-cancer agents. Drug Discov Today 24:685–702
Emanuele S, Lauricella M, Carlisi D, Vassallo B, D'Anneo A, Di Fazio P, Vento R, Tesoriere G (2007) SAHA induces apoptosis in hepatoma cells and synergistically interacts with the proteasome inhibitor Bortezomib. Apoptosis 12:1327–1338
Fortunati N, Marano F, Bandino A, Frairia R, Catalano MG, Boccuzzi G (2014) The pan-histone deacetylase inhibitor LBH589 (panobinostat) alters the invasive breast cancer cell phenotype. Int J Oncol 44:700–708
Ganai SA (2015) Strategy for enhancing the therapeutic efficacy of histone deacetylase inhibitor dacinostat: the novel paradigm to tackle monotonous cancer chemoresistance. Arch Pharm Res
Ganai SA (2016a) Histone deacetylase inhibitor pracinostat in doublet therapy: a unique strategy to improve therapeutic efficacy and to tackle herculean cancer chemoresistance. Pharm Biol 54:1926–1935
Ganai SA (2016b) Histone deacetylase inhibitor sulforaphane: the phytochemical with vibrant activity against prostate cancer. Biomed Pharmacother 81:250–257
Ganai SA (2018) Designing isoform-selective inhibitors against classical HDACs for effective anticancer therapy: insight and perspectives from in silico. Curr Drug Targets 19:815–824
Ganai SA, Rashid R, Abdullah E, Altaf M (2017) Plant derived inhibitor sulforaphane in combinatorial therapy against therapeutically challenging pancreatic cancer. Anti Cancer Agents Med Chem 17:365–373
Gandesiri M, Chakilam S, Ivanovska J, Benderska N, Ocker M, Di Fazio P, Feoktistova M, Gali-Muhtasib H, Rave-Frank M, Prante O, Christiansen H, Leverkus M, Hartmann A, Schneider-Stock R (2012) DAPK plays an important role in panobinostat-induced autophagy and commits cells to apoptosis under autophagy deficient conditions. Apoptosis 17:1300–1315
Garcia-Manero G, Assouline S, Cortes J, Estrov Z, Kantarjian H, Yang H, Newsome WM, Miller WH Jr, Rousseau C, Kalita A, Bonfils C, Dubay M, Patterson TA, Li Z, Besterman JM, Reid G, Laille E, Martell RE, Minden M (2008) Phase 1 study of the oral isotype specific histone deacetylase inhibitor MGCD0103 in leukemia. Blood 112:981–989
Giannopoulou AF, Velentzas AD, Konstantakou EG, Avgeris M, Katarachia SA, Papandreou NC, Kalavros NI, Mpakou VE, Iconomidou V, Anastasiadou E, Kostakis IK, Papassideri IS, Voutsinas GE, Scorilas A, Stravopodis DJ (2019) Revisiting histone deacetylases in human tumorigenesis: the paradigm of urothelial bladder cancer. Int J Mol Sci 20:1291
Gibbs A, Schwartzman J, Deng V, Alumkal J (2009) Sulforaphane destabilizes the androgen receptor in prostate cancer cells by inactivating histone deacetylase 6. Proc Natl Acad Sci 106:16663–16668
Gold PJ, Smith DA, Iriarte D, Boatman B, Kaplan HG (2012) Phase II trial of panobinostat (LBH589) in patients (pts) with refractory metastatic colorectal cancer (MCRC). J Clin Oncol 30:582–582

Gould JJ, Kenney PA, Rieger-Christ KM, Silva Neto B, Wszolek MF, LaVoie A, Holway AH, Spurrier B, Austin J, Cammarata BK, Canes D, Libertino JA, Summerhayes IC (2010) Identification of tumor and invasion suppressor gene modulators in bladder cancer by different classes of histone deacetylase inhibitors using reverse phase protein arrays. J Urol 183:2395–2402

Gravina GL, Marampon F, Giusti I, Carosa E, Di Sante S, Ricevuto E, Dolo V, Tombolini V, Jannini EA, Festuccia C (2012) Differential effects of PXD101 (belinostat) on androgen-dependent and androgen-independent prostate cancer models. Int J Oncol 40:711–720

Gryder BE, Sodji QH, Oyelere AK (2012) Targeted cancer therapy: giving histone deacetylase inhibitors all they need to succeed. Future Med Chem 4:505–524

Guimaraes I, Guimarães S, Daltoé R, Herlinger A, Klesia P, Madeira T, Ladislau I, Valadão P, Morais L, Junior S, Teixeira G, Amorim D, Zipinotti D, Santos K, Demuth L, Batista A, Rangel (2013) Conventional cancer treatment. https://doi.org/10.5772/55282

Gupta S, Albertson DJ, Parnell TJ, Butterfield A, Weston A, Pappas LM, Dalley B, O'Shea JM, Lowrance WT, Cairns BR, Schiffman JD, Sharma S (2019) Histone deacetylase inhibition has targeted clinical benefit in ARID1A-mutated advanced urothelial carcinoma. Mol Cancer Ther 18:185–195

Haefner M, Bluethner T, Niederhagen M, Moebius C, Wittekind C, Mossner J, Caca K, Wiedmann M (2008) Experimental treatment of pancreatic cancer with two novel histone deacetylase inhibitors. World J Gastroenterol 14:3681–3692

Howard SC, McCormick J, Pui C-H, Buddington RK, Harvey RD (2016) Preventing and managing toxicities of high-dose methotrexate. Oncologist 21:1471–1482

Hrabeta J, Stiborova M, Adam V, Kizek R, Eckschlager T (2014) Histone deacetylase inhibitors in cancer therapy. A review. Biomed Pap Med Fac Univ Palacky Olomouc Czech Repub 158:161–169

Jacobo E, Loening S, Schmidt JD, Culp DA (1977) Primary adenocarcinoma of the bladder: a retrospective study of 20 patients. J Urol 117:54–56

Jiang W, Guo Q, Wu J, Guo B, Wang Y, Zhao S, Lou H, Yu X, Mei X, Wu C, Qiao S, Wu Y (2012) Dual effects of sodium butyrate on hepatocellular carcinoma cells. Mol Biol Rep 39:6235–6242

Jin JS, Tsao TY, Sun PC, Yu CP, Tzao C (2012) SAHA inhibits the growth of colon tumors by decreasing histone deacetylase and the expression of cyclin D1 and survivin. Pathol Oncol Res 18:713–720

Kaletsch A, Pinkerneil M, Hoffmann MJ, Jaguva Vasudevan AA, Wang C, Hansen FK, Wiek C, Hanenberg H, Gertzen C, Gohlke H, Kassack MU, Kurz T, Schulz WA, Niegisch G (2018) Effects of novel HDAC inhibitors on urothelial carcinoma cells. Clin Epigenetics 10:100–100

Kim HJ, Bae SC (2011) Histone deacetylase inhibitors: molecular mechanisms of action and clinical trials as anti-cancer drugs. Am J Transl Res 3:166–179

Kim SW, Hooker JM, Otto N, Win K, Muench L, Shea C, Carter P, King P, Reid AE, Volkow ND, Fowler JS (2013) Whole-body pharmacokinetics of HDAC inhibitor drugs, butyric acid, valproic acid and 4-phenylbutyric acid measured with carbon-11 labeled analogs by PET. Nucl Med Biol 40:912–918

Kohler J, Erlenkamp G, Eberlin A, Rumpf T, Slynko I, Metzger E, Schule R, Sippl W, Jung M (2012) Lestaurtinib inhibits histone phosphorylation and androgen-dependent gene expression in prostate cancer cells. PLoS One 7:e34973

Kong D, Ahmad A, Bao B, Li Y, Banerjee S, Sarkar FH (2012) Histone deacetylase inhibitors induce epithelial-to-mesenchymal transition in prostate cancer cells. PLoS One 7:e45045

Kong LR, Tan TZ, Ong WR, Bi C, Huynh H, Lee SC, Chng WJ, Eichhorn PJA, Goh BC (2017) Belinostat exerts antitumor cytotoxicity through the ubiquitin-proteasome pathway in lung squamous cell carcinoma. Mol Oncol 11:965–980

Konsoula Z, Cao H, Velena A, Jung M (2009) Pharmacokinetics-pharmacodynamics and antitumor activity of mercaptoacetamide-based histone deacetylase inhibitors. Mol Cancer Ther 8:2844–2851

Kunnimalaiyaan S, Sokolowski K, Gamblin TC, Kunnimalaiyaan M (2017) Suberoylanilide hydroxamic acid, a histone deacetylase inhibitor, alters multiple signaling pathways in hepatocellular carcinoma cell lines. Am J Surg 213:645–651

LaBonte MJ, Wilson PM, Fazzone W, Groshen S, Lenz HJ, Ladner RD (2009) DNA microarray profiling of genes differentially regulated by the histone deacetylase inhibitors vorinostat and LBH589 in colon cancer cell lines. BMC Med Genet 2:67

LaBonte MJ, Wilson PM, Fazzone W, Russell J, Louie SG, El-Khoueiry A, Lenz HJ, Ladner RD (2011) The dual EGFR/HER2 inhibitor lapatinib synergistically enhances the antitumor activity of the histone deacetylase inhibitor panobinostat in colorectal cancer models. Cancer Res 71:3635–3648

Lakshmaiah KC, Jacob LA, Aparna S, Lokanatha D, Saldanha SC (2014) Epigenetic therapy of cancer with histone deacetylase inhibitors. J Cancer Res Ther 10:469–478

Lee JE, Kim JH (2015) Valproic acid inhibits the invasion of PC3 prostate cancer cells by upregulating the metastasis suppressor protein NDRG1. Genet Mol Biol 38:527–533

Lee HS, Park SB, Kim SA, Kwon SK, Cha H, Lee DY, Ro S, Cho JM, Song SY (2017) A novel HDAC inhibitor, CG200745, inhibits pancreatic cancer cell growth and overcomes gemcitabine resistance. Sci Rep 7:41615

Li Y, Seto E (2016) HDACs and HDAC inhibitors in cancer development and therapy. Cold Spring Harb Perspect Med 6:a026831

Li GC, Zhang X, Pan TJ, Chen Z, Ye ZQ (2006) Histone deacetylase inhibitor trichostatin A inhibits the growth of bladder cancer cells through induction of p21WAF1 and G1 cell cycle arrest. Int J Urol 13:581–586

Li Y, Karagöz GE, Seo YH, Zhang T, Jiang Y, Yu Y, Duarte AMS, Schwartz SJ, Boelens R, Carroll K, Rüdiger SGD, Sun D (2012) Sulforaphane inhibits pancreatic cancer through disrupting Hsp90-p50(Cdc37) complex and direct interactions with amino acids residues of Hsp90. J Nutr Biochem 23:1617–1626

Li QQ, Hao JJ, Zhang Z, Hsu I, Liu Y, Tao Z, Lewi K, Metwalli AR, Agarwal PK (2016) Histone deacetylase inhibitor-induced cell death in bladder cancer is associated with chromatin modification and modifying protein expression: a proteomic approach. Int J Oncol 48:2591–2607

Li Q, Ding C, Meng T, Lu W, Liu W, Hao H, Cao L (2017) Butyrate suppresses motility of colorectal cancer cells via deactivating Akt/ERK signaling in histone deacetylase dependent manner. J Pharmacol Sci 135:148–155

Liao B, Sun Q, Yuan Y, Yin Y, Qiao J, Jiang P (2020) Histone deacetylase inhibitor MGCD0103 causes cell cycle arrest, apoptosis, and autophagy in liver cancer cells. J Cancer 11:1915–1926

Liu P, Atkinson SJ, Akbareian SE, Zhou Z, Munsterberg A, Robinson SD, Bao Y (2017) Sulforaphane exerts anti-angiogenesis effects against hepatocellular carcinoma through inhibition of STAT3/HIF-1α/VEGF signalling. Sci Rep 7:12651

Lombardi PM, Cole KE, Dowling DP, Christianson DW (2011) Structure, mechanism, and inhibition of histone deacetylases and related metalloenzymes. Curr Opin Struct Biol 21:735–743

Loprevite M, Tiseo M, Grossi F, Scolaro T, Semino C, Pandolfi A, Favoni R, Ardizzoni A (2005) In vitro study of CI-994, a histone deacetylase inhibitor, in non-small cell lung cancer cell lines. Oncol Res 15:39–48

Lu X, Ning Z, Li Z, Cao H, Wang X (2016) Development of chidamide for peripheral T-cell lymphoma, the first orphan drug approved in China. Intractable Rare Dis Res 5:185–191

Ma Y, Yue Y, Pan M, Sun J, Chu J, Lin X, Xu W, Feng L, Chen Y, Chen D, Shin VY, Wang X, Jin H (2015) Histone deacetylase 3 inhibits new tumor suppressor gene DTWD1 in gastric cancer. Am J Cancer Res 5:663–673

Machado MC, Bellodi-Privato M, Kubrusly MS, Molan NA, Tharcisio T Jr, de Oliveira ER, D'Albuquerque LA (2011) Valproic acid inhibits human hepatocellular cancer cells growth in vitro and in vivo. J Exp Ther Oncol 9:85–92

Mariadason JM (2008) HDACs and HDAC inhibitors in colon cancer. Epigenetics 3:28–37

Mármol I, Sánchez-de-Diego C, Pradilla Dieste A, Cerrada E, Rodriguez Yoldi MJ (2017) Colorectal carcinoma: a general overview and future perspectives in colorectal cancer. Int J Mol Sci 18:197

Martin DT, Hoimes CJ, Kaimakliotis HZ, Cheng CJ, Zhang K, Liu J, Wheeler MA, Kelly WK, Tew GN, Saltzman WM, Weiss RM (2013) Nanoparticles for urothelium penetration and delivery of the histone deacetylase inhibitor belinostat for treatment of bladder cancer. Nanomedicine 9:1124–1134

Martinez-Zamudio R, Ha HC (2012) Histone ADP-ribosylation facilitates gene transcription by directly remodeling nucleosomes. Mol Cell Biol 32:2490–2502

Mawatari T, Ninomiya I, Inokuchi M, Harada S, Hayashi H, Oyama K, Makino I, Nakagawara H, Miyashita T, Tajima H, Takamura H, Fushida S, Ohta T (2015) Valproic acid inhibits proliferation of HER2-expressing breast cancer cells by inducing cell cycle arrest and apoptosis through Hsp70 acetylation. Int J Oncol 47:2073–2081

McClure JJ, Li X, Chou CJ (2018) Advances and challenges of HDAC inhibitors in cancer therapeutics. Adv Cancer Res 138:183–211

Mehdi O, Francoise S, Sofia CL, Urs G, Kevin Z, Bernard S, Igor S, Anabela CD, Dominique L, Eric M, Ali O (2012) HDAC gene expression in pancreatic tumor cell lines following treatment with the HDAC inhibitors panobinostat (LBH589) and trichostatine (TSA). Pancreatology 12:146–155

Mickley LA, Bates SE, Richert ND, Currier S, Tanaka S, Foss F, Rosen N, Fojo AT (1989) Modulation of the expression of a multidrug resistance gene (mdr-1/P-glycoprotein) by differentiating agents. J Biol Chem 264:18031–18040

Miyanaga A, Gemma A, Noro R, Kataoka K, Matsuda K, Nara M, Okano T, Seike M, Yoshimura A, Kawakami A, Uesaka H, Nakae H, Kudoh S (2008) Antitumor activity of histone deacetylase inhibitors in non-small cell lung cancer cells: development of a molecular predictive model. Mol Cancer Ther 7:1923–1930

Moriya K, Sato S, Furukawa M, Saikawa S, Namisaki T, Kitade M, Kawaratani H, Kaji K, Takaya H, Shimozato N, Sawada Y, Seki K, Kitagawa K, Akahane T, Mitoro A, Okura Y, Yoshiji H, Yamao J (2018) Sulforaphane inhibits liver cancer cell growth and angiogenesis. Arch Can Res 6(4):23. https://doi.org/10.21767/2254-6081.100189

Mostofi FK (1968) Pathological aspects and spread of carcinoma of the bladder. JAMA 206:1764–1769 passim

Niegisch G, Knievel J, Koch A, Hader C, Fischer U, Albers P, Schulz WA (2013) Changes in histone deacetylase (HDAC) expression patterns and activity of HDAC inhibitors in urothelial cancers. Urol Oncol 31:1770–1779

Niknamian S (2019) Valproic acid inhibit non-small-cell lung cancer (A549 cell line) metastasis via inhibition of CD44v6 and Nm23H1. https://doi.org/10.31219/osf.io/j6tcq

Noh H, Park J, Shim M, Lee Y (2016) Trichostatin A enhances estrogen receptor-alpha repression in MCF-7 breast cancer cells under hypoxia. Biochem Biophys Res Commun 470:748–752

Novotny-Diermayr V, Sangthongpitag K, Hu CY, Wu X, Sausgruber N, Yeo P, Greicius G, Pettersson S, Liang AL, Loh YK, Bonday Z, Goh KC, Hentze H, Hart S, Wang H, Ethirajulu K, Wood JM (2010) SB939, a novel potent and orally active histone deacetylase inhibitor with high tumor exposure and efficacy in mouse models of colorectal cancer. Mol Cancer Ther 9:642–652

Ota S, Zhou ZQ, Hurlin PJ (2018) Suppression of FGFR3- and MYC-dependent oncogenesis by tubacin: association with HDAC6-dependent and independent activities. Oncotarget 9:3172–3187

Ozawa A, Tanji N, Kikugawa T, Sasaki T, Yanagihara Y, Miura N, Yokoyama M (2010) Inhibition of bladder tumour growth by histone deacetylase inhibitor. BJU Int 105:1181–1186

Paillas S, Then CK, Kilgas S, Ruan J-L, Thompson J, Elliott A, Smart S, Kiltie AE (2020) The histone deacetylase inhibitor romidepsin spares normal tissues while acting as an effective radiosensitizer in bladder tumors in vivo. Int J Radiat Oncol*Biol*Phys 107:212–221

Pandey M, Kaur P, Shukla S, Abbas A, Fu P, Gupta S (2012) Plant flavone apigenin inhibits HDAC and remodels chromatin to induce growth arrest and apoptosis in human prostate cancer cells: in vitro and in vivo study. Mol Carcinog 51:952–962

Pant K, Yadav AK, Gupta P, Islam R, Saraya A, Venugopal SK (2017) Butyrate induces ROS-mediated apoptosis by modulating miR-22/SIRT-1 pathway in hepatic cancer cells. Redox Biol 12:340–349

Parizadeh SM, Jafarzadeh-Esfehani R, Ghandehari M, Seifi S, Parizadeh SMR, Moetamani-Ahmadi M, Hassanian SM, Khazaei M, Ghayour-Mobarhan M, Ferns GA, Avan A (2018) Epigenetic drug therapy in the treatment of colorectal cancer. Curr Pharm Des 24:2701–2709

Pellizzaro C, Coradini D, Daniotti A, Abolafio G, Daidone MG (2001) Modulation of cell cycle-related protein expression by sodium butyrate in human non-small cell lung cancer cell lines. Int J Cancer 91:654–657

Peulen O, Gonzalez A, Peixoto P, Turtoi A, Mottet D, Delvenne P, Castronovo V (2013) The anti-tumor effect of HDAC inhibition in a human pancreas cancer model is significantly improved by the simultaneous inhibition of cyclooxygenase 2. PLoS One 8:e75102

Pinkerneil M, Hoffmann MJ, Deenen R, Köhrer K, Arent T, Schulz WA, Niegisch G (2016) Inhibition of class I histone deacetylases 1 and 2 promotes urothelial carcinoma cell death by various mechanisms. Mol Cancer Ther 15:299–312

Platta CS, Greenblatt DY, Kunnimalaiyaan M, Chen H (2008) Valproic acid induces Notch1 signaling in small cell lung cancer cells. J Surg Res 148:31–37

Qian DZ, Wei YF, Wang X, Kato Y, Cheng L, Pili R (2007) Antitumor activity of the histone deacetylase inhibitor MS-275 in prostate cancer models. Prostate 67:1182–1193

Rana Z, Diermeier S, Hanif M, Rosengren RJ (2020) Understanding failure and improving treatment using HDAC inhibitors for prostate cancer. Biomedicine 8:22

Ray A, Alalem M, Ray BK (2013) Loss of epigenetic Kruppel-like factor 4 histone deacetylase (KLF-4-HDAC)-mediated transcriptional suppression is crucial in increasing vascular endothelial growth factor (VEGF) expression in breast cancer. J Biol Chem 288:27232–27242

Richon VM, Sandhoff TW, Rifkind RA, Marks PA (2000) Histone deacetylase inhibitor selectively induces p21WAF1 expression and gene-associated histone acetylation. Proc Natl Acad Sci U S A 97:10014–10019

Saha RN, Pahan K (2006) HATs and HDACs in neurodegeneration: a tale of disconcerted acetylation homeostasis. Cell Death Differ 13:539–550

Salimi V, Shahsavari Z, Safizadeh B, Hosseini A, Khademian N, Tavakoli-Yaraki M (2017) Sodium butyrate promotes apoptosis in breast cancer cells through reactive oxygen species (ROS) formation and mitochondrial impairment. Lipids Health Dis 16:208–208

Sanaei M, Kavoosi F (2019) Histone deacetylases and histone deacetylase inhibitors: molecular mechanisms of action in various cancers. Adv Biomed Res 8:63–63

Sarkar R, Chakraborty Mukherjee S (2015) Sulforaphane inhibits metastatic events in breast cancer cells through genetic and epigenetic regulation. J Carcinog Mutagen 06

Sharma S, Witteveen PO, Lolkema MP, Hess D, Gelderblom H, Hussain SA, Porro MG, Waldron E, Valera SZ, Mu S (2015) A phase I, open-label, multicenter study to evaluate the pharmacokinetics and safety of oral panobinostat in patients with advanced solid tumors and varying degrees of renal function. Cancer Chemother Pharmacol 75:87–95

Shi P, Yin T, Zhou F, Cui P, Gou S, Wang C (2014) Valproic acid sensitizes pancreatic cancer cells to natural killer cell-mediated lysis by upregulating MICA and MICB via the PI3K/Akt signaling pathway. BMC Cancer 14:370

Shi X-Y, Ding W, Li T-Q, Zhang Y-X, Zhao S-C (2017) Histone deacetylase (HDAC) inhibitor, suberoylanilide hydroxamic acid (SAHA), induces apoptosis in prostate cancer cell lines via the Akt/FOXO3a signaling pathway. Med Sci Monit 23:5793–5802

Siegel RL, Miller KD, Jemal A (2020) Cancer statistics, 2020. CA Cancer J Clin 70:7–30

Song Q, Li M, Fan C, Liu Y, Zheng L, Bao Y, Sun L, Yu C, Song Z, Sun Y, Wang G, Huang Y, Li Y (2019) A novel benzamine lead compound of histone deacetylase inhibitor ZINC24469384 can suppresses HepG2 cells proliferation by upregulating NR1H4. Sci Rep 9:2350

Soukupova J, Bertran E, Peñuelas-Haro I, Urdiroz-Urricelqui U, Borgman M, Kohlhof H, Fabregat I (2017) Resminostat induces changes in epithelial plasticity of hepatocellular carcinoma cells and sensitizes them to sorafenib-induced apoptosis. Oncotarget 8:110367–110379

Subramanian S, Bates SE, Wright JJ, Espinoza-Delgado I, Piekarz RL (2010) Clinical toxicities of histone deacetylase inhibitors. Pharmaceuticals (Basel) 3:2751–2767

Sun P-C, Tzao C, Chen B-H, Liu C-W, Jin J-S (2010) Suberoylanilide hydroxamic acid induces apoptosis and sub-G1 arrest of 320 HSR colon cancer cells. J Biomed Sci 17:76

Sun S, Han Y, Liu J, Fang Y, Tian Y, Zhou J, Ma D, Wu P (2014) Trichostatin A targets the mitochondrial respiratory chain, increasing mitochondrial reactive oxygen species production to trigger apoptosis in human breast cancer cells. PLoS One 9:e91610

Sun J-Y, Wang J-D, Wang X, Liu H-C, Zhang M-M, Liu Y-C, Zhang C-H, Su Y, Shen Y-Y, Guo Y-W, Shen A-J, Geng M-Y (2017) Marine-derived chromopeptide A, a novel class I HDAC inhibitor, suppresses human prostate cancer cell proliferation and migration. Acta Pharmacol Sin 38:551–560

Tang Y-A, Wen W-L, Chang J-W, Wei T-T, Tan Y-HC, Salunke S, Chen C-T, Chen C-S, Wang Y-C (2010) A novel histone deacetylase inhibitor exhibits antitumor activity via apoptosis induction, F-actin disruption and gene acetylation in lung cancer. PLoS One 5:e12417

Tate CR, Rhodes LV, Segar HC, Driver JL, Pounder FN, Burow ME, Collins-Burow BM (2012) Targeting triple-negative breast cancer cells with the histone deacetylase inhibitor panobinostat. Breast Cancer Res 14:R79–R79

Tee YT, Chen GD, Lin LY, Ko JL, Wang PH (2006) Nm23-H1: a metastasis-associated gene. Taiwan J Obstet Gynecol 45:107–113

Thomas DG, Ward AM, Williams JL (1971) A study of 52 cases of adenocarcinoma of the bladder. Br J Urol 43:4–15

Thomas M, Kelly ED, Abraham J, Kruse M (2019) Invasive lobular breast cancer: a review of pathogenesis, diagnosis, management, and future directions of early stage disease. Semin Oncol 46:121–132

Vallo S, Xi W, Hudak L, Juengel E, Tsaur I, Wiesner C, Haferkamp A, Blaheta RA (2011) HDAC inhibition delays cell cycle progression of human bladder cancer cells in vitro. Anti-Cancer Drugs 22:1002–1009

Wang G, He J, Taub JW, Guo Y, Ge Y (2012) Abstract 1830: both class I and class II histone deacetylases are required for proliferation and survival of human pancreatic cancer cells. Cancer Res 72:1830–1830

Wang HG, Huang XD, Shen P, Li LR, Xue HT, Ji GZ (2013) Anticancer effects of sodium butyrate on hepatocellular carcinoma cells in vitro. Int J Mol Med 31:967–974

Wang H, Huang C, Zhao L, Zhang H, Yang JM, Luo P, Zhan BX, Pan Q, Li J, Wang BL (2016) Histone deacetylase inhibitors regulate P-gp expression in colorectal cancer via transcriptional activation and mRNA stabilization. Oncotarget 7:49848–49858

Wang D-X, Zou Y-J, Zhuang X-B, Chen S-X, Lin Y, Li W-L, Lin J-J, Lin Z-Q (2017) Sulforaphane suppresses EMT and metastasis in human lung cancer through miR-616-5p-mediated GSK3β/β-catenin signaling pathways. Acta Pharmacol Sin 38:241–251

Wang F, Wu H, Fan M, Yu R, Zhang Y, Liu J, Zhou X, Cai Y, Huang S, Hu Z, Jin X (2020) Sodium butyrate inhibits migration and induces AMPK-mTOR pathway-dependent autophagy and ROS-mediated apoptosis via the miR-139-5p/Bmi-1 axis in human bladder cancer cells. FASEB J 34:4266–4282

Whitehead RP, Rankin C, Hoff PM, Gold PJ, Billingsley KG, Chapman RA, Wong L, Ward JH, Abbruzzese JL, Blanke CD (2009) Phase II trial of romidepsin (NSC-630176) in previously treated colorectal cancer patients with advanced disease: a Southwest Oncology Group Study (S0336). Investig New Drugs 27:469–475

Wilting RH, Dannenberg J-H (2012) Epigenetic mechanisms in tumorigenesis, tumor cell heterogeneity and drug resistance. Drug Resist Updat 15:21–38

Xia Q, Sung J, Chowdhury W, Chen CL, Hoti N, Shabbeer S, Carducci M, Rodriguez R (2006) Chronic administration of valproic acid inhibits prostate cancer cell growth in vitro and in vivo. Cancer Res 66:7237–7244

Xu Y, Zheng S, Chen B, Wen Y, Zhu S (2016) Sodium phenylbutyrate antagonizes prostate cancer through the induction of apoptosis and attenuation of cell viability and migration. Onco Targets Ther 9:2825–2833

Xu Z, Tao J, Chen P, Chen L, Sharma S, Wang G, Dong Q (2018) Sodium butyrate inhibits colorectal cancer cell migration by downregulating Bmi-1 through enhanced miR-200c expression. Mol Nutr Food Res 62:e1700844

Yamamoto H, Fujimoto J, Okamoto E, Furuyama J, Tamaoki T, Hashimoto-Tamaoki T (1998) Suppression of growth of hepatocellular carcinoma by sodium butyrate in vitro and in vivo. Int J Cancer 76:897–902

Yang H, Zhong Y, Xie H, Lai X, Xu M, Nie Y, Liu S, Wan YJ (2013) Induction of the liver cancer-down-regulated long noncoding RNA uc002mbe.2 mediates trichostatin-induced apoptosis of liver cancer cells. Biochem Pharmacol 85:1761–1769

Yaxley JP (2016) Urinary tract cancers: an overview for general practice. J Family Med Prim Care 5:533–538

Zappa C, Mousa SA (2016) Non-small cell lung cancer: current treatment and future advances. Transl Lung Cancer Res 5:288–300

Zhang J, Yi M, Zha L, Chen S, Li Z, Li C, Gong M, Deng H, Chu X, Chen J, Zhang Z, Mao L, Sun S (2016a) Sodium butyrate induces endoplasmic reticulum stress and autophagy in colorectal cells: implications for apoptosis. PLoS One 11:e0147218

Zhang Q, Sun M, Zhou S, Guo B (2016b) Class I HDAC inhibitor mocetinostat induces apoptosis by activation of miR-31 expression and suppression of E2F6. Cell Death Discov 2:16036

Zhang H, Liu L, Liu C, Pan J, Lu G, Zhou Z, Chen Z, Qian C (2017) Notch3 overexpression enhances progression and chemoresistance of urothelial carcinoma. Oncotarget 8:34362–34373

Zhao J, Gray SG, Wabitsch M, Greene CM, Lawless MW (2018) The therapeutic properties of resminostat for hepatocellular carcinoma. Onco Targets Ther 5:196–208

Zhou SF (2008) Structure, function and regulation of P-glycoprotein and its clinical relevance in drug disposition. Xenobiotica 38:802–832

Zhou H, Cai Y, Liu D, Li M, Sha Y, Zhang W, Wang K, Gong J, Tang N, Huang A, Xia J (2018) Pharmacological or transcriptional inhibition of both HDAC1 and 2 leads to cell cycle blockage and apoptosis via p21(Waf1/Cip1) and p19(INK4d) upregulation in hepatocellular carcinoma. Cell Prolif 51:e12447–e12447

10 Combining Histone Deacetylase Inhibitors with Other Anticancer Agents as a Novel Strategy for Circumventing Limited Therapeutic Efficacy and Mitigating Toxicity

Although histone deacetylase inhibitors (HDACi) are emerging as prosperous therapeutics for cancer therapy yet they exert only modest cytotoxic effect when used singly (singlet therapy). Further the single agent use of these inhibitors escalates the probability of cancer cells for eliciting resistance mechanisms which in reciprocation hampers their therapeutic efficacy. Moreover, in singlet therapy the higher doses of HDACi used also sensitize the typical body cells and as thus impart toxicity to them. This restricted therapeutic efficacy and concurrent toxicity has been circumvented by using a novel therapeutic strategy where HDACi are used in concert with other conventional therapeutics or targeted agents. This therapy namely combined therapy involves the two or three drugs in combination either simultaneously or sequentially. While therapy involving two drugs in combination is referred as doublet therapy, triplet combination is the term used when three drugs are given in combination.

10.1 Combination of HDACi and Platinum Coordination Complexes

The combined effects of vorinostat or sirtinol (inhibits sirtuins) and cisplatin along with the underlying mechanism have been studied using HeLa cells as models. Cisplatin-vorinostat or cisplatin-sirtinol combination synergistically reduced cell viability. Potentiation of cisplatin induced cytotoxic effects by HDAC inhibitors has been related to depleted levels of XIAP and Bcl-2. Moreover, HDAC inhibitor induced chromatin decondensation may promote the cisplatin access to DNA thereby augmenting cytotoxicity (Jin et al. 2010). Synergistic effect of cisplatin while in conjunction with vorinostat or another hydroxamate TSA has been observed in two cholangiocarcinoma cell lines (KKU-M214, KKU-100). Growth inhibition in KKU-100 cells was accompanied with Bcl-2 and CDK4 downregulation besides the upregulation of p21 and p53 by individual treatment or by combination. However, the intensity of upregulation or downregulation was comparatively more in cooperative treatment. In KKU-M214 cells, the drug combination enhanced the p21

S. A. Ganai, *Histone Deacetylase Inhibitors in Combinatorial Anticancer Therapy*,
https://doi.org/10.1007/978-981-15-8179-3_10

expression but unlike KKU-100 cells no effect was noted on expression of p53 (Asgar et al. 2016). Further ovarian cancer cells resistant to cisplatin got sensitized to oncolytic adenoviruses on treatment with pan-HDAC inhibitor TSA (Hulin-Curtis et al. 2018). Two novel HDAC inhibitors namely 13a and 13d inhibiting HDAC2/HDAC6 have been synthesized. Both these compounds especially 13d potentiated the cytotoxic effect induced by cisplatin. Among the two cell lines Cal27CisR and Cal27 the effect was more noticeable in the former line which is resistant to cisplatin (tenfold) than the latter Cal27. This enhanced effect in cancer cell cytotoxicity on combined exposure was accompanied with activation of caspase-3/7 (Asfaha et al. 2020). Cisplatin resistance often noted in oesophageal squamous cell carcinoma was reversed by Class I selective HDAC inhibitor entinostat not only under in vitro condition but also under in vivo set-up. This reversal was associated with cell proliferation inhibition, apoptosis induction and suppression of multidrug resistance gene 1/MDR1, cyclin D1, Mcl-1, P-Src and enhanced cleavage of PARP (Huang et al. 2018). Bladder cancer is another cancer where the issue of cisplatin resistance is often observed. Three urothelial bladder cancer lines including J82CisR were pre-incubated with entinostat for 48 h following which the cisplatin treatment was given. This sequential combination of entinostat and cisplatin reversed cisplatin resistance in J82CisR only partly (Wang et al. 2020).

HDAC inhibitor induced anticancer effects are mediated through modulation of histone and non-histone targets. These inhibitors have demonstrated efficacy when used in associative manner with platinum-based anticancer agents (Diyabalanage et al. 2013; Suraweera et al. 2018). A clinical study has been performed involving vorinostat in combinatorial manner with carboplatin and paclitaxel in advanced-stage patients of NSCLC. The outcome of this trial was that this HDAC inhibitor boosted the effectiveness of these conventional agents (carboplatin/paclitaxel) in the aforesaid patients (Ramalingam et al. 2010). Vorinostat plus paclitaxel/carboplatin treatment when tested in patients of NSCLC induced a response in 53% of them. The mechanism for this encouraging result has been studied using four NSCLC cells as models. In these cells different drug combinations including vorinostat/paclitaxel, vorinostat/carboplatin and vorinostat plus paclitaxel/carboplatin were used and the underneath mechanism was studied at finer level. Substantial growth inhibitory effect was seen both on combining vorinostat/carboplatin or vorinostat/paclitaxel. The triple combination involving vorinostat plus carboplatin and paclitaxel inhibited growth of 128-88T cells synergistically. While vorinostat strengthened the carboplatin induced DNA double strand breaks and raised acetylation status of α-tubulin in the above cell line and A549, vorinostat/paclitaxel combination synergistically elevated α-tubulin acetylation (Owonikoko et al. 2010).

HDAC inhibitor vorinostat has been tested along with oxaliplatin in three hepatocellular carcinoma cell lines (HepG2, BEL7402 and SMMC7721). Vorinostat/oxaliplatin combination in all the three cell lines showed synergistic antiproliferative effect and this was quite evident from the combination index values indicating synergism. This combination through induction of cell cycle arrest and apoptosis restrained proliferation under both in vitro and in vivo conditions. Further oxaliplatin-induced BRCA1 was put down by vorinostat treatment (Liao et al.

2018). The promising effects thus observed suggest the further clinical trial based studies for vorinostat/oxaliplatin combination. ACY-1215 (ricolinostat), the HDAC inhibitor interfering HDAC6 as shown by various studies is not suitable as a single agent for solid cancers. Thus its anticancer activity has been tested in combination with other traditional agents including oxaliplatin using colorectal cancer models such as HT29 and HCT116. Among the various combinations tested, ACY-1215/ oxaliplatin treatment proved to be more effective in causing apoptosis that the combination components separately. The combined approach resulted in caspase-3 activation and depleted Bcl-xL protein. Additionally, this combination upregulated the programmed death-ligand 1 in HCT116 cell model (Lee et al. 2018). Promising results have been obtained when valproic acid was used in a combined manner with other agents including oxaliplatin against a gastric cancer cell line CRL 1739. VPA and oxaliplatin treatment simultaneously or first VPA followed by oxaliplatin showed synergistic anticancer effect in gastric cancer cells. Pretreatment with HDAC inhibitor facilitates the binding of oxaliplatin or cisplatin to chromatin through enhanced chromatin decondensation (Amnekar et al. 2020).

10.2 HDACi in Cooperation with Taxanes

Paclitaxel resistant non-small cell lung cancer (NSCLC) cells exhibit the overexpression of HDAC1 which in turn brings down the levels of p21. SNOH-3, a novel HDAC inhibitor capable of obstructing HDACs has shown anticancer effect. Combining this inhibitor with paclitaxel exerted synergistic antiproliferative effect on paclitaxel resistant NSCLC cell line (A549/T). This antiproliferative effect thus observed was due to induction of apoptosis as enhanced cleavage of caspase 3 and PARP was recorded compared to single agent intervention. In A549/3 xenograft model the dual treatment inhibited tumour growth strongly in comparison to models provided only either HDAC inhibitor or paclitaxel. This treatment was associated with elevated apoptosis, angiogenesis suppression, enhanced H3 acetylation, and most importantly HDAC1 downregulation and p21 induction (Wang et al. 2016). Study performed on the sequential combination of romidepsin (macrocyclic HDAC inhibitor) and gemcitabine or docetaxel has demonstrated promising results in hormone refractory prostate cancer cells (HRPC DU145) in vitro and in xenograft models. Romidepsin pretreatment heightened the gemcitabine or docetaxel induced cytotoxic effects, the results being more favourable with romidepsin/gemcitabine combination. Apart from this prior treatment with romidepsin followed by gemcitabine administration prolonged the tumour doubling time in comparison to single agent therapy (Kanzaki et al. 2007).

In phase I study on castration-resistant prostate cancer patients combined therapy with panobinostat and docetaxel showed 50% or greater decrease in prostate specific antigen in 63% of the patients (Rathkopf et al. 2010). It has been proved in mesenchyme-like TNBC (triple negative breast cancer)-cells that co-delivery of paclitaxel and vorinostat by using micelle system enhances the cytotoxic effect by 5.91-fold in comparison to free combination of the defined agents (Kutty et al. 2015).

Combination of paclitaxel and valproic acid was not encouraging in anaplastic thyroid cancer patients as no impact on general survival was recorded in the patients (Catalano et al. 2016). Promising results were obtained when TSA and paclitaxel were given in combination to endometrial cells (KLE, Ark2) and xenograft (Ark2) models. Cooperation of these therapeutics synergistically reduced cell growth through drastic induction of apoptosis. Replacing TSA with another oxamflatin (another HDAC inhibitor) also yielded similar results. This dual treatment enhanced the α-tubulin acetylation and consequently the microtubule stabilization. Moreover, TSA/paclitaxel in xenograft model decreased tumour weight by 50% and these findings provide the rationale for further clinical trial studies with the defined combination (Dowdy et al. 2006). Improved growth inhibition was reported in four NSCLC cell lines including A549 and 128-88T on addition of vorinostat to paclitaxel. This effect proved to be synergistic in case of 128-88T line and further it was proved that vorinostat/paclitaxel treatment augments α-tubulin acetylation in A549 cells (Owonikoko et al. 2010). Further studies have been performed in mice bearing serous ovarian cancer (human) xenografts using panobinostat in combination with carboplatin plus paclitaxel (P/C). In one of the three patient derived xenografts the above combination exerted enhanced tumour regression as compared to panobinostat (Garrett et al. 2016). Paclitaxel induced cytotoxicity in urothelial carcinoma cells (BFTC-909, BFTC-905) was raised by TSA addition. Combination of these two increased the cleaved PARP besides the cleaved caspase-3 when compared to sole paclitaxel exposure. Synergistic reduction as indicated by the combination index below 1 was observed after dual treatment. Experimental evidences certified that enhanced phospho-ERK1/2 levels on singlet treatment of paclitaxel hamper its cytotoxicity. Thus the synergistic effect acquired following the combined intervention has been attributed to depleted phospho-ERK1/2 levels (Hsu et al. 2019).

10.3 Triazenes and HDACi in Union

Triazene compounds such as dacarbazine and temozolomide due to their better pharmacokinetic properties have great clinical importance. While temozolomide has significance in treating melanoma and primary brain tumours, dacarbazine is effective against Hodgkin disease and melanoma (Marchesi et al. 2007; Neyns et al. 2010; Quirbt et al. 2007). Evidences suggest that HDAC inhibitor RGFP-109 (inhibits HDAC1/2) circumvents temozolomide resistance in glioblastoma cell line resistant to this triazene by inhibiting the expression of prosurvival genes controlled by NF-κB. This makes it clear that HDAC inhibitor and temozolomide combination may prove promising for temozolomide-resistant glioblastoma patients (Li et al. 2016). Valproate has been studied in association with dacarbazine/interferon-α. Patients were subjected to single agent valproate for 6 weeks. After this 18 patients were given the combination of valproate/dacarbazine/interferon-α and 29 were given only valproate. While complete response was observed in one, partial response was seen in two patients of the group who received combination regime.

VPA/chemo-immunotherapy combination engendered no superior results compared to standard melanoma therapy. Further the toxicity induced by triplet combination was not minimal thereby raising concerns regarding the introduction of VPA in the defined setting (Rocca et al. 2009). On evaluating the VPA and temozolomide combination on two temozolomide-resistant glioblastoma cell lines (U138, T98), substantial anticancer effect was observed. This enhanced anticancer effect on VPA addition has been ascribed to lowered MGMT (O6-methylguanine-DNA methyltransferase) expression. MGMT has a crucial role in making the cells resistant to therapeutic effects of alkylating agents (Sarkaria et al. 2008). Additionally increased cell death observed was found to occur not only through apoptosis but also through autophagic mechanism. Most importantly, VPA/temozolomide co-treatment markedly reduced the tumour growth under in vivo condition in mice model as compared to isolated treatment groups (Ryu et al. 2012). The conclusion that can be derived from these evidence based findings is that temozolomide resistance encountered in malignant glioblastoma can be overcome by administering VPA/temozolomide combined therapy. In Glioblastoma patients (freshly diagnosed) vorinostat plus chemo (temozolomide)-radiation therapy combination showed allowable tolerability (Galanis et al. 2017). Another study on newly diagnosed patients of glioblastoma revealed that valproic acid incorporation to radiation therapy plus temozolomide was quite bearable (Krauze et al. 2015).

10.4 HDACi in Concert with Hydroxyurea

Mounting evidences suggest that HDACi trigger apoptosis modulate cell cycle and sensitize neoplastic cells to other therapeutic agents. It has been reported that VPA when used as co-agent with hydroxyurea (ribonucleotide reductase) strengthen the pro-apoptotic effects of each other in a variety of cancer cell lines. Hydroxyurea induces the degradation of p21 and p27 (cyclin-dependent kinase inhibitors) through proteasome or by caspase-3. Activation of this caspase has key role in VPA-induced apoptosis. The above specified CDKI act as apoptosis inhibitors by interacting with caspase-3 and subsequently competing with its other substrates (Krämer et al. 2008). VPA/hydroxyurea treatment inhibited cancer cell survival in a cooperative manner. This effect has been related to obstruction of HU-invoked homologous recombination by VPA by way of inhibiting the activity of replication protein A2, the protein critical for homologous recombination repair (Tian et al. 2017). Combination of valproic acid and hydroxyurea as expected proved to be promising in inducing cytotoxicity in head and neck squamous cell carcinoma (HNSCC) cell lines. Cell lines apart from freshly derived tumour cells were subjected to combined regime. This treatment effectually blocked proliferation and induced apoptosis in HNSCC cells. This anticancer effect has been correlated to enhanced expression of BIM (pro-apoptotic protein). While overexpression of BIM was accompanied with induction of apoptosis, its knockdown rescued HNSCC cells from VPA/HU stimulated apoptosis (Stauber et al. 2012).

10.5 Co-Treatment with HDACi and Camptothecin Analogues

Just like other conventional agents, two camptothecin analogues namely topotecan and irinotecan, approved by FDA, have been studied with respect to anticancer activity in combination with HDACi (Sriram et al. 2005). Topotecan, derivative of camptothecin and topoisomerase I inhibitor is used for treating small cell lung cancer (relapsed). Camptothecin (pentacyclic alkaloid) or its derivative topotecan has been investigated at molecular level in sequential or concomitant mode with vorinostat both against topoisomerase I-resistant (H526) and the sensitive NSCLC (H209) models. While individual treatment proved cytotoxic towards the both cell lines, H526 cell line was highly resistant to topotecan or camptothecin exposure. Potent synergistic effect was observed in both cell lines at equipotent (50:50) doses of vorinostat/camptothecin or vorinostat/topotecan. This synergistic effect observed on concomitant or sequential treatment was similar clearly indicating that schedule of the combination is immaterial for the defined effect. Both these combinations induced apoptosis in these NSCLC cells through escalated ROS generation (Bruzzese et al. 2009). HDACi like vorinostat or PXD101 are well known for obstructing proliferation and facilitating apoptosis in various cancer cell or tumour models (Suraweera et al. 2018). PXD101 as co-agent with irinotecan has been explored for anticancer effect in colon cancer. While SN38, the active metabolite of camptothecin analogue irinotecan, and PXD101, the HDAC inhibitor, as independent agents induced antiproliferative effect in colon cancer cell models (HT29, HCT116) dose dependently; the two agents in combination (PXD101 and SN38) demonstrated synergistic effect in these cell lines. This combination in comparison to single treatment modulated the expression of p21 and XIAP more effectively in HCT117 as compared to HT29. Moreover, this combination strikingly reduced the tumour growth in xenograft mice, like cell lines the effect being relatively more marked in HCT116 xenograft bearing mice, further strengthening the in vitro studies (Na et al. 2011).

HDAC inhibitor CG2 in combined form with other agents like SN38 or oxaliplatin or 5-fluorouracil has been evaluated against colon cancer cells (HCT116) and xenografts. Compared to other combinations, CG2/SN38 combination proved to be more promising. Although increased histone H3 acetylation and p21 expression was noted both in individually treated cells and combined ones, marked reduction in the levels of XIAP (antiapoptotic protein) was recorded in cells treated with CG2/SN38 compared to CG2 or SN38 alone. Interestingly, CG2/SN38 combination was more successful in inhibiting tumour growth in HCT116 xenografts (subcutaneous) in comparison to individual agents. While at third day CG2 and irinotecan independently afforded 10.4 and 11.8% tumour growth inhibition respectively, combined administration exhibited 39.9% inhibition. At day twenty-first (21) the defined inhibition for CG2 was 54.3%, for irinotecan 74.3% and for combination (CG2/irinotecan) 87.8% (Na et al. 2010). TSA, a pan-HDAC inhibitor potentiated the anticancer effects of various therapeutic including irinotecan in a pair of gastric cancer cell lines (MKN-74, OCUM-8). Combination of TSA and SN38 demonstrated synergistic antiproliferative effect in both these gastric cancer lines. TSA triggered synergistic effect might be due to increased protein levels of p21, p53 as well as DAPK1/2 (Zhang et al. 2006).

10.6 HDACi and Podophyllotoxin Analogues

HDACi have been tested in combination with other cytotoxic agents like podophyllotoxin analogues. Etoposide and teniposide, the two less toxic podophyllotoxin analogues, certified for clinical use, have been tested in combination with HDACi against a variety of cancers. NSCLC are associated with innate chemotherapy resistance. Etoposide co-treatment with TSA (pan-HDAC inhibitor) instigated apoptosis in NSCLC (drug-resistant) cells. This effect was not observed with structurally different inhibitor valproic acid. Mechanistically not only caspase-reliant but also caspase-independent apoptotic pathways participate in TSA/etoposide driven cytotoxicity of drug-resistant (NSCLC) cells. Combined exposure declined expression of Bcl-xL, activated Bax and aroused death pathway driven by apoptosis inducing factor (AIF). Further it was confirmed that TSA/etoposide provoked apoptosis in the above-mentioned resistant cells is mediated through induction of AIF as the death by predefined mechanism was fully rescinded by AIF knockdown (Hajji et al. 2008). Current therapies against medulloblastoma are not highly effective, thus targeting epigenetic regulators is tried for promising therapeutic effect. HDAC inhibitor and the antiepilepsy drug valproic acid have shown synergistic effects when used in two medulloblastoma cell lines D283 and Daoy. Although the effect of this HDAC inhibitor was limited as an isolated agent its combination with etoposide markedly enhanced the anticancer activity. While in presence of VPA, enhanced chromatin decompaction was found, the combination of the two led to increase in the DNA double strand breaks. VPA intervention increased the mRNA levels of 53BP1, acetylated p53 in addition to mRNA levels of p21 only in Daoy cells suggesting that cytotoxic effect in D283 may operate via p53 independent mechanism (Gopalakrishnan et al. 2008).

Triplet combination involving vorinostat/etoposide/cisplatin was studied using NSCLC (H146 and H209) cell models. The triplet combination in comparison to singlet therapy with vorinostat or doublet therapy with etoposide/cisplatin proved to be immensely effective in engendering cytotoxic effect. Adjusting the concentrations the most effective concentrations of the combination proved to be vorinostat (0.4 μM), cisplatin (0.2 μM) and etoposide (0.6 μM). While inhibition of cell viability was found to be 49.19% in H146, 37.74% inhibition was measured in H209 cells. Compared to vorinostat or etoposide/cisplatin, the increased reduction in viability in cells treated with triple combination has been allocated to relatively elevated PARP cleavage (Pan et al. 2016). Thus the triplet combination shows differential effect in lessening viability in NSCLC models. Moreover, valproic acid as a combined agent with DNA damaging agents like etoposide or cisplatin has proved to be promising against high-risk neuroblastoma. Using UKF-NB-4 cells as models of defined neuroblastoma, synergistic anticancer effect was manifested on VPA/cisplatin or VPA/etoposide combination. Unlike VPA, this effect was not reproduced by another inhibitor of HDACs, TSA and also by valpromide, the VPA derivative having no HDAC obstructing activity. This synergistic effect of VPA was restricted only to DNA damaging agents as no such effect was observed while using VPA with mechanistically distinct anticancer agent vincristine. VPA-enhanced cytotoxic effect of this cell line was seen to be dependent on the schedule of the drug treatment as the increased benefit was recorded only on either simultaneous combination or by first using DNA damaging agents followed by VPA. The synergistic effect has been related to enhanced caspase-3-driven apoptotic signalling (Hrabeta et al. 2014). Various drug combinations involving tubacin in combination with etoposide or SAHA or doxorubicin were evaluated against two pancreatic cancer (MCF-7, LNCaP) and one normal (HFS) cell lines. Tubacin and etoposide combination substantially increased cell death in LNCaP cells. This effect was also seen when this HDAC inhibitor was used in combinatorial way with doxorubicin or SAHA. MCF-7 cells were also sensitized by tubacin to etoposide invoked cell death. Importantly, no effect on drug treatment was seen in normal cell line. Nil-tubacin lacking HDAC6 inhibitory activity proved futile in reproducing tubacin-like effects suggesting that inhibition of HDAC6 activity is crucial mediating the effect. HDAC6 knockdown cells showed high sensitivity to etoposide, vorinostat or doxorubicin intervention (Namdar et al. 2010). This along with above experiment clearly indicates that either pharmacological inhibition of HDAC6 or genetically reduced levels of HDAC6 are sufficient for sensitizing pancreatic cancer cells to above cytotoxic agents (Namdar et al. 2010). Tubacin and etoposide combination enhanced the cleaved levels of PARP, the cell death proved to be partially dependent on caspase activation as pan-caspase inhibitor treatment prior to addition of tubacin/etoposide failed to fully rescue the cell death. Downstream analysis further revealed that tubacin potentiates etoposide-induced DNA damage and this was evident from the increased γH2AX and checkpoint kinase (Chk2) activation (Namdar et al. 2010). Speaking concisely, tubacin potentiates the anticancer effects of etoposide by multiple ways including through HDAC6 inhibition, enhanced PARP cleavage, and by augmenting DNA double strand breaks.

10.7 HDACi and Vinca Alkaloids

The detailed account of these alkaloids has been provided in Chap. 6 and here these alkaloids will be discussed as combinatorial agents of HDACi. Vincristine administration is accompanied with peripheral neuropathies termed as vincristine-induced peripheral neuropathies (Mora et al. 2016). Co-treatment effect of vincristine and HDAC6 specific inhibitor ACY-738 or tubastatin A has been studied in acute lymphoblastic leukaemia mouse model. Although vincristine demonstrated high efficacy as single agent in this model, animals given co-treatment of ACY-738 or another HDAC inhibitor tubastatin A exhibited almost full repression of cancer advancement. Importantly, the co-treated animals showed total absence of vincristine-driven neurotoxicity compared to animals treated with vincristine only (Van Helleputte et al. 2018). Taken together HDAC6 specific inhibitors not only potentiate the anticancer effects of vincristine but also bridle its associated peripheral neuropathies. It has been reported that diffuse large B cell lymphoma (DBCL) lines sensitive to HDACi, prior to programmed cell death undergo early arrest during mitosis. The DBCL resistant cell lines on the other hand complete the process of mitotic process despite the brief hold-up and arrest in the interphase G1. Combining vincristine or paclitaxel in low doses with HDAC inhibitor belinostat-induced cytotoxic effect in HDACi-resistant cell models synergistically. This greater than additive cytotoxicity has been linked to MCL-1 downregulation, BIM upregulation and effectual elimination of polyploid cells. From these results it can be inferred that vincristine elevates the sensitivity of DLBCL cells to belinostat-driven cytotoxicity while in reciprocation belinostat restrains vincristine-induced polyploidy (cause of resistance to vincristine) (Havas et al. 2016). Mitigating drug resistance through combinatorial approach has revolutionized the anticancer therapy. Although vincristine is globally used for treating acute leukaemia, the toxicity and chemoresistance issues raise serious concerns. A study has been performed where vincristine has been used in association with first approved pan-HDAC inhibitor vorinostat, the aim being whether the combination soothes toxicity and overcomes vincristine resistance. This drug combination heightened the cytotoxicity in human leukaemia (MOLT-4) cells as compared to the treatment where these agents were used alone. Enhanced cytotoxic effect has been credited to strong induction of G2/M arrest and caspase activation (Chao et al. 2015).

10.8 Anticancer Antibiotics and HDACi in Combination

HDACi have demonstrated prosperous effects when used in combination with antibiotics which may be chemically glycopeptides and aromatic polyketides/anthracyclines. Several HDACi including OSU-HDAC42, TSA, entinostat and SAHA have been reported to sensitize prostate cancer cells to DNA damaging agents such as bleomycin, etoposide and doxorubicin. This effect was more pronounced when HDACi and the DNA damaging agents were used sequentially. In other words prior treatment with HDACi followed by bleomycin or doxorubicin has proved to be more effective. This enhanced cytotoxic effect in prostate cancer cells on combination has been attributed to enhanced acetylation of KU70 which weakens its DNA binding inclination and in the long run DNA double strand breaks induced by these antitumour antibiotics remain unrepaired (Chen et al. 2007). ITF2357, a pan-HDAC inhibitor along with doxorubicin (anthracycline antibiotic) induced cytotoxicity in patient derived and established sarcoma cell lines (U2OS, SW872, HT1080) in a synergistic manner. Combination treatment was highly effective in reducing tumour volume in SW982 xenograft model. Contrarily no substantial reduction in this volume was reported on individual administration of doxorubicin or ITF2357 (Di Martile et al. 2018). These encouraging results in in vivo model support the clinical level studies of the aforementioned combination for fostering its use in clinic. Quisinostat, a strong inhibitor targeting Class I as well as Class II HDACs has been found to raise the doxorubicin-evoked cytotoxic effects not only in breast cancer stem cells but also in non-stem breast cancer models including HCC38, MCF-7, and MDA-MB-231 (Hii et al. 2020). This suggests that combining HDACi with conventional agents may prove as a promising approach for refractory breast cancer.

Sequential combination of valproic acid and doxorubicin, the latter inhibiting the resealing step of topoisomerase II, has been evaluated against immensely lethal anaplastic thyroid carcinoma. With two cell models of anaplastic thyroid carcinoma (ARO and CAL-62) the combination introduced above was checked. Pretreatment with valproic acid followed by doxorubicin addition enhanced the cytotoxic effect of latter (doxorubicin) by about twofold and threefold in ARO and CAL-62 cell lines, respectively. This increased effect on valproic acid/doxorubicin combination has been credited to elevated histone acetylation, enhanced caspase-3 activation and apoptosis and escalated doxorubicin-triggered G2 arrest (Catalano et al. 2006). Extensive study has been performed on rat model of acute myeloid leukaemia (AML) and in BNML model representing acute myelocytic leukaemia. HDAC inhibitor tacedinaline has been investigated in these models in combination with traditional anticancer agents. Three combinations of tacedinaline namely tacedinaline/cytarabine, tacedinaline/daunorubicin and tacedinaline/mitoxantrone were tested in the above models and in BCLO cells in vitro. While modest synergism was noted with all the above three combinations of tacedinaline in BCLO cells, tacedinaline/cytarabine proved to be the most effective against BNML. The remaining two combinations were also effective but to a lesser degree than tacedinaline/cytarabine combination (Hubeek et al. 2008). The crux taken from this study suggests that tacedinaline/cytarabine combination being highly active should be tested in clinical trials on patients of acute myelocytic leukaemia. As single entities, romidepsin and liposomal doxorubicin are only moderately successful against peripheral T-cell lymphoma (PTCL) and cutaneous T-cell lymphoma (CTCL). These two agents along with each other were tested against CTCL as well as in primary CTCL cells. Besides the phase I study of this combination has been evaluated in patients of relapsed/refractory lymphoma (CTCL and PTCL). Enhanced cytotoxicity was found in the cell models after intervention with this combination. This combination demonstrated 70% overall response rate, admissible safety profile as well as encouraging clinical efficacy (Vu et al. 2020). Phase I study, in patients with solid malignancies, has been performed using the valproic acid in association with epirubicin. This study revealed that valproic acid is quite tolerable. While 22% of the patients manifested partial responses, stable disease was observed in 39% of the patients. Notably, patients bearing resistant tumours (anthracycline-resistant) and excessively pre-treated ones also showed response to this combination (Münster et al. 2007). Another combination study involving valproic acid and other drugs including epirubicin, 5-fluorouracil and other has been performed. This study proved that this combination is not only safe but also feasible and tolerable. Objective responses were observed in 64% of the breast cancer patients of the dose expansion group (Munster et al. 2009a). Further phase I study involving vorinostat and conventional therapeutic doxorubicin has been done on patients with metastatic or advanced tumours (solid). The combination offered moderate clinical benefit and emphasized that HDAC2 selective inhibitors may offer greater degree of inhibition and safety (Munster et al. 2009b).

10.9 HDACi and Antimetabolites in Combinatorial Manner

As discussed in the Chap. 6 antimetabolites is an umbrella term and thus includes not only folate analogues but also purine and pyrimidine analogues. HDACi in association with these conventional anticancer agents have offered enhanced therapeutic benefit. Methotrexate, a folate analogue has been tested in combination with a variety of HDACi and it has been found that the final results are highly dependent on the sequence of treatment followed. In acute lymphoblastic leukaemia cell lines (NALM6, REH) methotrexate treatment for some hours followed by vorinostat proved to be synergistic in inducing cytotoxic effects (Leclerc et al. 2010). When cells were first treated with vorinostat or valproic acid and then with methotrexate or vorinostat plus methotrexate simultaneously or valproic acid and methotrexate concomitantly the effect produced was robustly antagonistic (Bastian et al. 2011). Cladribine, the purine analogue (synthetic) due to its anticancer properties has been used along with certain HDACi like entinostat (Leist and Weissert 2011). Individual effects of cladribine and entinostat were examined on various multiple myeloma cell lines including MM1.R and RPMI8226. Besides, the combination of these two drugs was also tested on the defined models. Although as single agents these inhibitors dose dependently restrained the proliferation of multiple myeloma cells, the combined chemotherapy manifested the synergistic antiproliferative and anti-survival effect. This dual treatment elicited mitotic catastrophe, upregulated p21, and intensely triggered apoptosis in addition to DNA damage response (Wang et al. 2018a). These findings support the further extensive evaluations of cladribine/entinostat combination towards its promising clinical use. Relapsed, refractory as well as poor-risk lymphoma patients from many years were treated by high dose chemotherapy (HDC) and subsequent ASCT (autologous stem cell transplantation). A novel combination involving cladribine plus gemcitabine and busulfan (CGB) has been found to be robust in inducing cytotoxicity to lymphoma lines synergistically (Ji et al. 2016). The efficacy of this combination got further augmented on addition of vorinostat or benzamide derivative group HDAC inhibitor that is chidamide (ChiCGB) (Ji et al. 2016). Phase II clinical trial results suggest that this combination can serve as effective HDC prior to ASCT for relapsed/refractory lymphoma patients or those with high-risk lymphomas (Ji et al. 2017).

Nucleoside analogue fludarabine in combination with other therapeutics serves as treatment for B-chronic lymphocytic leukaemia (B-CLL). Combined therapeutic intervention with fludarabine and valproic acid (HDAC inhibitor) in B-CLL cells synergistically triggered apoptosis. The apoptosis thus induced was proved to be caspase-dependent and executed through extrinsic pathway. Moreover, this combination triggered ROS production apart from overexpression of Fas and bax (Bouzar et al. 2009). Further studies in CLL cells (primary ones) proved the downregulation of Mcl-1 and XIAP (anti-apoptosis proteins) by this combination. The synergistic apoptotic death following the administration of fludarabine/vorinostat has been accredited to enhanced levels of cathepsin B, the protease from lysosomes (Yoon et al. 2013). Sequential and simultaneous treatment of entinostat and fludarabine evoked synergistic cytotoxic effects in leukaemia cells (myeloid and lymphoid) including CCRF-CEM and U937. These effects were comparatively more profound in case of sequential combination. Enhanced therapeutic effect of this combination was accompanied by increase in production of ROS, antiapoptotic proteins (XIAP/Mcl-1) downregulation, release of mitochondria localized pro-apoptotic proteins to cytosol and finally apoptosis. Pretreatment of leukaemia cells with entinostat restrained the activation of MEK1/2 and AKT, the key players hampering the fludarabine induced cytotoxicity (Maggio et al. 2004). Valproic acid-fludarabine combination has been explored in multiple cell lines including BJAB and I-83. This cell death was found to be mediated by elevated ROS production and this death was suppressed by ROS scavenger incorporation. Fludarabine-driven cytotoxic effect was found to have cross-talk with AKT or ATM (ataxia-telangiectasia mutated) as inhibition of either of them facilitated the apoptosis induced by this purine analogue. Analysis of four samples taken from patients having relapsed CLL and who were on valproic acid treatment for 30 days showed depleted ATM levels in 75% cases and AKT reduction in 25%. This all suggest that valproic acid by way of diminishing the levels of ATM (key kinase in DNA damage response) escalates ROS-reliant cytotoxic effect when used in conjunction with fludarabine (Yoon et al. 2014).

Like purine analogues, the enhanced anticancer effect has been achieved by using pyrimidine analogues in combination with HDACi. Heightened anticancer effect was seen when valproic acid was used in unity with 5-fluorouracil, the pyrimidine analogue. This effect was recorded both in cholangiocarcinoma (HuCCT1) cell line as well as in pancreatic cancer cell model (SUIT-2). While the dual combination (1 μM 5-fluorouracil and 5 mM VPA) exerted 19% inhibitory effect in SUIT-2, 30% decline in viability by the defined concentration was quantified in cholangiocarcinoma line (Iwahashi et al. 2011a). Overexpression of thymidylate synthase in cancer cells governs 5-fluorouracil resistance. Induction of this enzyme occurs following the single agent treatment with 5-fluorouracil and thus the addition of those therapeutic agents that may revert the 5-flurouracil-induced expression of thymidylate synthase may prove fruitful in cancer therapy. TSA, a pan-HDAC inhibitor has been certified to lower the expression of thymidylate synthase at both message and protein level. TSA intervention through promotion of Hsp70 and thymidylate synthase binding leads to the proteasomal degradation of latter. Most importantly, the low dose TSA potentiated the 5-fluorouracil provoked cytotoxicity in 5-fluorouracil resistance cancer cells. This heightened effect of TSA combination in the resistant cells has been allocated to the lowering of protein levels of the defined synthase (Lee et al. 2006).

In cell lines of colon cancer (HT29/SW48/HCT-116) 5-fluorouracil elicited anticancer effect was potentiated by depsipeptide (HDAC inhibitor). While as single agent this pyrimidine analogue was not able to trigger caspase-3/7 activation, the combination proved to be effective enough for activating these caspases. Genes related to apoptotic process and cell death were upregulated by the dual combination as evident from the caspase activation and colony formation inhibition. Robust induction in the MHC class II gene expression was also seen after the combination treatment (Okada et al. 2016). Fimepinostat (CUDC-907) which inhibits HDACs and phosphatidylinositide 3-kinase (dual inhibitor) have been inspected in multiple colorectal cancer cells including HT-29, RKO and HCT1116. This study proved the increased cytotoxicity of CUDC-907-5-flurouracil against all these models. Moreover, HCT116 colony forming ability was also impeded by the predefined combination. Among the cell death mechanisms apoptosis and necrosis were evident and the CUDC-907 treated cells exhibited elevated acetylation of histone H3 lysine 9 besides the decline in AKT phosphorylation (Hamam et al. 2017).

HDAC inhibitors were also tested in combined form against breast cancer cells resistant (MDA468/FU) or sensitive (MDA-MB-468) to 5-fluorouracil intervention. The resistant cells differed from the sensitive line in many aspects including the thymidylate synthetase transcriptional levels which are higher in the former. Sequential addition of valproic acid or SAHA to these cell lines followed by 5-fluorouracil raised the sensitivity of these models to latter (Minegaki et al. 2018). From this finding it is clear that pretreatment with structurally distinct inhibitors of HDACs sensitizes breast cancer cells to 5-fluorouracil irrespective of their resistant/sensitive status. Vorinostat in Cal27 cells (cisplatin resistant) in combination with 5-fluorouracil plus cisplatin greatly enhanced (synergistically) their antiproliferative and pro-apoptotic efficacy. This effect has also been demonstrated not only in orthotopic but also in heterotopic Cal27 xenograft mouse models. Vorinostat addition through inhibition of 5-flurouracil-cisplatin driven EGFR-nuclear translocation, and by upregulating the channel mediating cisplatin influx namely copper transporter 1 (CTR1) strengthened the effects of 5-flurouracil plus cisplatin combination (Piro et al. 2019). Another HDAC inhibitor belonging to benzamide derivatives group (tacedinaline or CI-994) has been explored in union with 5-fluorouracil against several neuroendocrine tumour (NET) cells. Tacedinaline individually and dose dependently inhibited the growth of H727, BON1 and QGP1 (NET cells). Combined strategy involving 5-fluorouracil-tacedinaline declined cell survival synergistically in all the named NET cells (Jin et al. 2019).

Gemcitabine, the predominantly used drug for pancreatic cancer has proved more effective when used in concert with other therapeutics. HDAC inhibitor (CG200745) addition to gemcitabine-erlotinib combination substantially enhanced the anticancer effect induced by this combination in BxPC3 pancreatic cancer cells. Enhancement of anticancer effect on CG200745 incorporation has been assigned to caspase-3 activation. Triplet combination involving CG200745-gemcitabine-erlotinib showed drastic reduction (50%) in tumour volume in xenograft (BxPC3) model (Lee et al. 2017). Combination of two HDACi mocetinostat (inhibits HDACs 1–3) and LMK-235 (inhibits HDACs 4–6) with gemcitabine induced cytotoxic effects in a series of pancreatic cancer lines some of them being T3M-4, MiaPaCa-2 and Capan-1. Majority of the cell lines on triplet combination exhibited synergistic cell death. Compared to other combinations, mocetinostat/LMK-235/gemcitabine showed more pronounced PARP cleavage. The cause of this enhanced cell death on the above-mentioned combination has been accredited to inhibition of two Class I

(HDAC1, 2) and one Class IIb HDAC by name HDAC6 (Laschanzky et al. 2019). These findings suggest that multiple HDACs may be involved in cell survival mechanisms and thus targeting the implicated HDACs only may yield better therapeutic outcome.

Another study showed the heightened antiproliferative effect in a pair of pancreatic cancer cells on addition of valproic acid to combination of gemcitabine plus pegylated interferon-α2b. This effect was more inclined towards SUIT-2 cells in comparison to BxPC3 as 88% and 67% suppression in proliferation was respectively observed following exposure to above combination (Iwahashi et al. 2011b). From these findings it is highly evident that the combined treatment sensitizes different pancreatic cancer cell lines to different extent. Impact of HDAC inhibitor combination with the antimetabolite gemcitabine has been tested on the pancreatic ductal adenocarcinoma cells (MiaPaCa-2 and PANC-1). While the individual use of gemcitabine slightly impeded their tumour sphere forming tendency, its combination with vorinostat or TSA potentiated the mentioned effect. Reduction in the number of tumour spheres was more profound in case of gemcitabine-TSA combination. TSA potentiated gemcitabine effects may be ascribed to downregulation of stem cell markers (Nanog, Oct-4, SOX2) and HDAC1, 7–8 (Cai et al. 2018). Beneficial effects of HDAC inhibitor and gemcitabine have also been evaluated in a couple of leiomyosarcoma (LMS1 and SKLMS1)-cells. Mocetinostat and gemcitabine unitedly demonstrated synergistic anti-leiomyosarcoma effects not only under in vitro set-up but also under in vivo conditions. While the markers indicating gemcitabine resistance were downregulated, the hallmarks of gemcitabine-sensitivity were lowered on mocetinostat treatment (Lopez et al. 2017). The promising effects of this combination justify the further clinical trial studies of this combination for promoting its clinical use.

Using paediatric AML (THP-1) cells as models, it has been revealed that Class I selective HDAC inhibitor entinostat as well as pan-HDAC inhibitor vorinostat individually (entinostat plus cytarabine or vorinostat along with cytarabine) heighten the cytarabine-driven apoptosis. Enhanced apoptotic effect noted on HDAC inhibitor cytarabine co-treatment has been ascribed to concomitant inhibition of HDAC1 and HDAC6 (Xu et al. 2011). Conclusively, it is obvious that raised cytotoxic effect following the above combination is through modulation of epigenetic mechanism. While cytotoxic antagonism was observed on concurrent use of vorinostat and cytarabine in leukaemia cell (K562, HL60) lines, sequential addition of vorinostat first, then cytarabine after a drug free interval between the two induced higher than additive (synergistic) cytotoxic effect. Vorinostat intervention depleted the quantity of cells in S-phase by triggering arrest at G1-phase (Shiozawa et al. 2009). This indicates that mode of combination of two therapeutic agents greatly impacts the end results of treatment. In AML or MDS patients the safety and efficacy of vorinostat in union with two other therapeutics idarubicin and cytarabine have been evaluated. As an induction therapy patients were orally given 500 mg vorinostat thrice a single day for 3 days. From the fourth day idarubin 12 mg/m^2 was administered daily for 3 days intravenously (4th to 6th day) and cytarabine 1.5 g/m^2 (also from 4th day) was given as a continuous infusion (intravenously) for maximum 4 days (day 4th to 7th).

Promising results were obtained as indicated by the all-inclusive response rate of 85%. Complete response was recorded in 76% cases. On the whole, combination of these three agents proved to be active and safe in AML (Garcia-Manero et al. 2012).

10.10 HDACi and Radiation Therapy

Radiation therapy like surgery and classical chemotherapy comes under conventional anticancer therapies. Radiation therapy through induction of breaks in genetic material of cancer cells invokes death signalling in these cells (Pinar et al. 2010; Wang et al. 2018b). Experiments on two prostate cancer cell lines (LNCaP, PC-3,) and one typical prostate epithelial cell line (RWPE-1) have been done using histone deacetylase inhibitors panobinostat. This inhibitor singly restrained the proliferation of all the three lines dose dependently. The normal cell line was showed less sensitivity compared to atypical cells. Panobinostat-induced apoptosis in prostate cancer models was associated with hyperacetylation of histone H3/H4 and enhanced caspase-3 cleavage. Panobinostat plus radiation therapy exposed cells showed heightened apoptosis compared to prostate cancer cells subjected to radiation therapy only. Combined exposure was accompanied with prolongation of DNA double strand breaks as γH2AX, the hallmarks of these breaks persisted even after 72 h of exposure (Xiao et al. 2013). Radiosensitizing ability of vorinostat was tested on three melanoma cell lines including A549 and MeWo by taking the advantage of clonogenic assays. While ionizing radiation and vorinostat separately failed to cause apoptosis in melanoma models, combination of the two resulted in elevated apoptosis indicating that vorinostat exposure powers radiation-driven cancer cell death. The combined treatment was not provided concurrently but sequentially by pre-treating the melanoma cells with vorinostat for 24 h followed by radiation. Mechanistically, vorinostat suppressed strongly the non-homologous end joining, one of the predominant DNA double strand break correcting pathway in A549 cells (Munshi et al. 2006). Using vorinostat and ionizing radiations in a scheduled way has great impact on melanoma cells apoptosis.

Two NSCLC lines (H460, H23) underwent apoptosis synergistically on sequential therapy of panobinostat and ionizing radiation. When radiation exposure was given to these cells alone, γ-H2AX foci sustained as far as 6 h. Prior treatment with this inhibitor followed by radiation exposure prolonged these foci over 24 h clearly suggesting that panobinostat somehow hampers repair of double strand breaks in DNA. Further studies proved that radiation therapy alone induces translocation of HDAC4 from cytoplasm to nucleus where it augments DNA damage repair thereby lowering cell death (Geng et al. 2006). Panobinostat pretreatment confined this HDAC to cytoplasm or in other words prevented its nuclear translocation and as such the DNA double strand breaks induced by subsequent radiation treatment are not repaired thereby driving the cells to apoptosis (Geng et al. 2006). The favourable effect of vorinostat on radiation induced death of rhabdomyosarcoma (A-204, RD) and osteosarcoma cell lines (SAOS2, KHOS-24OS) has been elucidated. Proliferation of all cell lines was inhibited by vorinostat treatment. While this HDAC inhibitor markedly enhanced the photon radiation treatment-induced apoptosis in osteosarcoma lines, the effect was not statistically significant in rhabdomyosarcoma lines. DNA repair protein inhibition (Rad51/Ku80) due to vorinostat exposure may be responsible for this effect (Blattmann et al. 2010). Moreover, radio-sensitization effects of HDAC inhibitor PCI-24781 were evaluated on two paediatric glioblastoma cell lines (KNS42, SF188). This inhibitor on individual use reduced the proliferation of these lines by inducing apoptosis and colony formation. Although PCI-24781 sensitized both cell models to ionizing radiation, the effect was more marked in KNS42 line (de Andrade et al. 2016). Radiation treatment alone was not able to reduce colony formation in SF188 line; the combined treatment induced 45% reduction. While in KNS42 only 20% reduction in colony formation was recorded, combination of both demonstrated 70% reduction. The expression of proteins involved in homologous recombination (Rasd51) or non-homologous end joining repair pathway (Ku86, Ku70, DNA-PKcs) was relatively more attenuated compared to radiotherapy alone (Table 10.1) (de Andrade et al. 2016). Increased antiproliferative effect on combined use of HDAC inhibitor and radiotherapy may thus be correlated to inhibition of DNA damage repair potential. Also in xenograft model of bladder cancer, panobinostat, the approved HDAC inhibitor prove to be very effectual in inducing radiosensitivity (Groselj et al. 2018).

10.11 Using HDACi in Association with Proteasome Inhibitors

Agglomeration of ubiquitinated proteins induces cell stress which in turn triggers cytotoxic effects in multiple myeloma cells. It is well accepted that ubiquitinated proteins (misfolded or unfolded) are degraded both by proteasomes and aggresomes, the latter being reliant on the activity of HDAC6 (Kawaguchi et al. 2003; Kopito 2000). Due to this fact it was presumed that preventing the degradation of ubiquitinated proteins through inhibition of both mechanisms may offer enhanced therapeutic benefit. A study has been performed where bortezomib, a proteasome inhibitor has been used in combination with tubacin, a HDAC6-specific inhibitor (Haggarty et al. 2003; Kane et al. 2003). Following the combined treatment statistically substantial amassing of polyubiquitinated proteins occurred in the experimental multiple myeloma cell models (RPMI8226 and MM.1S). As expected bortezomib increased bortezomib triggered cytotoxic effects in the defined cell lines. While bortezomib intervention at concentration of 5 μM induced 26% death in RPMI8226 cells, 66% death was recorded at 10 μM concentration of this inhibitor. Percentage of cell death increased to 87% and 91%, respectively, on combined treatment with tubacin (5 μM). Thus bortezomib and tubacin combination synergistically induced apoptotic cell death as evidenced from the combination index below 1. Further it was observed that bortezomib driven $p21^{Cip1}$ induction was supressed by tubacin treatment (Hideshima et al. 2005). Panobinostat in combination with bortezomib and dexamethasone has been approved to treat multiple myeloma patients who have undergone at least two initial standard therapies with proteasome inhibitor

Table 10.1 Summary of various molecular players influenced by histone deacetylase inhibitors in combination with other anticancer agents for inducing cytotoxic signalling in various cancer models

Combination used	Underlying molecular players modulated		Literature evidence
	Activated/ upregulated	Inhibited/ downregulated	
Vorinostat + cisplatin		XIAP. Bcl-2	Jin et al. (2010)
Vorinostat or TSA + Cisplatin	p21, p53	Bcl-2, CDK4	Asgar et al. (2016)
13a or 13d + cisplatin	Caspase-3/7		Asfaha et al. (2020)
Entinostat + cisplatin		MDR1, Mcl-1, P-Src cyclin D1	Huang et al. (2018)
Vorinostat + paclitaxel + carboplatin	α-Tubulin acetylation		Owonikoko et al. (2010)
Vorinostat + oxaliplatin		BRCA1	Liao et al. (2018)
ACY-1215 + oxaliplatin	Caspase-3 activation	Bcl-xL	Lee et al. (2018)
SNOH-3 + paclitaxel	H3 acetylation, p21	HDAC1	Wang et al. (2016)
TSA + paclitaxel	α-Tubulin acetylation		Dowdy et al. (2006)
Vorinostat + paclitaxel	α-Tubulin acetylation		Owonikoko et al. (2010)
TSA + paclitaxel		Phospho-ERK1/2	Hsu et al. (2019)
VPA + temozolomide		MGMT	Ryu et al. (2012)
VPA + hydroxyurea		Replication protein A2	Tian et al. (2017)
	BIM		Stauber et al. (2012)
Vorinostat + topotecan	ROS		Bruzzese et al. (2009)
CG2 + SN38 or 5-fluorouracil		XIAP	Na et al. (2010)
TSA + SN38	p21, p53, DAPK1/2		Zhang et al. (2006)
TSA + etoposide	Bax, AIF	Bcl-xL	Hajji et al. (2008)
VPA + etoposide	53BP1, acetylated p53		Gopalakrishnan et al. (2008)
VPA + CISPLATIN OR ETOPOSIDE	Caspase-3		Hrabeta et al. (2014)
Tubacin + etoposide or doxorubicin	γH2AX, Chk2	HDAC6	Namdar et al. (2010)
Belinostat + vincristine or paclitaxel	BIM	MCL-1	Havas et al. (2016)

(continued)

Table 10.1 (continued)

Combination used	Underlying molecular players modulated		Literature evidence
	Activated/ upregulated	Inhibited/ downregulated	
Valproic acid + doxorubicin	Caspase-3		Catalano et al. (2006)
Entinostat + cladribine	p21		Wang et al. (2018a)
Valproic acid + fludarabine	Fas, Bax		Bouzar et al. (2009)
	Cathepsin B	Mcl-1, XIAP	Yoon et al. (2013)
Entinostat + fludarabine		XIAP, mcl-1, MEK1/2, AKT	Maggio et al. (2004)
Valproic acid + fludarabine	ATM		Yoon et al. (2014)
TSA + 5-flurouracil		Thymidylate synthase	Lee et al. (2006)
Depsipeptide + 5-flurouracil	MHC class II		Okada et al. (2016)
Fimepinostat + 5-flurouracil	H3K9ac	AKT phosphorylation	Hamam et al. (2017)
Vorinostat + 5-flurouracil	CTR1		Piro et al. (2019)
CG200745 + gemcitabine + erlotinib	Caspase-3		Lee et al. (2017)
Mocetinostat + LMK-235 + gemcitabine		HDAC1, 2, 6	Laschanzky et al. (2019)
TSA + gemcitabine		Oct-4, SOX2, Nanog	Cai et al. (2018)
Entinostat + cytarabine		HDAC1, HDAC6	Xu et al. (2011)
Vorinostat + ionizing radiation	γH2AX		Xiao et al. (2013)
		Non-homologous end joining	Munshi et al. (2006)
Panobinostat + ionizing radiation	γ-H2AX	HDAC4 translocation to nucleus	Geng et al. (2006)
Vorinostat + XRT		Rad51/Ku80	Blattmann et al. (2010)
PCI-24781 + radiation		Ku86, Ku70	de Andrade et al. (2016)

bortezomib and with immunomodulatory drug for instance lenalidomide (Eleutherakis-Papaiakovou et al. 2020; Ganai 2016; San-Miguel et al. 2014).

Another proteasome inhibitor namely carfilzomib in cooperation with vorinostat or entinostat has been tested on mantle cell lymphoma (MCL) cells not only under in vitro conditions but also using in vivo system (Dasmahapatra et al. 2011; McBride et al. 2015). Synergistic cell death was observed in several MCL lines (HF-4B, REC-1, MINO and JVM-2) on co-administration. Carfilzomib/vorinostat combination markedly enhanced caspase activation, cleavage of PARP, DNA damage and activation of JNK. Importantly the defined regimen proved to be effective in inducing apoptosis in MCL cells resistant to bortezomib treatment. Most notably this combination suppressed MCL growth in xenograft model thereby providing a rationale for higher order clinical studies with the carfilzomib/vorinostat combination (Dasmahapatra et al. 2011). L-carnitine has been proved to be endogenous inhibitor of HDACs (Huang et al. 2012a). This has been examined in association with bortezomib in liver cancer (hepatoma) cells and in mice models with induced HepG2 tumour. L-carnitine/bortezomib combination invoked cytotoxic effect in cultured cells and also reduced tumour growth synergistically. This effect has been linked to synergistic elevation of $p21^{cip1}$, histone acetylation and potentiation of bortezomib-mediated proteasome inhibition by L-carnitine (Huang et al. 2012b). Proteasome inhibitor in conjunction with HDAC inhibitor-approach has also been investigated in synovial carcinoma. SS18-SSX, a fusion oncoprotein that drives synovial sarcoma is not addressed by conventional therapies thereby offering only moderate benefit in circumventing this cancer. This synovial sarcoma driving complex is disrupted by HDAC inhibitor Quisinostat which in turn reinstates the expression of tumour suppressor genes namely *CDKN2A and EGR1*. Quisinostat in combination with bortezomib reduced the viability of six synovial carcinoma cell lines. The defined combination showed synergistic antiproliferative effect in all the tested cancer cell models. Proteasome inhibition and subsequent heaping of misfolded proteins facilitate the aggresome formation (Johnston et al. 1998). The hallmark of aggresome formation is LC3B protein whose levels are depleted by obstruction of HDAC6. Quisinostat plus bortezomib treatment-induced synergistic effect is due to prevention of aggresome formation by the former by way of suppressing HDA6 activity (Laporte et al. 2017). Combination of these mechanistically distinct therapeutic agents highly enhanced endoplasmic reticulum stress, ROS, BIM and BIK (pro-apoptotic proteins) and Bcl-2 phosphorylation (Laporte et al. 2017). It is well established that phosphorylation of Bcl-2 is inversely related to its antiapoptotic function (Tamura et al. 2004).

10.12 Heat Shock Protein 90 Inhibitors and HDACi

Activity of HSP90 is crucial for survival and proliferation of tumour cells. Inhibition of this molecular chaperone not only attenuates tumour growth but also restrains inflammation and metastasis (Costa et al. 2020). In response to HDAC inhibitor treatment cell survival mechanisms reliant on HSP90 activity limit the cytotoxic potential of these inhibitors. This clearly indicates that using the inhibitors of HSP90 and HDACi in combinatorial mode may offer enhanced therapeutic benefits. Entinostat (HDAC inhibitor) and HSP90 inhibitor (17-AAG) have been in combination against two preclinical models (SYO1, Fuji) of synovial sarcoma. This combination proved to be synergistic in inducing cytotoxicity in these cells.

17-AAG inhibited entinostat-induced NF-κB activation thereby boosting its therapeutic effect. Entinostat effect was found to be mediated by inhibition of HDAC3 as its knockdown reproduced the entinostat driven effects. The involvement of NF-κB in the synergism was also validated by another experiment where its inhibitor BAY 11-7085 along with entinostat demonstrated synergistic effects (Nguyen et al. 2009). Vorinostat in combination with 17-NN-dimethyl ethylenediamine geldanamycin (17-DMAG) induced apoptosis both in cultured (MO2058, MCL Z138, JeKo-1) and primary mantle cell lymphoma cells synergistically. On elucidating the underlying mechanism it has been proved that the enhanced effect is due to significantly depleted levels of CDK4, c-Myc, cyclin D1, AKT as well as c-RAF (Rao et al. 2009). The promising results noted in in vitro study favour the examination of this combination in in vivo models.

Promising results have been obtained when the combination of panobinostat and 17-AAG was tested on acute leukaemia (MV4–11) and K562, the chronic myeloid leukaemia blast crisis (CML-BC) cells. Panobinostat/17-AAG, in both these leukaemia models induced apoptosis that proved to be synergistic on analysing the combination index. Leukaemia cells refractory to imatinib mesylate, expressing mutant Bcr-Abl also succumbed to combined therapy. While the joint treatment depleted the mutant Bcr-Abl in K562 cells, the expression of FLT-3 was attenuated in MV4–11. Moreover, this combination compared to single treatments diminished the protein expression of phosphorylated-STAT5 and p-AKT in both the models (George et al. 2005). The prosperous results observed with this combination provide justification for the further in vivo and patient based studies. Co-treatment with tubacin, a selective inhibitor of HDAC6, and 17-AAG resulted in cooperative reduction of K562 cell survival. Among the HSP90 client proteins HDAC6 is prominent and hyperacetylation of former intensifies not only the anti-HSP90 effects of 17-AAG but also its leukaemia-lessening effects (Rao et al. 2008). NVP-AUY922, a novel HSP90 inhibitor has been tested lonely as well as jointly with PXD101 (HDAC inhibitor) against the CAL62 and 8505C, the model cell lines for representing anaplastic thyroid carcinoma. NVP-AUY922 compared to other inhibitors demonstrated more potency in inducing death in these cell lines. Simultaneous treatment with both anticancer agents induced cytotoxicity in these cells which proved to be synergistic on further analysis. Further finer studies proved that concomitant treated was accompanied with declined expression of both total AKT and p-AKT. Although total ERK1/2 levels were not altered by this combination, depleted level of pERK1/2 was quite profound. Moreover combined approach manifested depleted levels of survivin, enhanced PARP cleavage and DNA double strand breaks as evident from enhanced γ-H2AX foci in these cells (Kim et al. 2015). HSP90 inhibitor 17-AAG was further tested in combination with belinostat which targets tubulin deacetylase HDAC6. 17AAG/belinostat combined treatment in triple negative breast cancer (MDA-MB-231) cells showed synergistic effect in inducing apoptosis. This combination was chosen as the expression profile of HDAC6 and HSP90 is high in triple negative breast cancer cells. Further, more suppression was observed in the migration and invasion of this TNBC cell line. While the belinostat-induced acetylation was supplemented by 17-AAG, belinostat in reciprocation potentiated the 17-AAG mediated acetylation of α-tubulin (Zuo et al. 2020). Speaking in combined form, HDACi and HSP90 inhibitors highly cooperate with each other to remove the therapeutic limitations when these agents are used as singlet therapy.

10.13 HDACi and mTOR Inhibitors in Combination Mode

Anomalous mammalian target of rapamycin (mTOR) signalling has involvement in various cancers (Pópulo et al. 2012). This serine/threonine kinase forms mTORC1 and mTORC2, the two discrete multiprotein complexes. These complexes show differential sensitivity and it has been proved that mTORC1 and mTORC2 are sensitive to nutrients and PI3K/growth factor signalling, respectively (Jhanwar-Uniyal et al. 2019; Watanabe et al. 2011). MLN0128, the second generation, ATP-competitive dual mTOR inhibitor, along with panobinostat has been studied in various B-ALL cell models (Graham et al. 2018). While in murine pre-B cell model (p190 cells), this inhibitor induced cell death, human Philadelphia chromosome positive (Ph+) and Ph− cells were resistant to certain level to MLN0128 triggered cytotoxicity. Combining this inhibitor with vorinostat synergistically increased cytotoxic effect against SUP-B15 and Nalm-6 which represent Ph+ and Ph− lines, respectively (Beagle et al. 2015). The cytotoxic effect induced by this combination was found to be due to apoptosis as the enhanced PARP cleavage was noted in these two cell lines and furthermore concentration dependent inhibition of apoptosis was seen on administration of pan-caspase inhibitor. The synergistic effect of the above combination has been linked to transcriptional downregulation of FOXO target genes including those encoding death receptor-4 (DR4), TRAIL, Fas ligand (Beagle et al. 2015).

Rapamycin also termed as sirolimus, relatively specific for mTOR1, on prolonged exposure targets mTOR2 as well (Sarbassov et al. 2006). Combination of this inhibitor with entinostat has shown promising effects in a series of multiple myeloma cell lines with genetic distinction. Greater than additive effects were observed with this combination in various cell lines including Burkitt's lymphoma, multiple myeloma and mantle cell lymphoma cell lines. Synergistic effect was also found in murine plasmacytoma cells. Association of these inhibitors increased cell cycle arrest and programmed cell death or apoptosis. Mechanism-wise the combination of the two downregulated the various molecular players such as antiapoptotic BCL-XL, BIRC5 and cyclin D. Besides, MAPK inhibition was sharper in cells subjected to dual inhibition in comparison to individual use of entinostat or rapamycin (Simmons et al. 2014). Only some years back a dual inhibitor of PI3K/mTOR namely BEZ235 along with TSA (pan-HDAC inhibitor) has been investigated against six breast cancer cell lines including MCF-7 and MB-453. Synergistic cytotoxic effect on treatment with the aforesaid combination was engendered in three cell lines (MCF-7, T47D and MDA-MB-231) only among the six tested lines. Additionally this combination-induced cytotoxicity occurred both by apoptosis and autophagy. More importantly, BEZ235 plus TSA substantially reduced tumour growth in breast cancer (MDA-MB-231) xenograft model compared to single agent treatment. Further the combined treatment induced no discernible toxicity in these models which further supports the clinical applicability of this combination (Chen et al. 2017).

10.14 Collaboration of HDACi and Growth Factor Receptor/ Growth Factor Inhibitors

7 are associated with increased expression of epidermal growth factor receptor (EGFR) and in regulating its expression HDACs play a critical role. Erlotinib, the inhibitor of EGFR tyrosine kinase has been investigated with various HDACi for various cancers including glioblastomas either sensitive or resistant to erlotinib. HDACi markedly declined the proliferation of glioblastoma cells resistant to erlotinib and partly reinstated their erlotinib-sensitivity. Besides combining erlotinib with HDACi prevented the resistance development (Liffers et al. 2016). Combination of vorinostat and erlotinib used on NSCLC patients bearing EGFR mutations seemed to be well tolerable (Reguart et al. 2009).

Another (EGFR)-tyrosine kinase inhibitor in union with vorinostat has been extensively studied in hepatocarcinoma and lung adenocarcinoma cells not only under conditions of in vitro but also by employing in vivo condition. By obstructing the insulin like growth factor-1 (IGF-1R)/protein kinase B (PKB)-driven signalling route gefitinib/vorinostat combination potentially induced apoptosis in these cells. Moreover strong growth inhibitory effects of this combination were seen in mice bearing hepatocarcinoma or lung adenocarcinoma xenografts. However, no induction of apoptosis was demonstrated when sorafenib (multikinase inhibitor) was coupled with vorinostat (Jeannot et al. 2016; Wilhelm et al. 2008). Speaking in few words vorinostat plus gemcitabine induce synergistic cytotoxic effect in lung adenocarcinoma as well as in hepatocarcinoma cells and replacing gemcitabine with sorafenib offers no therapeutic advantage. EGFR has implications in the head and neck cancer development and as such is emerging as a candidate target for this malady. The limited efficacy of EGFR inhibitors is circumvented by using them in combination with HDACi. Vorinostat in various tumours without the exception of head and neck reverses the epithelial to mesenchymal transition. The collaborative role of vorinostat and gefitinib has been explored in human papilloma virus positive and also in human papilloma virus negative head and neck cancer cells. As individual agents vorinostat or gefitinib showed antitumour effect in HPV-negative and HPV-negative cell lines but in association these inhibitors exhibited synergism in suppressing cell growth. Vorinostat through downregulation of ΔNp63α (transcription factor) lowers the proteins levels of EGFR, declines cell proliferation and TGFβ-mediated migration of head and neck cancer cell lines (Citro et al. 2019).

EGFR inhibitors require further strengthening for improved therapeutic index in order to be more successful for tackling head and neck squamous cell carcinoma. Vorinostat collaboratively with gefitinib demonstrated synergism in restraining proliferation and apoptosis induction in squamous cell carcinoma of head and neck models. This was the case even with cells that were resistant to gefitinib. In CAL27 cells vorinostat downregulated the expression of various receptors including ErbB3, EGFR and ErbB2 and attenuated their signalling. Gefitinib-resistant (KB and Hep-2) cells associated with very low ErbB3 expression and that have gone through metastasis regressed to mesenchymal phenotype by vorinostat exposure. This effect has been ascribed to vorinostat induced enhanced expression of ErbB3, E-cadherin and reduction in the expression profile of EGFR/ErbB2 (Bruzzese et al. 2011).

NSCLC bearing KRAS mutation does not show a good response to treatment with EGFR inhibitors and thus therapeutic approaches attenuating KRAS signalling may prove beneficial. HDAC inhibitor panobinostat, the FDA approved therapeutic for multiple myeloma has also shown good results against NSCLC. Taking this rationale into consideration, this inhibitor was checked for its therapeutic effects in NSCLC cells (H460, H441, A549) harbouring KRAS mutation. Panobinostat intervention circumvented gemcitabine resistance in these cells. Combination of this HDAC inhibitor with gefitinib synergistically diminished tumour growth under in vivo conditions as well (Jeannot et al. 2016). From mechanistic perspective panobinostat inhibited the transcriptional co-activator with PDZ-binding motif (TAZ) transcription and synergistic effect on its downregulation was noted on coupling this inhibitor with gefitinib. Panobinostat-induced TAZ (oncogene) inhibition sensitized NSCLC cell lines (KRAS mutant but EGFR canonical) to gefitinib by way of negating AKT/mTOR signalling (Jeannot et al. 2016; Zhou et al. 2011). Thus panobinostat/gefitinib coupling proved to be fruitful against NSCLC with mutated KRAS and this was observed even under in vivo set-up. These findings justify the further studies of this combination on patient models towards its clinical benefit.

It is well obeyed that tumour angiogenesis has key role in the development and advancement of renal cell carcinoma. Although the HDACi possess antiangiogenesis activity to certain extent, this activity gets potentiated when these inhibitors are given collectively with vascular endothelial growth factor (VEGF) inhibitors (Deng et al. 2020). Vorinostat and bevacizumab combination was evaluated in patients harbouring metastatic conventional renal cell carcinoma and were previously treated. Among the 36 patients selected only 33 were evaluable and among the six objective responses observed five were partial while one was complete. In 48% patient's progression-free survival (6 months) was recorded. This combination was comparatively well tolerated and clinical activity was observed with this combination as evident from the response rate besides the survival (progression-free) (Pili et al. 2017).

10.15 HDACi and DNMT Inhibitors in Combination

Decitabine, a DNMT inhibitor tested along with HDACi panobinostat or mocetinostat exerted synergistic anticancer effect in majority of small cell lung cancer (SCLC) cells. However, this synergy has been attributed to modulation of non-epigenetic mechanisms as this effect neither correlated with inhibition of DNMT1 nor with HDACs. Mocetinostat and decitabine co-treatment elevated

DNA damage and this was quite evident from the enhanced phospho-H2A.X levels (Luszczek et al. 2010). Panobinostat and 5-azacitidine (DNA dimethylating agent) combination proved to be safe and was tolerated by patients having myelodysplastic syndrome or chronic myelomonocytic leukaemia (Kobayashi et al. 2018).

A variety of HDACi including belinostat, romidepsin, scriptaid and panobinostat were tested in diffuse large B-cell lymphoma (DLBCL) models. Finally panobinostat in conjunction with decitabine was also studied in these cancer models. The combination proved to be synergistic in suppression of cell growth and in inducing cytotoxic effect. While panobinostat as single agent invoked apoptotic death in 9.95% of the OCI-Ly1 cells, decitabine singly proved to be more effective than panobinostat and induced apoptosis in 39.5% of defined cells. The combination of the two proved to be highly effective as 61.4% of the cells succumbed to apoptosis. The combined therapy specifically altered the expression profile of von Hippel Lindau (VHL), WT1 (encodes Wilms tumour protein), in addition to transcription elongation factor B polypeptide 1-encoding gene TCEB1 and DIRAS3 which encodes GTP-binding protein Di-Ras3 (Kalac et al. 2011).

Combined treatment involving vorinostat and decitabine has also been evaluated on a pair of acute myeloid leukaemia cell lines (HL-60 and OCI-AML3). This combination proved to be more effective in HL-60 line and synergistic decline in proliferation was noted. Further this strategy boosted apoptosis and increased the acetylation status of histone proteins. Comparatively lesser sensitivity in OCI-AML3 following the doublet therapy was ascribed to upregulation of AXL, a gene encoding tyrosine-protein kinase receptor UFO. For circumventing AXL-elicited resistance triplet combination therapy involving the above-mentioned anticancer agents with AXL inhibitor BGB324 was tested. This strategy proved to be effective in sensitizing OCI-AML3 cells to decitabine-vorinostat treatment (Young et al. 2017). Efficacy of combinatorial treatment involving decitabine and mocetinostat has been explored on several chondrosarcoma cell models (SW1353, CH2879 and JJ012). Chondrosarcomas (bone malignancy type) although being resistant to radiotherapy and conventional chemotherapy, respond better to agents targeting epigenetic route. Mocetinostat and decitabine reduced the viability of all the above-mentioned chondrosarcoma lines and the combination of these two agents proves to be additive. This treatment was accompanied by enhanced DNA damage, increased Bim levels, enhanced acetylation status of core histones H3 and H4 and induction of E-cadherin (Sheikh et al. 2018).

In AML pathogenesis epigenetic aberrations contribute significantly and in its preclinical models pharmacological intervention of epigenetic modification regulators have offered a ray of hope. Study involving HDAC1/HDAC2 selective inhibitors namely ACY-957 or ACY-1035, in combination with azacitidine, in two cell (MOLM-13, MV-4-11) models showed substantial increase in apoptosis as compared to single agent exposure. In 73.7% patient samples (primary AML) ACY-957 and azacytidine demonstrated synergistic antiproliferative effect as evident from the combination index less than 1. Importantly, the defined combination proved to be quite effective in restraining tumour growth in AML xenograft (MOLM-13 AML) mice models. Certain genes with tumour suppressor role including CDKN1A, CDKN1C were found to be relatively more upregulated by co-treatment as compared to individual treatments. The expression profile of two transcription factor encoding genes (GATA2 and HES1) followed similar trend on the combined approach (Min et al. 2017). Another study certified the effectiveness of combined therapy of entinostat and azacytidine in oesophageal cancer cell models. These cell lines included a pair of oesophageal adenocarcinoma cells (SK-GT-4, OE33) and three oesophageal squamous cell carcinoma lines including Kyse-410, OE21 and Kyse-270. This approach resulted in selective cytotoxicity of above-mentioned cancer lines through DNA damage induction, reduced viability and programmed cell death/apoptosis. The cytotoxic effects induced by this combination as evident from transcriptome analyses were driven by downregulation of MLKL and FAIM besides the induction of Hes2 and BCL6 (Ahrens et al. 2015). In three AML cell lines including one with canonical p53 (MV4–11), another with mutated p53 (AML-193) and third with no discernible p53 messenger RNA (THP-1) the cytotoxic effects of inhibitors of HDACs and DNMTs in union have been studied. From the experimental outcome it was quite evident that azacytidine as single therapeutic decreases viability of MV4–11 line modestly with no such effect on other two lines. Reduction in viability proved to be synergistic when azacytidine and romidepsin/panobinostat were used jointly against MV4–11 and THP-1 cells. Cell viability depletion on the combined use of defined inhibitors was found to be reliant on p53 as AML-193 cell line (mutated p53) showed nearly no sensitivity to the treatment (Gopalakrishnapillai et al. 2013).

Decitabine and vorinostat combination when tested on a couple of ovarian cancer cell lines (SKOv3, Hey) induced synergistic growth inhibitory effects in both of them. While the antiproliferative effect of this combination in SKOv3 cells has been linked to induction of apoptosis and cell cycle arrest, the Hey cells unlike the former undergo autophagy as well (Chen et al. 2011). PEG3 (paternally expressed 3) and ARHI are key players in the pathogenesis of ovarian cancer. While 88% of the ovarian cancers are associated with downregulation of the latter, 75% show PEG3 downregulation (Feng et al. 2008). Growth inhibition of ovarian cancer cells on co-treatment inversely correlated with the upregulation of the transcription levels of PEG3 and ARHI (Chen et al. 2011). DNMT inhibitor aza-deoxycytidine exerts great impact on cell cycle regulation whereas TSA strongly induces apoptosis in endometrioid cancer cell lines. The combined procedure resulted in synergistic effect in the endometrial cancer cells (Xu et al. 2014). Pracinostat, a hydroxamate group HDAC inhibitor in conjunction with azacytidine has shown synergistic anticancer activity. Phase II study involving these drugs in combination in AML patients (minimum age 65) proved to be promising. This combination was finely tolerated and completed remission was recorded in 52% of the patients (Garcia-Manero et al. 2019). Selective and potent inhibitor of HDAC6, nexturastat A in association with 5-Azacytidine has been tested against ovarian tumour. Together these agents lessened tumour load and enhanced the immune signalling unfavourable for tumour (Moufarrij et al. 2020).

10.16 Histone Methyltransferase Inhibitors Plus HDACi in Antineoplastic Therapy

Like histone deacetylase inhibitors, histone methyltransferases are also involved in regulating gene expression programs epigenetically. These enzymes may be components of a single complex and may function in concert in regulating the transcription of various genes. As an example, histone methyltransferase EZH2 (enhancer of zeste homolog 2) in association with HDACs promotes gene silencing by enhancing histone H3 lysine 27 trimethylation (H3K79me3) (Ganai et al. 2015). Overexpression of this methyltransferase has implications in several cancer types one of them being triple negative breast cancer. From the above explanation it is quite rational to speculate that EZH2 inhibitors may prove highly effective when used together with inhibitors of HDACs. Co-treatment effects of GSK126 (EZH2 inhibitor) and panobinostat (HDAC inhibitor) have been studied on two cell lines (MDA-MB-436, MDA-MB-231) of triple negative breast cancer. The combination proved to be additive in inducing the apoptosis in these lines. This enhanced apoptotic effect has been linked to enhanced expression of BIM at both levels (mRNA and protein). While GSK126 increased the H3K27 acetylation status in the promoter of BIM, panobinostat enhanced the defined mark at the two BIM enhancers (Huang and Ling 2017).

Another study explored the combined effects of 3-Deazaneplanocin A also known as DZNep (indirect EZH2 inhibitor) and panobinostat (HDAC inhibitor) in AML cell models (HL60, OCI-AML3) HL-60 leukaemia mice model (Fiskus et al. 2009; Girard et al. 2014). Individual treatment with DZNep induced apoptosis in both AML lines but to a varying extent. Further this indirect inhibitor depleted protein expression of EZH2 both in primary and cultured AML cells. Moreover, exposure to DZNep resulted in 40% reduction in H3K27me3 in cultured AML lines which in turn induced the expression of various genes regulating cell cycle including p21, p16, p27 and cell death modulator FBXO32. DZNep and panobinostat combination synergistically induced apoptosis in AML cells and this joint treatment was associated with relatively more declines in EZH2 levels and relatively raised induction of above cell cycle modulating genes and FBXO32 (Table 10.2). This effect was also observed when panobinostat in the above combination was substituted by entinostat. HL-60 implanted mice models demonstrated prolonged survival on co-treatment DZNep/panobinostat (Fiskus et al. 2009). In nutshell these findings strongly favour the higher order studies of this combination to promote its clinical use in the near future. Study involving inhibition of EZH2 and HDACs has also been performed in lymphoma cell lines. GSK126 and romidepsin combination showed strong synergy in these lines. The cooperative therapeutic effect has been credited to PRC2 complex disruption (Lue et al. 2019).

Up to now I have shed light on the combined use of HDACi along with other therapeutic agents. First I have explained the benefits of combining HDACi with platinum coordination complexes following which these inhibitors were also discussed in concert with taxanes, triazenes, hydroxyurea, camptothecin analogues, podophyllotoxin analogues, vinca alkaloids, anticancer antibiotics and

Table 10.2 Distinct molecular targets modulated by combined therapy involving histone deacetylase inhibitors as one of the cytotoxic agents

Drug combination used	Cellular players upregulated	Downstream players inhibited	Literature evidence
Vorinostat + carfilzomib	Cleaved PARP, JNK		Dasmahapatra et al. (2011)
L-carnitine + bortezomib	$p21^{cip1}$	HDAC6	Huang et al. (2012b)
Quisinostat + bortezomib	ROS, BIK, BIM		Laporte et al. (2017)
Entinostat + 17-AAG		HDAC3	Nguyen et al. (2009)
Vorinostat + 17-DMAG		CDK4, cyclin D1, AKT	Rao et al. (2009)
Panobinostat + 17-AAG		Mutant Bcr-Abl, FLT-3	George et al. (2005)
PXD101 + NVP-AUY922		AKT, p-AKT, pERK1/2, survivin,	Kim et al. (2015)
Belinostat + 17-AAG	α-Tubulin		Zuo et al. (2020)
Vorinostat + MLN0128		DR4, TRAIL	Beagle et al. (2015)
Entinostat + rapamycin		Cyclin D, BCL-XL, BIRC5, MAPK	Simmons et al. (2014)
Vorinostat + gefitinib		ΔNp63α, EGFR,	Citro et al. (2019)
Vorinostat + gefitinib	ErbB3, E-cadherin	EGFR/ErbB2	Bruzzese et al. (2011)
Panobinostat + gefitinib		TAZ, AKT/mTOR	Jeannot et al. (2016)
Mocetinostat + decitabine	Bim, E-cadherin		Sheikh et al. (2018)
ACY-957 + azacytidine	CDKN1A, CDKN1C		Min et al. (2017)
Entinostat + azacytidine	*Hes2, BCL6*	*MLKL, FAIM*	Ahrens et al. (2015)
Panobinostat + GSK126	BIM,H3K27 acetylation		Huang and Ling (2017)
Panobinostat + DZNep	FBXO32	EZH2	Fiskus et al. (2009)

antimetabolites. Further the benefits of sequential combination of HDACi and radiation therapy were rigorously explained. Apart from this, promising benefits of HDACi when used in collaboration with proteasome inhibitors or HSP90 inhibitors were thoroughly taken into account. Moreover, the combinatorial advantages of HDACi in association with mTOR, growth factor receptor and growth factor inhibitors were potentially elaborated. Lastly, the therapeutic advantages of collective use of HDACi with other epi-drugs such as DNMT and histone methyltransferase inhibitors were also described. Combination of HDACi with the above agents was taken into consideration not only at preclinical level but clinical trial based studies wherever available were also included. Speaking briefly, through combinatorial approach, enhanced therapeutic benefit is achieved from HDACi that too even using low dose combinations. Cooperative use of these inhibitors overcomes chemo-radioresistance issues encountered while treating different cancer models. In addition to this while using HDACi in combination with other therapeutics logic and sequence of treatment should not be ignored. While synergistic cytotoxic effect is achieved both by concurrent and sequential combinations in some cases, in others sequential mode may prove only effective as is the case with HDAC inhibitor/radiation therapy.

References

Ahrens TD, Timme S, Hoeppner J, Ostendorp J, Hembach S, Follo M, Hopt UT, Werner M, Busch H, Boerries M, Lassmann S (2015) Selective inhibition of esophageal cancer cells by combination of HDAC inhibitors and Azacytidine. Epigenetics 10:431–445

Amnekar RV, Khan SA, Rashid M, Khade B, Thorat R, Gera P, Shrikhande SV, Smoot DT, Ashktorab H, Gupta S (2020) Histone deacetylase inhibitor pre-treatment enhances the efficacy of DNA-interacting chemotherapeutic drugs in gastric cancer. World J Gastroenterol 26:598–613

Asfaha Y, Schrenk C, Alves Avelar LA, Lange F, Wang C, Bandolik JJ, Hamacher A, Kassack MU, Kurz T (2020) Novel alkoxyamide-based histone deacetylase inhibitors reverse cisplatin resistance in chemoresistant cancer cells. Bioorg Med Chem 28:115108

Asgar MA, Senawong G, Sripa B, Senawong T (2016) Synergistic anticancer effects of cisplatin and histone deacetylase inhibitors (SAHA and TSA) on cholangiocarcinoma cell lines. Int J Oncol 48:409–420

Bastian L, Einsiedel HG, Henze G, Seeger K, Shalapour S (2011) The sequence of application of methotrexate and histone deacetylase inhibitors determines either a synergistic or an antagonistic response in childhood acute lymphoblastic leukemia cells. Leukemia 25:359–361

Beagle BR, Nguyen DM, Mallya S, Tang SS, Lu M, Zeng Z, Konopleva M, Vo T-T, Fruman DA (2015) mTOR kinase inhibitors synergize with histone deacetylase inhibitors to kill B-cell acute lymphoblastic leukemia cells. Oncotarget 6:2088–2100

Blattmann C, Oertel S, Ehemann V, Thiemann M, Huber PE, Bischof M, Witt O, Deubzer HE, Kulozik AE, Debus J, Weber KJ (2010) Enhancement of radiation response in osteosarcoma and rhabdomyosarcoma cell lines by histone deacetylase inhibition. Int J Radiat Oncol Biol Phys 78:237–245

Bouzar AB, Boxus M, Defoiche J, Berchem G, Macallan D, Pettengell R, Willis F, Burny A, Lagneaux L, Bron D, Chatelain B, Chatelain C, Willems L (2009) Valproate synergizes with purine nucleoside analogues to induce apoptosis of B-chronic lymphocytic leukaemia cells. Br J Haematol 144:41–52

Bruzzese F, Rocco M, Castelli S, Di Gennaro E, Desideri A, Budillon A (2009) Synergistic antitumor effect between vorinostat and topotecan in small cell lung cancer cells is mediated by generation of reactive oxygen species and DNA damage-induced apoptosis. Mol Cancer Ther 8:3075–3087

Bruzzese F, Leone A, Rocco M, Carbone C, Piro G, Caraglia M, Di Gennaro E, Budillon A (2011) HDAC inhibitor vorinostat enhances the antitumor effect of gefitinib in squamous cell carcinoma of head and neck by modulating ErbB receptor expression and reverting EMT. J Cell Physiol 226:2378–2390

Cai M-H, Xu X-G, Yan S-L, Sun Z, Ying Y, Wang B-K, Tu Y-X (2018) Depletion of HDAC1, 7 and 8 by histone deacetylase inhibition confers elimination of pancreatic cancer stem cells in combination with gemcitabine. Sci Rep 8:1621

Catalano MG, Fortunati N, Pugliese M, Poli R, Bosco O, Mastrocola R, Aragno M, Boccuzzi G (2006) Valproic acid, a histone deacetylase inhibitor, enhances sensitivity to doxorubicin in anaplastic thyroid cancer cells. J Endocrinol 191:465–472

Catalano MG, Pugliese M, Gallo M, Brignardello E, Milla P, Orlandi F, Limone PP, Arvat E, Boccuzzi G, Piovesan A (2016) Valproic acid, a histone deacetylase inhibitor, in combination with paclitaxel for anaplastic thyroid cancer: results of a multicenter randomized controlled phase II/III trial. Int J Endocrinol 2016:2930414

Chao M-W, Lai M-J, Liou J-P, Chang Y-L, Wang J-C, Pan S-L, Teng C-M (2015) The synergic effect of vincristine and vorinostat in leukemia in vitro and in vivo. J Hematol Oncol 8:82

Chen C-S, Wang Y-C, Yang H-C, Huang P-H, Kulp SK, Yang C-C, Lu Y-S, Matsuyama S, Chen C-Y, Chen C-S (2007) Histone deacetylase inhibitors sensitize prostate cancer cells to agents that produce DNA double-strand breaks by targeting Ku70 acetylation. Cancer Res 67:5318–5327

Chen M-Y, Liao WSL, Lu Z, Bornmann WG, Hennessey V, Washington MN, Rosner GL, Yu Y, Ahmed AA, Bast RC Jr (2011) Decitabine and suberoylanilide hydroxamic acid (SAHA) inhibit growth of ovarian cancer cell lines and xenografts while inducing expression of imprinted tumor suppressor genes, apoptosis, G2/M arrest, and autophagy. Cancer 117:4424–4438

Chen L, Jin T, Zhu K, Piao Y, Quan T, Quan C, Lin Z (2017) PI3K/mTOR dual inhibitor BEZ235 and histone deacetylase inhibitor Trichostatin A synergistically exert anti-tumor activity in breast cancer. Oncotarget 8:11937–11949

Citro S, Bellini A, Miccolo C, Ghiani L, Carey TE, Chiocca S (2019) Synergistic antitumour activity of HDAC inhibitor SAHA and EGFR inhibitor gefitinib in head and neck cancer: a key role for ΔNp63α. Br J Cancer 120:658–667

Costa TEMM, Raghavendra NM, Penido C (2020) Natural heat shock protein 90 inhibitors in cancer and inflammation. Eur J Med Chem 189:112063

Dasmahapatra G, Lembersky D, Son MP, Attkisson E, Dent P, Fisher RI, Friedberg JW, Grant S (2011) Carfilzomib interacts synergistically with histone deacetylase inhibitors in mantle cell lymphoma cells in vitro and in vivo. Mol Cancer Ther 10:1686–1697

de Andrade PV, Andrade AF, de Paula Queiroz RG, Scrideli CA, Tone LG, Valera ET (2016) The histone deacetylase inhibitor PCI-24781 as a putative radiosensitizer in pediatric glioblastoma cell lines. Cancer Cell Int 16:31

Deng B, Luo Q, Halim A, Liu Q, Zhang B, Song G (2020) The antiangiogenesis role of histone deacetylase inhibitors: their potential application to tumor therapy and tissue repair. DNA Cell Biol 39:167–176

Di Martile M, Desideri M, Tupone MG, Buglioni S, Antoniani B, Mastroiorio C, Falcioni R, Ferraresi V, Baldini N, Biagini R, Milella M, Trisciuoglio D, Del Bufalo D (2018) Histone deacetylase inhibitor ITF2357 leads to apoptosis and enhances doxorubicin cytotoxicity in preclinical models of human sarcoma. Oncogenesis 7:20

Diyabalanage HVK, Granda ML, Hooker JM (2013) Combination therapy: histone deacetylase inhibitors and platinum-based chemotherapeutics for cancer. Cancer Lett 329:1–8

Dowdy SC, Jiang S, Zhou XC, Hou X, Jin F, Podratz KC, Jiang SW (2006) Histone deacetylase inhibitors and paclitaxel cause synergistic effects on apoptosis and microtubule stabilization in papillary serous endometrial cancer cells. Mol Cancer Ther 5:2767–2776

Eleutherakis-Papaiakovou E, Kanellias N, Kastritis E, Gavriatopoulou M, Terpos E, Dimopoulos MA (2020) Efficacy of panobinostat for the treatment of multiple myeloma. J Oncol 2020:7131802

Feng W, Marquez RT, Lu Z, Liu J, Lu KH, Issa JP, Fishman DM, Yu Y, Bast RC Jr (2008) Imprinted tumor suppressor genes ARHI and PEG3 are the most frequently down-regulated in human ovarian cancers by loss of heterozygosity and promoter methylation. Cancer 112:1489–1502

Fiskus W, Wang Y, Sreekumar A, Buckley KM, Shi H, Jillella A, Ustun C, Rao R, Fernandez P, Chen J, Balusu R, Koul S, Atadja P, Marquez VE, Bhalla KN (2009) Combined epigenetic therapy with the histone methyltransferase EZH2 inhibitor 3-deazaneplanocin A and the histone deacetylase inhibitor panobinostat against human AML cells. Blood 114:2733–2743

Galanis E, Anderson SK, Miller CR, Sarkaria JN, Jaeckle K, Buckner JC, Ligon KL, Ballman KV, Moore DF Jr, Nebozhyn M, Loboda A, Schiff D, Ahluwalia MS, Lee EQ, Gerstner ER, Lesser GJ, Prados M, Grossman SA, Cerhan J, Giannini C, Wen PY, Alliance for Clinical Trials in Oncology and ABTC (2017) Phase I/II trial of vorinostat combined with temozolomide and radiation therapy for newly diagnosed glioblastoma: results of Alliance N0874/ABTC 02. Neuro-Oncology 20:546–556

Ganai SA (2016) Panobinostat: the small molecule metalloenzyme inhibitor with marvelous anticancer activity. Curr Top Med Chem 16:427–434

Ganai SA, Kalladi SM, Mahadevan V (2015) HDAC inhibition through valproic acid modulates the methylation profiles in human embryonic kidney cells. J Biomol Struct Dyn 33:1185–1197

Garcia-Manero G, Tambaro FP, Bekele NB, Yang H, Ravandi F, Jabbour E, Borthakur G, Kadia TM, Konopleva MY, Faderl S, Cortes JE, Brandt M, Hu Y, McCue D, Newsome WM, Pierce SR, de Lima M, Kantarjian HM (2012) Phase II trial of vorinostat with idarubicin and cytarabine for patients with newly diagnosed acute myelogenous leukemia or myelodysplastic syndrome. J Clin Oncol 30:2204–2210

Garcia-Manero G, Abaza Y, Takahashi K, Medeiros BC, Arellano M, Khaled SK, Patnaik M, Odenike O, Sayar H, Tummala M, Patel P, Maness-Harris L, Stuart R, Traer E, Karamlou K, Yacoub A, Ghalie R, Giorgino R, Atallah E (2019) Pracinostat plus azacitidine in older patients with newly diagnosed acute myeloid leukemia: results of a phase 2 study. Blood Adv 3:508–518

Garrett LA, Growdon WB, Rueda BR, Foster R (2016) Influence of a novel histone deacetylase inhibitor panobinostat (LBH589) on the growth of ovarian cancer. J Ovarian Res 9:58

Geng L, Cuneo KC, Fu A, Tu T, Atadja PW, Hallahan DE (2006) Histone deacetylase (HDAC) inhibitor LBH589 increases duration of gamma-H2AX foci and confines HDAC4 to the cytoplasm in irradiated non-small cell lung cancer. Cancer Res 66:11298–11304

George P, Bali P, Annavarapu S, Scuto A, Fiskus W, Guo F, Sigua C, Sondarva G, Moscinski L, Atadja P, Bhalla K (2005) Combination of the histone deacetylase inhibitor LBH589 and the hsp90 inhibitor 17-AAG is highly active against human CML-BC cells and AML cells with activating mutation of FLT-3. Blood 105:1768–1776

Girard N, Bazille C, Lhuissier E, Benateau H, Llombart-Bosch A, Boumediene K, Bauge C (2014) 3-Deazaneplanocin A (DZNep), an inhibitor of the histone methyltransferase EZH2, induces apoptosis and reduces cell migration in chondrosarcoma cells. PLoS One 9:e98176

Gopalakrishnan V, Das C, Aguilera D, Thomas A, Joyce C, Wolff J (2008) Synergy between HDAC inhibitors and etoposide in medulloblastoma cells: p53-dependent and independent mechanisms. Cancer Res 68:3746–3746

Gopalakrishnapillai A, Barwe SP, Kolb EA (2013) Combination of inhibitors against DNA methyltransferases and histone deacetylases synergistically suppresses cell viability in acute myeloid cells in a p53-dependent manner. Blood 122:4894–4894

Graham L, Banda K, Torres A, Carver BS, Chen Y, Pisano K, Shelkey G, Curley T, Scher HI, Lotan TL, Hsieh AC, Rathkopf DE (2018) A phase II study of the dual mTOR inhibitor MLN0128 in patients with metastatic castration resistant prostate cancer. Investig New Drugs 36:458–467

Groselj B, Ruan J-L, Scott H, Gorrill J, Nicholson J, Kelly J, Anbalagan S, Thompson J, Stratford MRL, Jevons SJ, Hammond EM, Scudamore CL, Kerr M, Kiltie AE (2018) Radiosensitization in vivo by histone deacetylase inhibition with no increase in early Normal tissue radiation toxicity. Mol Cancer Ther 17:381–392

Haggarty SJ, Koeller KM, Wong JC, Grozinger CM, Schreiber SL (2003) Domain-selective small-molecule inhibitor of histone deacetylase 6 (HDAC6)-mediated tubulin deacetylation. Proc Natl Acad Sci U S A 100:4389–4394

Hajji N, Wallenborg K, Vlachos P, Nyman U, Hermanson O, Joseph B (2008) Combinatorial action of the HDAC inhibitor trichostatin A and etoposide induces caspase-mediated AIF-dependent apoptotic cell death in non-small cell lung carcinoma cells. Oncogene 27:3134–3144

Hamam R, Ali D, Vishnubalaji R, Alsaaran Z, Chalisserry E, Alfayez M, Aldahmash A, Alajez N (2017) Enhanced efficacy of 5-fluorouracil in combination with a dual histone deacetylase and phosphatidylinositide 3-kinase inhibitor (CUDC-907) in colorectal cancer cells. Saudi J Gastroenterol 23:34

Havas AP, Rodrigues KB, Bhakta A, Demirjian JA, Hahn S, Tran J, Scavello M, Tula-Sanchez AA, Zeng Y, Schmelz M, Smith CL (2016) Belinostat and vincristine demonstrate mutually synergistic cytotoxicity associated with mitotic arrest and inhibition of polyploidy in a preclinical model of aggressive diffuse large B cell lymphoma. Cancer Biol Ther 17:1240–1252

Hideshima T, Bradner JE, Wong J, Chauhan D, Richardson P, Schreiber SL, Anderson KC (2005) Small-molecule inhibition of proteasome and aggresome function induces synergistic antitumor activity in multiple myeloma. Proc Natl Acad Sci U S A 102:8567–8572

Hii L-W, Chung FF-L, Soo JS-S, Tan BS, Mai C-W, Leong C-O (2020) Histone deacetylase (HDAC) inhibitors and doxorubicin combinations target both breast cancer stem cells and non-stem breast cancer cells simultaneously. Breast Cancer Res Treat 179:615–629

Hrabeta J, Stiborova M, Adam V, Kizek R, Eckschlager T (2014) Histone deacetylase inhibitors in cancer therapy. A review. Biomed Pap Med Fac Univ Palacky Olomouc Czech Repub 158:161–169

Hsu F-S, Wu J-T, Lin J-Y, Yang S-P, Kuo K-L, Lin W-C, Shi C-S, Chow P-M, Liao S-M, Pan C-I, Hong J-Y, Chang H-C, Huang K-H (2019) Histone deacetylase inhibitor, Trichostatin A, synergistically enhances paclitaxel-induced cytotoxicity in urothelial carcinoma cells by suppressing the ERK pathway. Int J Mol Sci 20:1162

Huang JP, Ling K (2017) EZH2 and histone deacetylase inhibitors induce apoptosis in triple negative breast cancer cells by differentially increasing H3 Lys(27) acetylation in the BIM gene promoter and enhancers. Oncol Lett 14:5735–5742

Huang H, Liu N, Guo H, Liao S, Li X, Yang C, Liu S, Song W, Liu C, Guan L, Li B, Xu L, Zhang C, Wang X, Dou Q, Liu J (2012a) L-Carnitine is an endogenous HDAC inhibitor selectively inhibiting cancer cell growth in vivo and in vitro. PLoS One 7:e49062

Huang H, Liu N, Yang C, Liao S, Guo H, Zhao K, Li X, Liu S, Guan L, Liu C, Xu L, Zhang C, Song W, Li B, Tang P, Dou QP, Liu J (2012b) HDAC inhibitor L-carnitine and proteasome inhibitor bortezomib synergistically exert anti-tumor activity in vitro and in vivo. PLoS One 7: e52576

Huang XP, Li X, Situ MY, Huang LY, Wang JY, He TC, Yan QH, Xie XY, Zhang YJ, Gao YH, Li YH, Rong TH, Wang MR, Cai QQ, Fu JH (2018) Entinostat reverses cisplatin resistance in esophageal squamous cell carcinoma via down-regulation of multidrug resistance gene 1. Cancer Lett 414:294–300

Hubeek I, Comijn EM, Van der Wilt CL, Merriman RL, Padron JM, Kaspers GJ, Peters GJ (2008) CI-994 (N-acetyl-dinaline) in combination with conventional anti-cancer agents is effective against acute myeloid leukemia in vitro and in vivo. Oncol Rep 19:1517–1523

Hulin-Curtis SL, Davies JA, Jones R, Hudson E, Hanna L, Chester JD, Parker AL (2018) Histone deacetylase inhibitor trichostatin A sensitises cisplatin-resistant ovarian cancer cells to oncolytic adenovirus. Oncotarget 9:26328–26341

Iwahashi S, Ishibashi H, Utsunomiya T, Morine Y, Ochir TL, Hanaoka J, Mori H, Ikemoto T, Imura S, Shimada M (2011a) Effect of histone deacetylase inhibitor in combination with 5-fluorouracil on pancreas cancer and cholangiocarcinoma cell lines. J Med Investig 58:106–109

Iwahashi S, Shimada M, Utsunomiya T, Morine Y, Imura S, Ikemoto T, Mori H, Hanaoka J, Sugimoto K, Saito Y (2011b) Histone deacetylase inhibitor augments anti-tumor effect of gemcitabine and pegylated interferon-α on pancreatic cancer cells. Int J Clin Oncol 16:671–678

Jeannot V, Busser B, Vanwonterghem L, Michallet S, Ferroudj S, Cokol M, Coll J-L, Ozturk M, Hurbin A (2016) Synergistic activity of vorinostat combined with gefitinib but not with sorafenib in mutant KRAS human non-small cell lung cancers and hepatocarcinoma. Onco Targets Ther 9:6843–6855

Jhanwar-Uniyal M, Wainwright JV, Mohan AL, Tobias ME, Murali R, Gandhi CD, Schmidt MH (2019) Diverse signaling mechanisms of mTOR complexes: mTORC1 and mTORC2 in forming a formidable relationship. Adv Biol Regul 72:51–62

Ji J, Valdez BC, Li Y, Liu Y, Teo EC, Nieto Y, Champlin RE, Andersson BS (2016) Cladribine, gemcitabine, busulfan, and SAHA combination as a potential pretransplant conditioning regimen for lymphomas: a preclinical study. Exp Hematol 44:458–465

Ji J, Liu T, Kuang P, Liu Z, Dong T, Liu J (2017) Chidamide, a HDAC inhibitor, combined with Cladribine, high dose gemcitabine and Busulfan with autologous stem cell transplantation in patients with relapsed/refractory and poor-risk lymphoma. Blood 130:3288–3288

Jin KL, Park J-Y, Noh EJ, Hoe KL, Lee JH, Kim J-H, Nam J-H (2010) The effect of combined treatment with cisplatin and histone deacetylase inhibitors on HeLa cells. J Gynecol Oncol 21:262–268

Jin XF, Auernhammer CJ, Ilhan H, Lindner S, Nölting S, Maurer J, Spöttl G, Orth M (2019) Combination of 5-fluorouracil with epigenetic modifiers induces radiosensitization, somatostatin receptor 2 expression, and radioligand binding in neuroendocrine tumor cells in vitro. J Nucl Med 60:1240–1246

Johnston JA, Ward CL, Kopito RR (1998) Aggresomes: a cellular response to misfolded proteins. J Cell Biol 143:1883–1898

Kalac M, Scotto L, Marchi E, Amengual J, Seshan VE, Bhagat G, Ulahannan N, Leshchenko VV, Temkin AM, Parekh S, Tycko B, O'Connor OA (2011) HDAC inhibitors and decitabine are highly synergistic and associated with unique gene-expression and epigenetic profiles in models of DLBCL. Blood 118:5506–5516

Kane RC, Bross PF, Farrell AT, Pazdur R (2003) Velcade: U.S. FDA approval for the treatment of multiple myeloma progressing on prior therapy. Oncologist 8:508–513

Kanzaki M, Kakinuma H, Kumazawa T, Inoue T, Saito M, Narita S, Yuasa T, Tsuchiya N, Habuchi T (2007) Low concentrations of the histone deacetylase inhibitor, depsipeptide, enhance the effects of gemcitabine and docetaxel in hormone refractory prostate cancer cells. Oncol Rep 17:761–767

Kawaguchi Y, Kovacs JJ, McLaurin A, Vance JM, Ito A, Yao TP (2003) The deacetylase HDAC6 regulates aggresome formation and cell viability in response to misfolded protein stress. Cell 115:727–738

Kim SH, Kang JG, Kim CS, Ihm S-H, Choi MG, Yoo HJ, Lee SJ (2015) Novel heat shock protein 90 inhibitor NVP-AUY922 synergizes with the histone deacetylase inhibitor PXD101 in induction of death of anaplastic thyroid carcinoma cells. J Clin Endocrinol Metabol 100: E253–E261

Kobayashi Y, Munakata W, Ogura M, Uchida T, Taniwaki M, Kobayashi T, Shimada F, Yonemura M, Matsuoka F, Tajima T, Yakushijin K, Minami H (2018) Phase I study of panobinostat and 5-azacitidine in Japanese patients with myelodysplastic syndrome or chronic myelomonocytic leukemia. Int J Hematol 107:83–91

Kopito RR (2000) Aggresomes, inclusion bodies and protein aggregation. Trends Cell Biol 10:524–530

Krämer OH, Knauer SK, Zimmermann D, Stauber RH, Heinzel T (2008) Histone deacetylase inhibitors and hydroxyurea modulate the cell cycle and cooperatively induce apoptosis. Oncogene 27:732–740

Krauze AV, Myrehaug SD, Chang MG, Holdford DJ, Smith S, Shih J, Tofilon PJ, Fine HA, Camphausen K (2015) A phase 2 study of concurrent radiation therapy, temozolomide, and the histone deacetylase inhibitor Valproic acid for patients with glioblastoma. Int J Radiat Oncol Biol Phys 92:986–992

Kutty RV, Tay CY, Lim CS, Feng S-S, Leong DT (2015) Anti-migratory and increased cytotoxic effects of novel dual drug-loaded complex hybrid micelles in triple negative breast cancer cells. Nano Res 8:2533–2547

Laporte A, Barrott J, Yao RJ, Poulin N, Brodin B, Jones K, Underhill T, Nielsen T (2017) HDAC and proteasome inhibitors synergize to activate pro-apoptotic factors in synovial sarcoma. PLoS One 12:e0169407

Laschanzky RS, Humphrey LE, Ma J, Smith LM, Enke TJ, Shukla SK, Dasgupta A, Singh PK, Howell GM, Brattain MG, Ly QP, Black AR, Black JD (2019) Selective inhibition of histone deacetylases 1/2/6 in combination with gemcitabine: a promising combination for pancreatic Cancer therapy. Cancers 11:1327

Leclerc GJ, Mou C, Leclerc GM, Mian AM, Barredo JC (2010) Histone deacetylase inhibitors induce FPGS mRNA expression and intracellular accumulation of long-chain methotrexate polyglutamates in childhood acute lymphoblastic leukemia: implications for combination therapy. Leukemia 24:552–562

Lee J-H, Park J-H, Jung Y, Kim J-H, Jong H-S, Kim T-Y, Bang Y-J (2006) Histone deacetylase inhibitor enhances 5-fluorouracil cytotoxicity by down-regulating thymidylate synthase in human cancer cells. Mol Cancer Ther 5:3085–3095

Lee HS, Park SB, Kim SA, Kwon SK, Cha H, Lee DY, Ro S, Cho JM, Song SY (2017) A novel HDAC inhibitor, CG200745, inhibits pancreatic cancer cell growth and overcomes gemcitabine resistance. Sci Rep 7:41615

Lee DH, Won HR, Ryu HW, Han JM, Kwon SH (2018) The HDAC6 inhibitor ACY-1215 enhances the anticancer activity of oxaliplatin in colorectal cancer cells. Int J Oncol 53:844–854

Leist TP, Weissert R (2011) Cladribine: mode of action and implications for treatment of multiple sclerosis. Clin Neuropharmacol 34:28–35

Li Z-Y, Li Q-Z, Chen L, Chen B-D, Wang B, Zhang X-J, Li W-P (2016) Histone deacetylase inhibitor RGFP109 overcomes temozolomide resistance by blocking NF-κB-dependent transcription in glioblastoma cell lines. Neurochem Res 41:3192–3205

Liao B, Zhang Y, Sun Q, Jiang P (2018) Vorinostat enhances the anticancer effect of oxaliplatin on hepatocellular carcinoma cells. Cancer Med 7:196–207

Liffers K, Kolbe K, Westphal M, Lamszus K, Schulte A (2016) Histone deacetylase inhibitors resensitize EGFR/EGFRvIII-overexpressing, erlotinib-resistant glioblastoma cells to tyrosine kinase inhibition. Target Oncol 11:29–40

Lopez G, Braggio D, Zewdu A, Casadei L, Batte K, Bid HK, Koller D, Yu P, Iwenofu OH, Strohecker A, Choy E, Lev D, Pollock R (2017) Mocetinostat combined with gemcitabine for the treatment of leiomyosarcoma: preclinical correlates. PLoS One 12:e0188859

Lue JK, Prabhu SA, Liu Y, Gonzalez Y, Verma A, Mundi PS, Abshiru N, Camarillo JM, Mehta S, Chen EI, Qiao C, Nandakumar R, Cremers S, Kelleher NL, Elemento O, Amengual JE (2019) Precision targeting with EZH2 and HDAC inhibitors in epigenetically dysregulated lymphomas. Clin Cancer Res 25:5271–5283

Luszczek W, Cheriyath V, Mekhail TM, Borden EC (2010) Combinations of DNA methyltransferase and histone deacetylase inhibitors induce DNA damage in small cell lung cancer cells: correlation of resistance with IFN-stimulated gene expression. Mol Cancer Ther 9:2309–2321

Maggio SC, Rosato RR, Kramer LB, Dai Y, Rahmani M, Paik DS, Czarnik AC, Payne SG, Spiegel S, Grant S (2004) The histone deacetylase inhibitor MS-275 interacts synergistically with fludarabine to induce apoptosis in human leukemia cells. Cancer Res 64:2590–2600

Marchesi F, Turriziani M, Tortorelli G, Avvisati G, Torino F, Vecchis L (2007) Triazene compounds: mechanism of action and related DNA repair systems. Pharmacol Res 56:275–287

McBride A, Klaus JO, Stockerl-Goldstein K (2015) Carfilzomib: a second-generation proteasome inhibitor for the treatment of multiple myeloma. Am J Health Syst Pharm 72:353–360

Min C, Moore N, Shearstone JR, Quayle SN, Huang P, van Duzer JH, Jarpe MB, Jones SS, Yang M (2017) Selective inhibitors of histone deacetylases 1 and 2 synergize with azacitidine in acute myeloid leukemia. PLoS One 12:e0169128–e0169128

Minegaki T, Suzuki A, Mori M, Tsuji S, Yamamoto S, Watanabe A, Tsuzuki T, Tsunoda T, Yamamoto A, Tsujimoto M, Nishiguchi K (2018) Histone deacetylase inhibitors sensitize

5-fluorouracil-resistant MDA-MB-468 breast cancer cells to 5-fluorouracil. Oncol Lett 16:6202–6208

Mora E, Smith EML, Donohoe C, Hertz DL (2016) Vincristine-induced peripheral neuropathy in pediatric cancer patients. Am J Cancer Res 6:2416–2430

Moufarrij S, Srivastava A, Gomez S, Hadley M, Palmer E, Austin PT, Chisholm S, Roche K, Yu A, Li J, Zhu W, Lopez-Acevedo M, Villagra A, Chiappinelli KB (2020) Combining DNMT and HDAC6 inhibitors increases anti-tumor immune signaling and decreases tumor burden in ovarian cancer. Sci Rep 10:3470

Munshi A, Tanaka T, Hobbs ML, Tucker SL, Richon VM, Meyn RE (2006) Vorinostat, a histone deacetylase inhibitor, enhances the response of human tumor cells to ionizing radiation through prolongation of gamma-H2AX foci. Mol Cancer Ther 5:1967–1974

Münster P, Marchion D, Bicaku E, Schmitt M, Lee JH, DeConti R, Simon G, Fishman M, Minton S, Garrett C, Chiappori A, Lush R, Sullivan D, Daud A (2007) Phase I trial of histone deacetylase inhibition by valproic acid followed by the topoisomerase II inhibitor epirubicin in advanced solid tumors: a clinical and translational study. J Clin Oncol 25:1979–1985

Munster P, Marchion D, Bicaku E, Lacevic M, Kim J, Centeno B, Daud A, Neuger A, Minton S, Sullivan D (2009a) Clinical and biological effects of valproic acid as a histone deacetylase inhibitor on tumor and surrogate tissues: phase I/II trial of valproic acid and epirubicin/FEC. Clin Cancer Res 15:2488–2496

Munster PN, Marchion D, Thomas S, Egorin M, Minton S, Springett G, Lee JH, Simon G, Chiappori A, Sullivan D, Daud A (2009b) Phase I trial of vorinostat and doxorubicin in solid tumours: histone deacetylase 2 expression as a predictive marker. Br J Cancer 101:1044–1050

Na YS, Kim SM, Jung KA, Yang SJ, Hong YS, Ryu MH, Ro S, Cho DH, Kim JC, Jin DH, Lee JS, Kim TW (2010) Effects of the HDAC inhibitor CG2 in combination with irinotecan, 5-fluorouracil, or oxaliplatin on HCT116 colon cancer cells and xenografts. Oncol Rep 24:1509–1514

Na YS, Jung KA, Kim SM, Hong YS, Ryu MH, Jang SJ, Moon DH, Cho DH, Kim JC, Lee JS, Kim TW (2011) The histone deacetylase inhibitor PXD101 increases the efficacy of irinotecan in in vitro and in vivo colon cancer models. Cancer Chemother Pharmacol 68:389–398

Namdar M, Perez G, Ngo L, Marks PA (2010) Selective inhibition of histone deacetylase 6 (HDAC6) induces DNA damage and sensitizes transformed cells to anticancer agents. Proc Natl Acad Sci U S A 107:20003–20008

Neyns B, Tosoni A, Hwu WJ, Reardon DA (2010) Dose-dense temozolomide regimens: antitumor activity, toxicity, and immunomodulatory effects. Cancer 116:2868–2877

Nguyen A, Su L, Campbell B, Poulin NM, Nielsen TO (2009) Synergism of heat shock protein 90 and histone deacetylase inhibitors in synovial sarcoma. Sarcoma 2009:794901

Okada K, Hakata S, Terashima J, Gamou T, Habano W, Ozawa S (2016) Combination of the histone deacetylase inhibitor depsipeptide and 5-fluorouracil upregulates major histocompatibility complex class II and p21 genes and activates caspase-3/7 in human colon cancer HCT-116 cells. Oncol Rep 36:1875–1885

Owonikoko TK, Ramalingam SS, Kanterewicz B, Balius TE, Belani CP, Hershberger PA (2010) Vorinostat increases carboplatin and paclitaxel activity in non-small-cell lung cancer cells. Int J Cancer 126:743–755

Pan C-H, Chang Y-F, Lee M-S, Wen BC, Ko J-C, Liang S-K, Liang M-C (2016) Vorinostat enhances the cisplatin-mediated anticancer effects in small cell lung cancer cells. BMC Cancer 16:857

Pili R, Liu G, Chintala S, Verheul H, Rehman S, Attwood K, Lodge MA, Wahl R, Martin JI, Miles KM, Paesante S, Adelaiye R, Godoy A, King S, Zwiebel J, Carducci MA (2017) Combination of the histone deacetylase inhibitor vorinostat with bevacizumab in patients with clear-cell renal cell carcinoma: a multicentre, single-arm phase I/II clinical trial. Br J Cancer 116:874–883

Pinar B, Henríquez-Hernández LA, Lara PC, Bordon E, Rodriguez-Gallego C, Lloret M, Nuñez MI, De Almodovar MR (2010) Radiation induced apoptosis and initial DNA damage are inversely related in locally advanced breast cancer patients. Radiat Oncol 5:85

Piro G, Roca MS, Bruzzese F, Carbone C, Iannelli F, Leone A, Volpe MG, Budillon A, Di Gennaro E (2019) Vorinostat potentiates 5-fluorouracil/cisplatin combination by inhibiting chemotherapy-induced EGFR nuclear translocation and increasing cisplatin uptake. Mol Cancer Ther 18(8):1405–1417

Pópulo H, Lopes JM, Soares P (2012) The mTOR signalling pathway in human cancer. Int J Mol Sci 13:1886–1918

Quirbt I, Verma S, Petrella T, Bak K, Charette M, Members of the Melanoma Disease Site Group of Cancer Care Ontario's Program in Evidence-Based Care (2007) Temozolomide for the treatment of metastatic melanoma. Curr Oncol 14:27–33

Ramalingam SS, Maitland ML, Frankel P, Argiris AE, Koczywas M, Gitlitz B, Thomas S, Espinoza-Delgado I, Vokes EE, Gandara DR, Belani CP (2010) Carboplatin and paclitaxel in combination with either vorinostat or placebo for first-line therapy of advanced non-small-cell lung cancer. J Clin Oncol 28:56–62

Rao R, Fiskus W, Yang Y, Lee P, Joshi R, Fernandez P, Mandawat A, Atadja P, Bradner JE, Bhalla K (2008) HDAC6 inhibition enhances 17-AAG--mediated abrogation of hsp90 chaperone function in human leukemia cells. Blood 112:1886–1893

Rao R, Lee P, Fiskus W, Yang Y, Joshi R, Wang Y, Buckley K, Balusu R, Chen J, Koul S, Joshi A, Upadhyay S, Tao J, Sotomayor E, Bhalla KN (2009) Co-treatment with heat shock protein 90 inhibitor 17-dimethylaminoethylamino-17-demethoxygeldanamycin (DMAG) and vorinostat: a highly active combination against human mantle cell lymphoma (MCL) cells. Cancer Biol Ther 8:1273–1280

Rathkopf D, Wong BY, Ross RW, Anand A, Tanaka E, Woo MM, Hu J, Dzik-Jurasz A, Yang W, Scher HI (2010) A phase I study of oral panobinostat alone and in combination with docetaxel in patients with castration-resistant prostate cancer. Cancer Chemother Pharmacol 66:181–189

Reguart N, Cardona AF, Isla D, Cardenal F, Palmero R, Carrasco-Chaumel E, Rolfo C, Massuti B, Moran T, Rosell R (2009) Phase I trial of vorinostat in combination with erlotinib in advanced non-small cell lung cancer (NSCLC) patients with EGFR mutations after erlotinib progression. J Clin Oncol 27:e19057–e19057

Rocca A, Minucci S, Tosti G, Croci D, Contegno F, Ballarini M, Nolè F, Munzone E, Salmaggi A, Goldhirsch A, Pelicci PG, Testori A (2009) A phase I-II study of the histone deacetylase inhibitor valproic acid plus chemoimmunotherapy in patients with advanced melanoma. Br J Cancer 100:28–36

Ryu CH, Yoon WS, Park KY, Kim SM, Lim JY, Woo JS, Jeong CH, Hou Y, Jeun SS (2012) Valproic acid downregulates the expression of MGMT and sensitizes temozolomide-resistant glioma cells. J Biomed Biotechnol 2012:987495

San-Miguel JF, Hungria VT, Yoon SS, Beksac M, Dimopoulos MA, Elghandour A, Jedrzejczak WW, Günther A, Nakorn TN, Siritanaratkul N, Corradini P, Chuncharunee S, Lee JJ, Schlossman RL, Shelekhova T, Yong K, Tan D, Numbenjapon T, Cavenagh JD, Hou J, LeBlanc R, Nahi H, Qiu L, Salwender H, Pulini S, Moreau P, Warzocha K, White D, Bladé J, Chen W, de la Rubia J, Gimsing P, Lonial S, Kaufman JL, Ocio EM, Veskovski L, Sohn SK, Wang MC, Lee JH, Einsele H, Sopala M, Corrado C, Bengoudifa BR, Binlich F, Richardson PG (2014) Panobinostat plus bortezomib and dexamethasone versus placebo plus bortezomib and dexamethasone in patients with relapsed or relapsed and refractory multiple myeloma: a multicentre, randomised, double-blind phase 3 trial. Lancet Oncol 15:1195–1206

Sarbassov DD, Ali SM, Sengupta S, Sheen J-H, Hsu PP, Bagley AF, Markhard AL, Sabatini DM (2006) Prolonged rapamycin treatment inhibits mTORC2 assembly and Akt/PKB. Mol Cell 22:159–168

Sarkaria JN, Kitange GJ, James CD, Plummer R, Calvert H, Weller M, Wick W (2008) Mechanisms of chemoresistance to alkylating agents in malignant glioma. Clin Cancer Res 14:2900–2908

Sheikh TN, Patwardhan P, Schwartz GK (2018) Abstract A30: preclinical study of a combination of mocetinostat (HDAC inhibitor) and 5-AZA-dC (decitabine) in chondrosarcoma. Clin Cancer Res 24:A30–A30

Shiozawa K, Nakanishi T, Tan M, Fang H-B, Wang W-C, Edelman MJ, Carlton D, Gojo I, Sausville EA, Ross DD (2009) Preclinical studies of vorinostat (suberoylanilide hydroxamic acid) combined with cytosine arabinoside and etoposide for treatment of acute leukemias. Clin Cancer Res 15:1698–1707

Simmons JK, Patel J, Michalowski A, Zhang S, Wei B-R, Sullivan P, Gamache B, Felsenstein K, Kuehl WM, Simpson RM, Zingone A, Landgren O, Mock BA (2014) TORC1 and class I HDAC inhibitors synergize to suppress mature B cell neoplasms. Mol Oncol 8:261–272

Sriram D, Yogeeswari P, Thirumurugan R, Ratan Bal T (2005) Camptothecin and its analogues: a review on their chemotherapeutic potential. Nat Prod Res 19:393–412

Stauber R, Knauer S, Habtemichael N, Bier C, Unruhe B, Weisheit S, Spange S, Nonnenmacher F, Fetz V, Ginter T, Reichardt S, Liebmann C, Schneider G, Krämer O (2012) A combination of a ribonucleotide reductase inhibitor and histone deacetylase inhibitors downregulates EGFR and triggers BIM-dependent apoptosis in head and neck cancer. Oncotarget 3:31–43

Suraweera A, O'Byrne KJ, Richard DJ (2018) Combination therapy with histone deacetylase inhibitors (HDACi) for the treatment of cancer: achieving the full therapeutic potential of HDACi. Front Oncol 8:92

Tamura Y, Simizu S, Osada H (2004) The phosphorylation status and anti-apoptotic activity of Bcl-2 are regulated by ERK and protein phosphatase 2A on the mitochondria. FEBS Lett 569:249–255

Tian Y, Liu G, Wang H, Tian Z, Cai Z, Zhang F, Luo Y, Wang S, Guo G, Wang X, Powell S, Feng Z (2017) Valproic acid sensitizes breast cancer cells to hydroxyurea through inhibiting RPA2 hyperphosphorylation-mediated DNA repair pathway. DNA Repair 58:1–12

Van Helleputte L, Kater M, Cook DP, Eykens C, Rossaert E, Haeck W, Jaspers T, Geens N, Vanden Berghe P, Gysemans C, Mathieu C, Robberecht W, Van Damme P, Cavaletti G, Jarpe M, Van Den Bosch L (2018) Inhibition of histone deacetylase 6 (HDAC6) protects against vincristine-induced peripheral neuropathies and inhibits tumor growth. Neurobiol Dis 111:59–69

Vu K, Wu CH, Yang CY, Zhan A, Cavallone E, Berry W, Heeter P, Pincus L, Wieduwilt MJ, William BM, Andreadis C, Kaplan LK, McCormick F, Porcu P, Brammer JE, Ai WZ (2020) Romidepsin plus liposomal doxorubicin is safe and effective in patients with relapsed or refractory T-cell lymphoma: results of a phase I dose-escalation study. Clin Cancer Res 26:1000–1008

Wang L, Li H, Ren Y, Zou S, Fang W, Jiang X, Jia L, Li M, Liu X, Yuan X, Chen G, Yang J, Wu C (2016) Targeting HDAC with a novel inhibitor effectively reverses paclitaxel resistance in non-small cell lung cancer via multiple mechanisms. Cell Death Dis 7:e2063–e2063

Wang B, Lyu H, Pei S, Song D, Ni J, Liu B (2018a) Cladribine in combination with entinostat synergistically elicits anti-proliferative/anti-survival effects on multiple myeloma cells. Cell Cycle 17:985–996

Wang J-S, Wang H-J, Qian H-L (2018b) Biological effects of radiation on cancer cells. Mil Med Res 5:20–20

Wang C, Hamacher A, Petzsch P, Köhrer K, Niegisch G, Hoffmann MJ, Schulz WA, Kassack MU (2020) Combination of decitabine and entinostat synergistically inhibits urothelial bladder cancer cells via activation of FoxO1. Cancers 12

Watanabe R, Wei L, Huang J (2011) mTOR signaling, function, novel inhibitors, and therapeutic targets. J Nucl Med 52:497–500

Wilhelm SM, Adnane L, Newell P, Villanueva A, Llovet JM, Lynch M (2008) Preclinical overview of sorafenib, a multikinase inhibitor that targets both Raf and VEGF and PDGF receptor tyrosine kinase signaling. Mol Cancer Ther 7:3129–3140

Xiao W, Graham PH, Hao J, Chang L, Ni J, Power CA, Dong Q, Kearsley JH, Li Y (2013) Combination therapy with the histone deacetylase inhibitor LBH589 and radiation is an effective regimen for prostate cancer cells. PLoS One 8:e74253

Xu X, Xie C, Edwards H, Zhou H, Buck SA, Ge Y (2011) Inhibition of histone deacetylases 1 and 6 enhances cytarabine-induced apoptosis in pediatric acute myeloid leukemia cells. PLoS One 6: e17138–e17138

Xu S, Ren J, Chen HB, Wang Y, Liu Q, Zhang R, Jiang SW, Li J (2014) Cytostatic and apoptotic effects of DNMT and HDAC inhibitors in endometrial cancer cells. Curr Pharm Des 20:1881–1887

Yoon JY, Szwajcer D, Ishdorj G, Benjaminson P, Xiao W, Kumar R, Johnston JB, Gibson SB (2013) Synergistic apoptotic response between valproic acid and fludarabine in chronic lymphocytic leukaemia (CLL) cells involves the lysosomal protease cathepsin B. Blood Cancer J 3: e153–e153

Yoon JY, Ishdorj G, Graham BA, Johnston JB, Gibson SB (2014) Valproic acid enhances fludarabine-induced apoptosis mediated by ROS and involving decreased AKT and ATM activation in B-cell-lymphoid neoplastic cells. Apoptosis 19:191–200

Young CS, Clarke KM, Kettyle LM, Thompson A, Mills KI (2017) Decitabine-Vorinostat combination treatment in acute myeloid leukemia activates pathways with potential for novel triple therapy. Oncotarget 8:51429–51446

Zhang X, Yashiro M, Ren J, Hirakawa K (2006) Histone deacetylase inhibitor, trichostatin A, increases the chemosensitivity of anticancer drugs in gastric cancer cell lines. Oncol Rep 16:563–568

Zhou Z, Hao Y, Ngai L, Raptis L, Tsao M-S, Yang X (2011) TAZ is a novel oncogene in non-small cell lung cancer. Oncogene 30:2181–2186

Zuo Y, Xu H, Chen Z, Xiong F, Zhang B, Chen K, Jiang H, Luo C, Zhang H (2020) 17-AAG synergizes with belinostat to exhibit a negative effect on the proliferation and invasion of MDA-MB-231 breast cancer cells. Oncol Rep 43:1928–1944

Futuristic Approaches Towards Designing of Isozyme-Selective Histone Deacetylase Inhibitors Against Zinc-Dependent Histone Deacetylases

11

The toxicity issue associated with histone deacetylase inhibitors (HDACi) has been soothed to a greater degree through combinatorial therapeutic strategy. Most of the HDACi being pan-inhibitors often target a broad range of classical HDACs thereby inducing off-target effects. One of the reasons for off-targeting is the high sequence identity at the active sites of classical HDACs. As of now among the four US FDA approved inhibitors three are pan and only one romidepsin is Class I selective. Certain noticeable side effects like thrombocytopenia and fatigue have been reported with pan-HDACi. Due to various aspects of HDACs including the role of specific isozymes in different cancer types it has been hypothesized that intervention with isozyme-selective HDACi may show superior therapeutic index and lesser toxicity. The studies with isozyme-selective inhibitors are ongoing and their enhanced therapeutic benefit is yet to be proved in clinical models. Here I will discuss the different strategies that have been employed for designing isozyme-selective HDACi.

11.1 Many Cancers Are Associated with Anomalous Expression/Activity of Particular HDAC or HDACs

Not all HDACs are overexpressed in all cancers. While aberrant expression of HDAC5 has implications in hepatocellular carcinoma, elevated expression of HDAC6 has cross-talk with oral squamous cell carcinoma (Feng et al. 2014; Sakuma et al. 2006). Overactivity of HDAC1 has been related to prostate and gastric cancer. High Class IIa HDAC (HDAC7) levels have been identified from the patients of pancreatic cancer (Choi et al. 2001; Ouaïssi et al. 2008). Patients of high-risk medulloblastoma showed enhanced expression of two Class II HDACs (HDAC5 and HDAC9) (Milde et al. 2010). It has been reported that HDAC6 overactivity by way of stabilizing androgen receptor fuels prostate cancer signalling (Gibbs et al. 2009). In glioma cells HDAC1 overexpression facilitates invasion and knockdown of this HDAC hampered invasion and invoked apoptosis in these cells (Wang et al. 2017). Thus from these findings it is evident that not all the 11 classical HDACs are

S. A. Ganai, *Histone Deacetylase Inhibitors in Combinatorial Anticancer Therapy*,
https://doi.org/10.1007/978-981-15-8179-3_11

overexpressed in cancers. Certain cancers show aberrant expression of some HDACs but others are associated with overexpression of a single HDAC.

11.2 Current Concerns with Pan-HDACi

A variety of toxicities have been noticed during the clinical trial study of pan-HDAC inhibitor vorinostat. These harmful effects were dysgeusia, diarrhoea, nausea and thrombocytopenia. Pulmonary embolism, sepsis, anaemia and hypotension were also recorded as dose-limited toxicities. Serious dose-dependent side effects such as anaemia, infection, dehydration, sepsis, hypotension and pulmonary embolism were also observed (Duvic et al. 2007). The efficacy of vorinostat against various solid tumours including colorectal, breast and thyroid cancers was evaluated. Discouragingly none of the patients showed partial or complete response and the frequency of vorinostat-induced harmful effects was found to be quite high. Among the reported side effects are anorexia, fatigue, diarrhoea and thrombocytopenia (Vansteenkiste et al. 2008; Woyach et al. 2009). Vorinostat and romidepsin (Class I selective inhibitor) have been observed to cause cardiotoxicity (Gryder et al. 2012).

11.3 High Sequence Identity Among Isozymes Offers Impediment in Isozyme-Selective Inhibitor Design

Isozyme-selective inhibitors serve as chemical probes for delineating the function of isozymes and have the potential to serve as strong therapeutic agents with relatively lesser adverse effects compared to inhibitors targeting a wide range of HDAC isozymes (Gupta et al. 2012). The high degree of sequence identity at the active sites of isozymes makes the isozyme-selective inhibitor design herculean (Seto and Yoshida 2014). Among the identified HDACi only few have shown strong inhibitory activity and concurrent isozyme selectivity. Maximum sequence identity is seen among the isozymes of a given Class rather than the members of various Classes. For explanation I have performed multiple sequence alignment of Class I HDACs and Class IIa HDACs separately using the MultAlin software (Corpet 1988) (Figs. 11.1 and 11.2).

11.4 Distinct Structural Components of Typical HDACi

HDACi such as TSA and vorinostat contain three distinct regions. These include zinc binding group (ZBG), linker and cap region (Fig. 11.3a) (Ganai 2018). The cap region forms interactions with active site rim residues and macrocyclic HDACi have complex cap region. The linker region unites the ZBG with the cap region and interacts with the tunnel residues of active site (Fig. 11.3b). Moreover, correct positioning of ZBG is also governed by linker region (Yang et al. 2020). ZBG by

```
           1             15             30             45             60            75
           |-------------+--------------+--------------+--------------+-------------|
HDAC1         MAQTQGT-RRKVCYYYDGDVGNYYYGQGHPMKPHRIRMTHNLLLNYGLYRKMEIYRPHKANAEEMTKYHSD
HDAC2         MAYSQGGGKKKVCYYYDGDIGNYYYGQGHPMKPHRIRMTHNLLLNYGLYRKMEIYRPHKATAEEMTKYHSD
HDAC3                   MAKTVAYFYDPDVGNFHYGAGHPMKPHRLALTHSLVLHYGLYKKMIVFKPYQASQHDMCRFHSE
HDAC8      MEEPEEPADSGQSLVPVYIYSPEYVSMCDSLAKI--PKRASMVHSLIEAYALHKQMRIVKPKVASMEEMATFHTD
Consensus  ..........g....v.Y.Ydp#.gn..yg.ghpmkPhR..$tHsL.l.YgLykkM.!.kP..As.e#M..%Hs#

           76            90             105            120            135           150
           |-------------+--------------+--------------+--------------+-------------|
HDAC1      DYIKFLRSIRPDNMSEYSKQMQRFNVGEDCPVFDGLFEFCQLSTGGSVASAVKLNKQQTDIAVNWAGGLHHAKKS
HDAC2      EYIKFLRSIRPDNMSEYSKQMQRFNVGEDCPVFDGLFEFCQLSTGGSVAGAVKLNRQQTDMAVNWAGGLHHAKKS
HDAC3      DYIDFLQRVSPTNMQGFTKSLNAFNVGDDCPVFPGLFEFCSRYTGASLQGATQLNNKICDIAINWAGGLHHAKKF
HDAC8      AYLQHLQKVSQEGDDDHPDSIE-YGLGYDCPATEGIFDYAAAIGGATITAAQCLIDGMCKVAINWSGGWHHAKKD
Consensus  .Yi.fLq.!sp.nm....ks.#.%nvG.DCPvf.GlF#%c...tGas...A..Ln...cd.A!NWaGGlHHAKK.

           151           165            180            195            210           225
           |-------------+--------------+--------------+--------------+-------------|
HDAC1      EASGFCYVNDIVLAILELLKYHQRVLYIDIDIHHGDGVEEAFYTTDRVMTVSFHKYGEYF-PGTGDLRDIGAGKG
HDAC2      EASGFCYVNDIVLAILELLKYHQRVLYIDIDIHHGDGVEEAFYTTDRVMTVSFHKYGEYF-PGTGDLRDIGAGKG
HDAC3      EASGFCYVNDIVIGILELLKYHPRVLYIDIDIHHGDGVQEAFYLTDRVMTVSFHKYGNYFFPGTGDMYEVGAESG
HDAC8      EASGFCYLNDAVLGILRLRRKFERILYVDLDLHHGDGVEDAFSFTSKVMTVSLHKFSPGFFPGTGDVSDVGLGKG
Consensus  EASGFCYvNDiVlgILeLlkyh.R!LY!DiDiHHGDGV##AFy.TdrVMTVSfHK%g.yFfPGTGD..#!GagkG

           226           240            255            270            285           300
           |-------------+--------------+--------------+--------------+-------------|
HDAC1      KYYAVNYPLRDGIDDESYEAIFKPVMSKVMEMFQPSAVVLQCGSDSLSGDRLGCFNLTIKGHAKCVEFVKSFNLP
HDAC2      KYYAVNFPMRDGIDDESYGQIFKPIISKVMEMYQPSAVVLQCGADSLSGDRLGCFNLTVKGHAKCVEVVKTFNLP
HDAC3      RYYCLNVPLRDGIDDQSYKHLFQPVINQVVDFYQPTCIVLQCGADSLGCDRLGCFNLSIRGHGECVEYVKSFNIP
HDAC8      RYYSVNVPIQDGIQDEKYYQICESVLKEVYQAFNPKAVVLQLGADTIAGDPMCSFNMTPVGIGKCLKYILQWQLA
Consensus  rYY.vNvP.rDGI#D#sY.qif.p!...V.#.%#P.a!VLQcGaDsl.gDr$gcFN$t..GhgkCvey!k.f#lp

           301           315            330            345            360           375
           |-------------+--------------+--------------+--------------+-------------|
HDAC1      MLMLGGGGYTIRNVARCWTYETAVALDTEIPNELPYNDYFEYFGPDFKLHISPSN-MTNQNTNEYLEKIKQRLFE
HDAC2      LLMLGGGGYTIRNVARCWTYETAVALDCEIPNELPYNDYFEYFGPDFKLHISPSN-MTNQNTPEYMEKIKQRLFE
HDAC3      LLVLGGGGYTVRNVARCWTYETSLLVEEAISEELPYSEYFEYFAPDFTLHPDVSTRIENQNSRQYLDQIRQTIFE
HDAC8      TLILGGGGYNLANTARCWTYLTGVILGKTLSSEIPDHEFFTAYGPDYVLEITPSCR-PDRNEPHRIQQILNYIKG
Consensus  .L.LGGGGYt.rNvARCWTYeT.v.l...is.ElPy.#%Fey%gPD%.Lhi.pS.r..#qN.p.y.#qI.#.ife

           376           390            405            420            435           450
           |-------------+--------------+--------------+--------------+-------------|
HDAC1      NLRMLPHAPGVQMQAIPEDAIPEESGDEDEDDPDKRISICSSDKRIACEEEFSDSEEEGEGGRKNSSNFKK-AKR
HDAC2      NLRMLPHAPGVQMQAIPEDAVHEDSGDEDGEDPDKRISIRASDKRIACDEEFSDSEDEGEGGRRNVADHKKGAKK
HDAC3      NLKMLNHAPSVQIHDVPADLLTYDRTDE--ADAEERGPEENYSRPEAPNEFYDGDHDNDKESDVEI
HDAC8      NLKHVV
Consensus  NLkml.hap.vq....p.d.......de...d...r..........a..e.........................

           451           465            480          494
           |-------------+--------------+------------|
HDAC1      VKTEDEKEKDPEEKKEVTEEEKTKE---EKPEAKGVKEEVKLA
HDAC2      ARIEEDKKETEDKKTDVKEEDKSKDNSGEKTDTKGTKSEQLSNP
HDAC3
HDAC8
Consensus  ............................................
```

Fig. 11.1 Multiple sequence alignment of four Class I HDAC isozymes showing the similarity/identity of these enzymes. HDAC1, 2 and 3 share more identity than HDAC8. Sequences of human HDAC1/2/3 and 8 were retrieved from UniProt bearing primary accession number Q13547, Q92769, O15379 and Q9BY41, respectively. Alignment was performed by using MultAlin (version 5.4.1). High sequence identity at the active sites of isozymes makes it difficult to design isozyme-selective inhibitors

binding to zinc ion and other nearby residues plays a critical role in HDAC inhibitory function of HDACi (Somoza et al. 2004; Zhang et al. 2018). Modifications in the above-mentioned regions of HDACi have been employed for designing Class-selective and isozyme-selective HDACi.

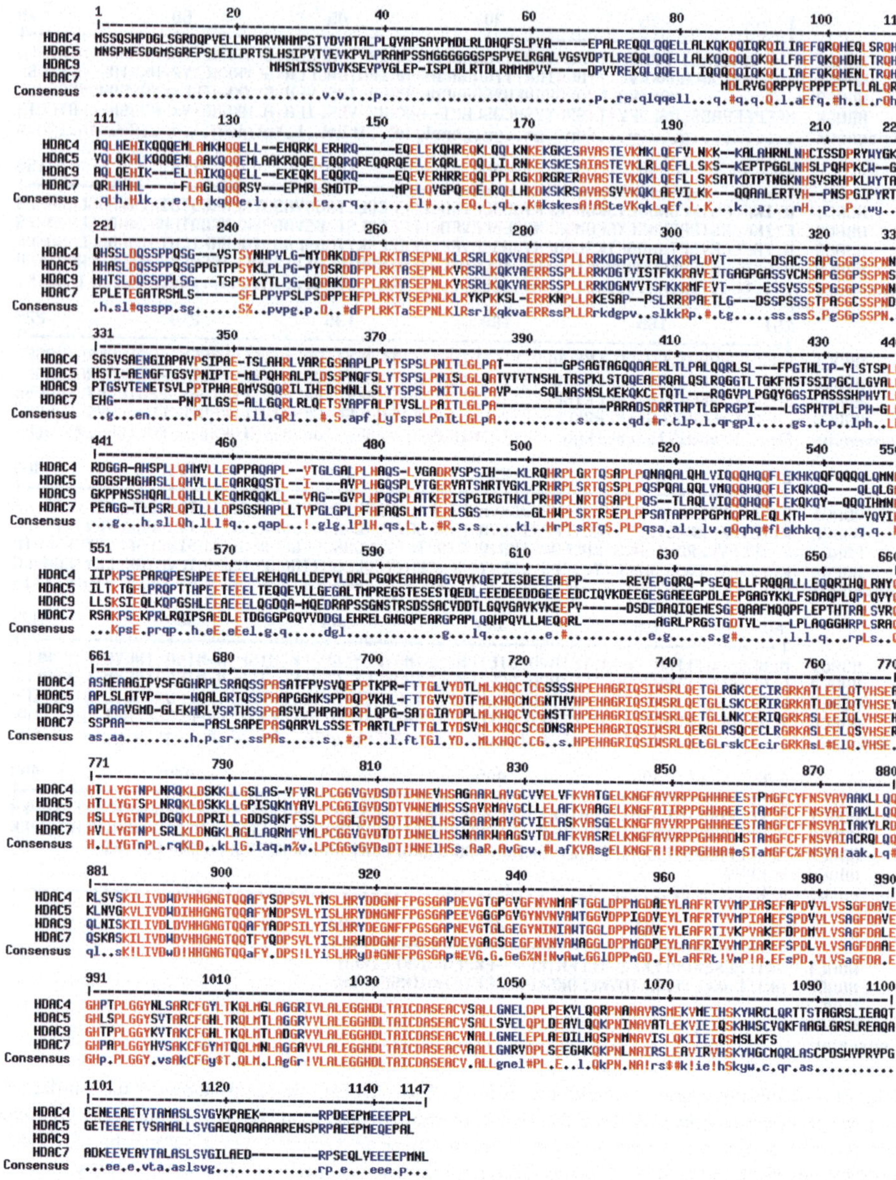

Fig. 11.2 Class IIa HDACs like Class I HDACs share sequence identity which makes the ground fertile for off-targeting. The sequences of human HDAC4/5/7 and 9 were fetched from UniProt bearing citable accession number P56524, Q9UQL6, Q8WUI4 and Q9UKV0, respectively. MultAlin software served the purpose of multiple sequence alignment

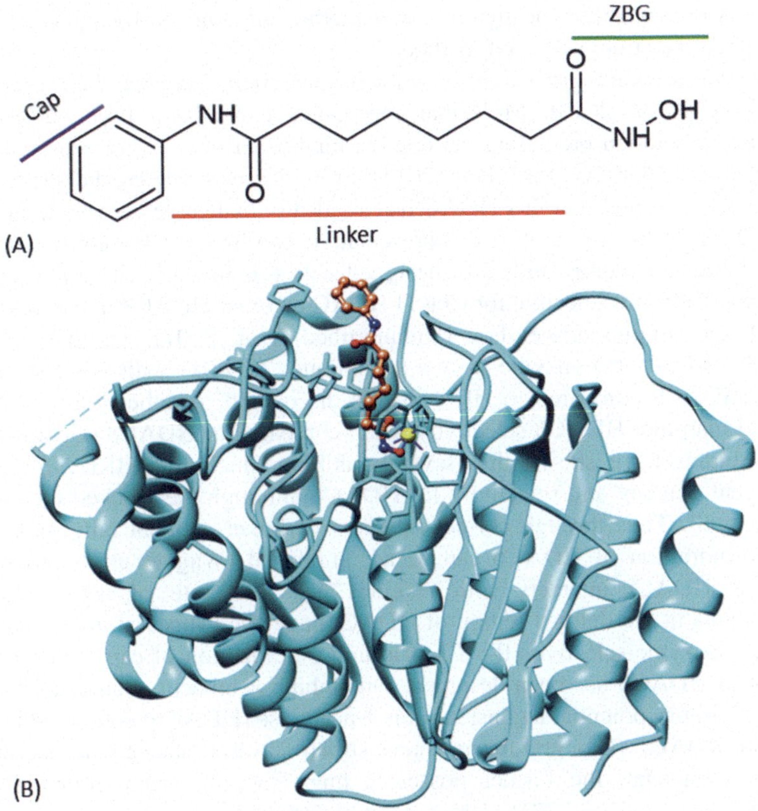

Fig. 11.3 Structural components of the typical pharmacophore of HDAC inhibitors. The structure of vorinostat has been shown and the various components of this inhibitor have been indicated with the help of lines of different colours. This canonical pharmacophore consists of a cap which is connected to zinc binding group (ZBG) with the help of linker (**a**). Modifications in these regions have been used for designing isozyme-selective HDAC inhibitors. (**b**) The Chimera rendered image of PDB structure retrieved from Protein Data Bank (PDB ID: 1T69). The structure is of human HDAC8 with co-crystallized inhibitor vorinostat (SAHA). HDAC8 protein is designated by cyan colour, zinc by yellow sphere and SAHA by orange red colour (ball and stick). The ZBG of SAHA chelates zinc, the linker fits in the active site tunnel and cap region interacts with rim residues of active site

11.5 Isozyme-Selective HDAC Inhibitor Design

It has been speculated that isozyme-selective HDACi may result in relatively better therapeutic efficacy. This has facilitated the designing of isozyme-selective inhibitors. Various approaches including in silico and synthetic ones have been used for designing such inhibitors by modifying the specific regions of HDACi (Ganai 2016). Computer-aided drug designing methods are either structure based where the receptor structure is available or ligand based where the structure of receptor is unavailable. The ligand based method of drug designing is relatively quicker than the structure based method.

Molecular docking sheds light on critical interactions between the receptor and ligands (Fu et al. 2018). Molecular mechanics generalized born surface area (MMGBSA) helps in predicting the relative binding affinity of congeneric ligands towards a particular receptor (Ganai 2015). From the available crystal structures of HDAC8 it was found that this HDAC has atypically malleable active site in which the inhibitors differing from the representative cap-linker-ZBG concept can fit. Certain inhibitors (compounds 5/6) designed with this novel scaffold have proved to be above 100-fold selective for Class I HDAC member HDAC8 in comparison to HDAC1 and tubulin deacetylase (Krennhrubec et al. 2007). Through in silico approach and in vitro enzyme assays an inhibitor (SD-01) with good selectivity towards HDAC8 compared to HDAC6 was discovered. Another inhibitor SD-02 exhibited marginal HDAC6 selectivity when compared to HDAC8 (Debnath et al. 2019). Studies have shown that several inhibitors including BRD9757 exhibit magnificent potency and selectivity for HDAC6 although they lacked the surface-binding motif. Thus through correct selection of linker element only, potent and selective inhibition of HDAC6 can be accomplished (Wagner et al. 2013). Cap region of HDACi interacts with residues near the gate of active site and modifications in the defined region of HDACi have been successfully exploited in designing isozyme-selective HDACi. Through modification of cap region specific inhibitor of HDAC6 has been designed. This inhibitor tubacin manifested fourfold greater inhibitory potential against tubulin deacetylase (HDAC6) compared to Class I isozyme HDAC1. Though tubacin shares structural resemblance with pan-HDAC inhibitor vorinostat, the former possesses huge capping group mimicking the HDAC6 substrate (Ganai 2018; Haggarty et al. 2003a, b).

PCI34051, the broadly used specific inhibitor of HDAC8 (IC50 10 nM) showed above 200-fold selectivity towards this HDAC. The reason for its selectivity has been put forward as its 4-methoxybenzyl part which gets accommodated in the subpocket of this isozyme (Chakrabarti et al. 2016). For designing HDAC8 selective inhibitors, cap region optimization has been the prime focus of researchers among the available designing strategies including ZBG and linker modifications. Evidence based study suggests that the residues making the subpocket of HDAC8, namely tyrosine 293 and methionine 261 should be specially considered while designing HDAC8 selective inhibitors (Zhang et al. 2020). Compound 14, the inhibitor with bulky branched cap showed 20- and 10-fold potencies towards HDAC1 as compared to HDAC3 and HDAC2, respectively (Siliphaivanh et al. 2007; Yang et al. 2019). Isozyme-selective inhibitor design has great medical importance and it is quite challenging to design such inhibitors where the high structural identity occurs among the isozymes. As an example, HDAC1 and HDAC2 share structural similarity to a greater extent and thus designing inhibitors targeting either of these isozymes is an arduous task. A novel and selective time-dependent HDAC2 inhibitor, namely β-hydroxymethyl chalcone has been discovered (Zhou et al. 2015). Two kinetically selective HDAC2 inhibitors (BRD6688 and BRD4884) of HDAC2 were also

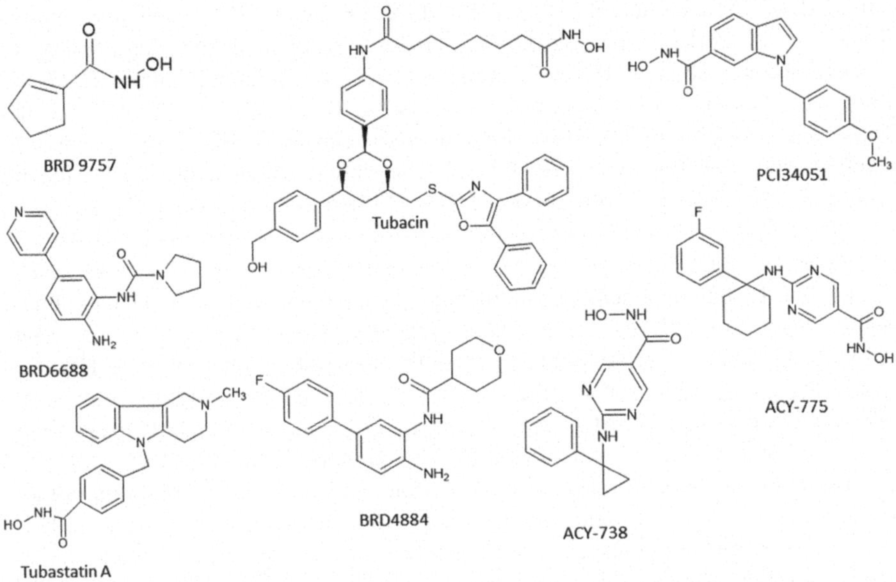

Fig. 11.4 Chemical structures of certain isozyme-selective inhibitors. While BRD 9757, tubacin, tubastatin A, ACY-738 and ACY-775 are selective inhibitors of HDAC6. PCI34051 selectively inhibits HDAC8. BRD6688 and BRD4884 are kinetically selective HDAC2 inhibitors. The help of freeware, namely ACD/ChemSketch was taken for production of these structures

developed (Fig. 11.4). These kinetically selective inhibitors exhibited biased residence time for HDAC2 in comparison to structurally similar HDAC1 (Wagner et al. 2015).

Through click chemistry approach, a novel lead compound 5 g (NSC746457) has been discovered. During preliminary testing this compound showed HDAC1 inhibitory activity with no substantial HDAC8 inhibition. Its inhibitory activity was comparable to that of first approved HDAC inhibitor vorinostat (Shen et al. 2008). Among the novel hybrid molecules synthesized, 8n showed better HDAC1 selectivity over HDAC8/HDAC6 compared to vorinostat. This inhibitor also invoked cancer cell cycle arrest in a dose-dependent fashion (Cai et al. 2015). A library of compounds were designed for finding HDAC8 selective inhibitors. These compounds synthesized by way of click chemistry were composed of a zinc binding group which was connected to a capping structure through triazole moiety. Screening of this library resulted in the identification of a HDAC8 selective inhibitor C149 which proved to be more potent than the known HDAC8 inhibitor PCI-34058 (IC50 = 0.31 μM) (Suzuki et al. 2012). In another study compound 12a has been synthesized by combining the BG45 and tacedinaline scaffolds. Further the structure of this compound was optimized and various 2-aminobenzamide derivatives were generated. These compounds on subsequent testing proved to be effective HDACi. Compound 26c showed HDAC3 selectivity even better than the BG45, the prototype inhibitor of HDAC3. This compound (26c) showed superior antitumour efficacy in cell models when compared to two prototype inhibitors BG45 and tacedinaline (Trivedi et al. 2018). A triazole library screening resulted in the identification of two potent and selective HDAC3 inhibitors. These compounds, namely T247 and T326 were found to inhibit HDAC3 potentially where other isozymes were not obstructed strongly. In HCT116 (human colon cancer) cells, these compounds induced selective escalation of NF-κB acetylation dose dependently clearly suggesting the selective obstruction of HDAC3 (Suzuki et al. 2013).

Ligand T247 exhibited more stability in docked condition with HDAC3 in comparison to HDAC2 as revealed by simulation output (Tsukamoto et al. 2018). Compound 17 having phenyl ring (meta-position) compared to SAHA exhibited better selectivity for HDAC1 over Class IIa member HDAC7. The anticancer effect of this compound was relatively better than the above-mentioned drug SAHA (Sun et al. 2013). Combinatorial in silico approach was used for designing potent and selective inhibitors of HDAC6. An optimized compound with HDAC6 selectivity, designated as tubastatin A was screened. This compound enhanced the acetylated alpha-tubulin levels in primary cortical neurons under in vitro conditions without altering the histone acetylation strongly validating its selectivity towards tubulin deacetylase HDAC6 (Butler et al. 2010). HDAC6 overexpression/overactivity has implications in a variety of disorders ranging from cancer to neurodegeneration. This HDAC deacetylates several non-histone targets including cortactin, HSP90, Foxp3 and tubulin. Selective inhibition of this isozyme does not induce cytotoxic effects in typical (healthy) cells, the condition often seen with intervention of Class I isozymes. Potent and selective inhibitors of this Class IIb HDAC, with pentaheterocyclic cyclic core have been synthesized. These inhibitors under in vitro *and* in vivo conditions showed low toxicity and were found to enhance the regulatory T cells function at nicely bearable concentrations (Vergani et al. 2019). Two compounds of pyrimidine hydroxyl amide family (ACY-738 and ACY-775) were found to be selective towards HDAC6 (Fig. 11.4). These two small molecules, over Class I HDACs, demonstrated selectivity (60- to 1500-fold) for HDAC6 (Jochems et al. 2013). Another small molecule inhibitor selective to HDAC6 (WT161) has been developed and tested alone or in conjunction with proteasome inhibitors against multiple myeloma (Hideshima et al. 2016).

Designing isozyme-selective inhibitors through modifications in the zinc binding group is comparatively more difficult than designing such inhibitors through cap group modifications. This difficulty has been ascribed to high structural similarity near the zinc across HDAC isozymes. However, HDACs do contain different binding pockets in the zinc vicinity which have been utilized for isozyme-selective inhibitor design (Ganai 2016). Several chlamydocin derivatives bearing different metal (zinc) binding substituents were developed. One of the compounds designated as compound 24 showed 15- and 4-fold selectivity for HDAC4 (IC50 = 60 nM) over HDAC6 and HDAC1, respectively (Bhuiyan et al. 2006; Bieliauskas and Pflum 2008). The active site tunnel residues being highly conserved among the isozymes offer strong hindrance in designing isozyme-selective inhibitors. Crystal structure of HDAC8 in bound state with CRA-A (inhibitor with aryl linker) showed a subpocket that was not evident when this HDAC was in docked state with vorinostat which has non-aryl linker (Somoza et al. 2004). This large subpocket on CRA-A bound HDAC8 gets formed by substantial shifting of Phe152 of active site channel of HDAC8 from its usual position. Based on this analysis HDACi were designed for targeting this subpocket specifically and this resulted in the genesis of HDAC8 selective inhibitors. While compound 26 showed over 180- and 330-fold selectivity for HDAC8 over HDAC6 and HDAC1, respectively, compound 25 displayed greater than 115- and 140-fold selectivity for HDAC8 over the other two HDACs (HDAC6/1), respectively. It has been proposed that HDAC8 selectivity may be influenced by the existence of aromatic linker next to zinc binding moiety (Bieliauskas and Pflum 2008). PCI-34051, a rationally designed inhibitor, possessing an indole ring in the linker region exhibited more than 290-fold selectivity towards HDAC8 over HDAC1/2/3/6/10 (Balasubramanian et al. 2008). Another study also proved the importance of aromatic linker in determining HDAC8 selectivity (Bieliauskas and Pflum 2008; Hu et al. 2003). Thiolate analogues were designed keeping in view the structure of selective substrate of HDAC6. Aliphatic compounds (17b–20b) enhanced tubulin acetylation selectively over histone H4 acetylation. Compounds 17–19a demonstrated selective inhibition of tubulin deacetylase over HDAC1/HDAC4 (Ganai 2016; Suzuki et al. 2006).

As aforementioned isozyme-selective HDAC inhibitor plays a key role in exploring the biological functions of HDAC to which the inhibitor is specific. Class IIb member HDAC10 also known as polyamine deacetylase has implications in chemotherapy resistance (Islam et al. 2017; Oehme et al. 2013). Tubastatin A, the well-known inhibitor of HDAC6, was found to strongly bind HDAC10 (Géraldy et al. 2019). Polyamine deacetylase shares not only structural but also pharmacological features with tubulin deacetylase (Guardiola and Yao 2002). Derivatives of tubastatin A were synthesized and it was revealed that basic amine in its cap region promotes strong binding to HDAC10. In neuroblastoma cell line HDAC10 inhibitors mimicked the knockdown of this HDAC as evident from the assembling of acid vesicles in this cell line. Molecular docking studies using HDAC10 models showed that hydrogen bonding interaction between the cap group nitrogen and the Glu272 (gate keeper residue) is crucial for strong polyamine deacetylase binding (Géraldy et al. 2019). Recently the lone member of Class IV (HDAC11) has also been focussed. Isozyme-selective inhibitors will serve as scaffolds for designing better therapeutic agents. Activity-guided design was employed for development of HDAC11 specific inhibitors. Very little is known about the biological function and activity of this HDAC. Emerging evidences suggest that HDAC11 has effective defatty-acylation function and thus intervention of this enzyme may prove promising in tackling various human diseases such as multiple sclerosis and metabolic diseases (Cao et al. 2019). The best inhibitor SIS17 prevented the demyristoylation of serine hydroxymethyl transferase 2 (HDAC11 substrate) without targeting other HDACs suggesting that this inhibitor is active in cells (Cao et al. 2019; Son et al. 2019). From these results it is obvious that activity-guided design may prove fruitful for developing specific inhibitors for other HDACs as well.

Selectivity of tubulin deacetylase inhibitors is susceptible to physicochemical properties of their zinc binding group, linker apart from their cap group size thereby making the discovery of novel HDAC6 specific inhibitors arduous. Binding modes of trichostatin A enantiomers (two) were studied at atomic level, particularly in HDAC6 using various computational strategies. Two non-conservative residues of HDAC6 (M205 and F202) and four residues (conservative) located deeply inside the binding pocket (hydrophobic) were first discovered to be critical in determining the selectivity of selective *(S)*-TSA towards this HDAC. Following this a novel mechanism for *(S)*-TSA selectivity towards HDAC6 was put forward. This mechanism was thought to involve the trigger by non-conservative residues (F202/M205) and the better fitting of *(S)*-TSA in the hydrophobic binding cavity (pocket). Taken together, two enantiomers of TSA were employed to unravel the mechanism crucial for deciding selectivity of selective-HDAC6 inhibitors (Zhang et al. 2019). As two residues F202 and M205 were found to be decisive in determining HDAC6 selectivity of *(S)*-TSA, thus these residues should be given prime importance while designing novel and selective inhibitors of HDAC6. HDACs being epigenetic regulators of gene expression have been identified as potential drug targets for circumventing a variety of cancers. Three novel inhibitors of HDAC8 were discovered through pharmacophore-based virtual screening. Downstream studies showed that compound H8-A5 to be selective for HDAC8 over HDAC1 and HDAC4. This compound inhibited proliferation of cancer cells (MDA-MB-231) under in vitro conditions. Docking studies followed by molecular dynamics simulations proposed a possible binding mode of this novel inhibitor against HDAC8 (Hou et al. 2015). Energetically optimized pharmacophores approach has been used for identifying different pharmacophoric features of inhibitors against Class I and Class II HDACs. The same inhibitor in the active sites of different HDAC isozymes shows distinct e-pharmacophoric features. For instance, HDAC inhibitor NK57 showed 3, 5, and 6 features against HDAC4, HDAC6, and HDAC5, respectively (Ganai et al. 2015; Kalyaanamoorthy and Chen 2013). These e-pharmacophores may be used as queries in e-pharmacophores based virtual screening for identifying isozyme-selective hits which on enzymatic studies may turn to lead compounds.

Various biaryl indolyl benzamide compounds were synthesized and tested for selectivity towards HDAC1. In this design two distinct fragments from known HDAC1/3 and HDAC1/2 selective inhibitors were combined. These fragments may result in selectivity towards HDAC1. Molecular docking studies were done to certify the design of inhibitors. In vitro screening was used for evaluating the potency and isozyme selectivity of novel compounds. Modest selectivity towards HDAC1 was exhibited by all analogues over HDAC2. The best compound in terms of potency and selectivity proved to be Bnz-3. Speaking in terms of potency, Bnz-3 compared to other compounds demonstrated 10- to 100-fold greater potency. Bnz-3 over HDAC2 manifested 10.6-fold selectivity for another member of Class I (HDAC1). This inhibitor showed nearly no inhibitory activity against HDAC6 or HDAC3 (Negmeldin and Pflum 2019). Thus Bnz-3 represents the strong lead compound that may assist in designing more potent and selective inhibitors against HDAC1. Keeping in view the concerns of pan-HDACi strong emphasis has been given to designing and synthesis of isozyme-selective inhibitors. It has been found that selective inhibitors of HDAC8 adopt a conformation (L-shaped) due to which they can bind the HDAC8-specific pocket. This pocket is formed by catalytic tyrosine residue of HDAC8 and the L1/L6 loops of this HDAC. The L-shaped inhibitor binding in other isozymes is sterically obstructed by L1–L6 lock. Shielding of this specific HDAC8 pocket through protein engineering diminishes the potency of selective inhibitors of this HDAC (Marek et al. 2018). Among the classical HDACs, the active site of HDAC8 is highly conserved. From a panel of trifluoromethylketone inhibitors of Class I member HDAC8, compound 10 demonstrated kinetic selectivity against HDAC8. Further studies such as molecular docking and flexibility analysis of binding site revealed that this compound occupies the catalytic site (conserved) and a nearby subpocket (transient) of HDAC8 (Schweipert et al. 2019). For selective and potent HDAC6 inhibitors phenothiazine system was recognized as suitable cap group. The ability of phenothiazine-based benzhydroxamic acids to selectively inhibit HDAC6 was evaluated using recombinant enzyme assays and cell based assays followed by western blotting for tubulin and histone acetylation. As revealed by structure–activity relationship studies, introduction of nitrogen atom in phenothiazine framework culminated in enhanced potency and selectivity towards the tubulin deacetylase (HDAC6). Greater than 500-fold selectivity was observed for HDAC6 over HDAC1/4/8. The binding mode of the strong azaphenothiazine inhibitor with the zebra fish HDAC6 (catalytic domain 2) was assured through co-crystallization (Vögerl et al. 2019).

11.6 Future Directions

HDACi have proved effective and thus have been approved for treating certain haematological malignancies. However, their use against solid tumours is strongly impeded due to little treatment efficacy. Nowadays nanotechnology has gained importance in cancer therapy and studies have shown that through nanotechnology approaches drug stability can be highly improved, circulation half-life of the drug can be prolonged and enhanced intratumoural drug availability can be acquired. Thus the therapeutic efficacy of HDACi can be strongly enhanced by using nanotechnology approaches. As aforementioned the restricted success of HDACi against solid cancers may be attributed to poor pharmacokinetics (short half-life/clearance/fast metabolism), low solubility and low cell or tissue permeability, low specificity resulting in off-target effects and rapid induction of drug resistance. The strong solution to these concerns of HDAC inhibitor-based therapy is target delivery and controlled release of the drug/inhibitor. Anticancer nanomedicines show improved efficacy, lesser side toxicity by way of escalating drug solubility, better pharmacokinetic profiles and increased drug delivery to tumours. Through size and surface property fine-tuning nanoparticles which are rationally designed can increase drug accumulation and its uptake by tumour cells. Intratumoural infiltration can be enhanced by small size as nanoparticles in the size range between 20 and 60 nm are having high likeliness of penetrating tumour tissues (Tu et al. 2020).

By circumventing the limitations of HDACi, nanotechnology-based delivery can boost their antitumour efficacy. Many HDAC inhibitors show poor solubility. Due to its biocompatibility and biodegradable properties starch has wide usage in pharmaceutics. It has been found that the water solubility/cellular uptake of HDAC inhibitor (CG-1521) can be increased by encapsulating this inhibitor in starch nanoparticles. This encapsulation substantially enhanced its cytotoxic effect against breast cancer cells compared to free CG-1521 (Alp et al. 2019). Poor solubility of hydroxamate panobinostat hampers its efficacy against brain cancer. Panobinostat-loaded P407 micelles enhanced the intracranial concentration of this HDAC inhibitor culminating in glioma repression (Singleton et al. 2017). Due to bladder permeability barrier it is herculean for drugs to get penetrated in the bladder tumour. PLGA nanoparticles loaded with HDAC inhibitor belinostat were modified with poly (guanidinium oxanorbornene) for betterment of their blood permeability barrier permeability. Compared to nanoparticles without modification with PGON, the PGON-modified nanoparticles increased the penetration (mouse bladder) over 10-fold (Martin et al. 2013). Combinatorial therapy as discussed in the previous chapter not only improves efficacy but also reduces dose-limited toxicity as well as drug resistance. Despite this benefit, due to asynchronous distribution more than additive effect (synergistic effect) may not be achieved under in vivo conditions. This distinct fate of combined drugs under in vivo set-up may be responsible for the variability of results that is observed during transition of experimental conditions from in vitro *to* in vivo (Zhang et al. 2017). Synchronized delivery and synchronized pharmacokinetic profile of drug combination can be achieved by employing nanotechnology-based combinatorial therapy. It is well known that HDACi act in the cellular nucleus and thus nuclear targeted delivery tactics based on nanotechnology may possibly enhance the efficacy of treatment. Certain HDACs get recruited to genome by interacting with other proteins and thus facilitate gene repression. For instance, Class IIa HDACs require the interaction with Myocyte enhancer factor 2 (MEF2) for genomic targeting (Lu et al. 2000). Amphipathic helix in the amino terminal regulatory domains of these HDACs binds to hydrophobic groove on MADS-box/MEF2 domain, the region of MEF2 (Han et al. 2005). Thus small molecules having the ability to bind to the hydrophobic pocket (MEF2) may prevent the recruitment of these (Class IIa) HDACs to DNA thereby inhibiting their function. Thus targeting the interaction site between HDACs and the protein/proteins recruiting them to specific genes may indirectly inhibit HDACs. By this way many HDACs will be simultaneously but indirectly inhibited by a single small molecule.

HDAC inhibitors are also investigated against brain cancers and other central nervous system diseases including Alzheimer's disease (Kazantsev and Thompson 2008). HDACi like vorinostat, tubastatin A and entinostat due to low blood–brain barrier permeability show lesser than average brain uptake. This hampers their clinical usage against central nervous system disorders. Recently a panel of HDACi containing benzoheterocycle were designed followed by synthesis and biological evaluation. Compound 9b among the synthesized compounds demonstrated encouraging antiproliferative effect against cancer cell line (SH-SY5Y) dose and time dependently. This compound proved to be the efficient inhibitor of HDAC1/HDAC6 as evidenced from the enzyme and cell-based assays. Compound 9b compared to vorinostat demonstrated higher blood–brain barrier permeability as revealed by the in vivo pharmacokinetic studies (Choi et al. 2019). However, further studies are required to design highly brain penetrant isozyme-selective HDACi for effective therapeutic intervention against central nervous system maladies.

From combinatorial therapeutic strategy the attention has shifted towards designing of ligands targeting two or more therapeutic targets. Designing of such multi-target ligands has gained substantial scientific interest due to certain benefits offered by this approach (Rosini 2014; Talevi 2015). The advantages of using multi-target ligands/inhibitors over using mixtures of drugs are low risk of drug interactions, better predictable pharmacokinetics; lesser intellectual property rights hurdles (Morphy et al. 2004; Zimmermann et al. 2007). However, it is highly advisable that multi-target ligands should possess desired selectivity towards the receptors needed to be targeted (Skok et al. 2020). A new bendamustine derived therapeutic, namely CY190602 has been developed. This DNA/HDAC dual targeting molecule has shown markedly enhanced anticancer activity under both in vitro and in vivo set-up (Liu et al. 2015). The dual targeting inhibitors having HDAC/HDACs as one of the targets require further evaluation from clinical trial studies for promotion from lab to clinic.

Hydroxamate group HDACi under in vivo conditions are quickly metabolized and are not thus very stable. Taking vorinostat as an example its half-life is below 2 h and thus continuous administration of this HDAC inhibitor is required for achieving the desired therapeutic effect (Hamze 2020). These stability related issues can be solved either by designing suitable prodrugs or by employing novel and highly stable zinc binding groups. Substrate like peptide inhibitor inhibiting HDAC1 at nanomolar concentration was developed. This inhibitor was based on specific substrate of HDAC1 (H4K16), the hydroxamic acid functionality was substituted for K16 (Watson et al. 2016). Peptide based HDAC inhibitors derived from another HDAC1 preferential substrate H3K56 were designed. From the crystal structure of histone H3 it is quite evident that H3K56 occurs towards C terminus region of H3 which is alpha-helical consisting of residues from 45 to 56. Better HDAC1/2 inhibition was demonstrated by 16cyc-HxA. Peptide 16cyc-HxA like TSA exhibited pan-HDAC inhibition but only little effect was noted on SIRT1 (Class III HDAC). Further the relatively stronger potency of 16cyc-HxA compared to 16lin-HxA indicated the positive correlation between helical stabilization and target/receptor binding affinity. The designed peptides induced stronger antiproliferative effects in cancer stem like cells and imparted minimum toxicity to typical cells in comparison to small molecule HDAC inhibitor vorinostat (Wang et al. 2019). Thus, it tempts me to speculate that, isozyme-selective peptide inhibitors should be designed and tested. These inhibitors may replace the conventional HDAC inhibitors which are notorious for exerting cytotoxicity towards normal cells as well.

References

Alp E, Damkaci F, Guven E, Tenniswood M (2019) Starch nanoparticles for delivery of the histone deacetylase inhibitor CG-1521 in breast cancer treatment. Int J Nanomedicine 14:1335–1346

Balasubramanian S, Ramos J, Luo W, Sirisawad M, Verner E, Buggy JJ (2008) A novel histone deacetylase 8 (HDAC8)-specific inhibitor PCI-34051 induces apoptosis in T-cell lymphomas. Leukemia 22:1026–1034

Bhuiyan MP, Kato T, Okauchi T, Nishino N, Maeda S, Nishino TG, Yoshida M (2006) Chlamydocin analogs bearing carbonyl group as possible ligand toward zinc atom in histone deacetylases. Bioorg Med Chem 14:3438–3446

Bieliauskas AV, Pflum MKH (2008) Isoform-selective histone deacetylase inhibitors. Chem Soc Rev 37:1402–1413

Butler KV, Kalin J, Brochier C, Vistoli G, Langley B, Kozikowski AP (2010) Rational design and simple chemistry yield a superior, neuroprotective HDAC6 inhibitor, tubastatin A. J Am Chem Soc 132:10842–10846

Cai M, Hu J, Tian J-L, Yan H, Zheng C-G, Hu W-L (2015) Novel hybrids from N-hydroxyarylamide and indole ring through click chemistry as histone deacetylase inhibitors with potent antitumor activities. Chin Chem Lett 26:675–680

Cao J, Sun L, Aramsangtienchai P, Spiegelman N, Zhang X, Huang W, Seto E, Lin H (2019) HDAC11 regulates type I interferon signaling through defatty-acylation of SHMT2. Proc Natl Acad Sci 116:201815365

Chakrabarti A, Melesina J, Kolbinger FR, Oehme I, Senger J, Witt O, Sippl W, Jung M (2016) Targeting histone deacetylase 8 as a therapeutic approach to cancer and neurodegenerative diseases. Future Med Chem 8:1609–1634

Choi JH, Kwon HJ, Yoon BI, Kim JH, Han SU, Joo HJ, Kim DY (2001) Expression profile of histone deacetylase 1 in gastric cancer tissues. Japanese J Cancer Res 92:1300–1304

Choi MA, Park SY, Chae HY, Song Y, Sharma C, Seo YH (2019) Design, synthesis and biological evaluation of a series of CNS penetrant HDAC inhibitors structurally derived from amyloid-β probes. Sci Rep 9:13187

Corpet F (1988) Multiple sequence alignment with hierarchical clustering. Nucleic Acids Res 16:10881–10890

Debnath S, Debnath T, Bhaumik S, Majumdar S, Kalle AM, Aparna V (2019) Discovery of novel potential selective HDAC8 inhibitors by combine ligand-based, structure-based virtual screening and in-vitro biological evaluation. Sci Rep 9:17174

Duvic M, Talpur R, Ni X, Zhang C, Hazarika P, Kelly C, Chiao JH, Reilly JF, Ricker JL, Richon VM, Frankel SR (2007) Phase 2 trial of oral vorinostat (suberoylanilide hydroxamic acid, SAHA) for refractory cutaneous T-cell lymphoma (CTCL). Blood 109:31–39

Feng GW, Dong LD, Shang WJ, Pang XL, Li JF, Liu L, Wang Y (2014) HDAC5 promotes cell proliferation in human hepatocellular carcinoma by up-regulating Six1 expression. Eur Rev Med Pharmacol Sci 18:811–816

Fu Y, Zhao J, Chen Z (2018) Insights into the molecular mechanisms of protein-ligand interactions by molecular docking and molecular dynamics simulation: a case of oligopeptide binding protein. Comput Math Methods Med 2018:3502514

Ganai S (2015) In silico approaches towards safe targeting of class I histone deacetylases. https://doi.org/10.1007/978-1-4614-6436-5_459-1, pp 1–9

Ganai SA (2016) Novel approaches towards designing of isoform-selective inhibitors against class II histone deacetylases: the acute requirement for Targetted anticancer therapy. Curr Top Med Chem 16:2441–2452

Ganai SA (2018) Designing isoform-selective inhibitors against classical HDACs for effective anticancer therapy: insight and perspectives from in Silico. Curr Drug Targets 19:815–824

Ganai SA, Shanmugam K, Mahadevan V (2015) Energy-optimised pharmacophore approach to identify potential hotspots during inhibition of class II HDAC isoforms. J Biomol Struct Dyn 33:374–387

Géraldy M, Morgen M, Sehr P, Steimbach RR, Moi D, Ridinger J, Oehme I, Witt O, Malz M, Nogueira MS, Koch O, Gunkel N, Miller AK (2019) Selective inhibition of histone deacetylase 10: hydrogen bonding to the gatekeeper residue is implicated. J Med Chem 62:4426–4443

Gibbs A, Schwartzman J, Deng V, Alumkal J (2009) Sulforaphane destabilizes the androgen receptor in prostate cancer cells by inactivating histone deacetylase 6. Proc Natl Acad Sci 106:16663–16668

Gryder BE, Sodji QH, Oyelere AK (2012) Targeted cancer therapy: giving histone deacetylase inhibitors all they need to succeed. Future Med Chem 4:505–524

Guardiola AR, Yao TP (2002) Molecular cloning and characterization of a novel histone deacetylase HDAC10. J Biol Chem 277:3350–3356

Gupta P, Reid RC, Iyer A, Sweet MJ, Fairlie DP (2012) Towards isozyme-selective HDAC inhibitors for interrogating disease. Curr Top Med Chem 12:1479–1499

Haggarty SJ, Koeller KM, Wong JC, Butcher RA, Schreiber SL (2003a) Multidimensional chemical genetic analysis of diversity-oriented synthesis-derived deacetylase inhibitors using cell-based assays. Chem Biol 10:383–396

Haggarty SJ, Koeller KM, Wong JC, Grozinger CM, Schreiber SL (2003b) Domain-selective small-molecule inhibitor of histone deacetylase 6 (HDAC6)-mediated tubulin deacetylation. Proc Natl Acad Sci U S A 100:4389–4394

Hamze A (2020) How do we improve histone deacetylase inhibitor drug discovery? Expert Opin Drug Discovery 15:527–529

Han A, He J, Wu Y, Liu JO, Chen L (2005) Mechanism of recruitment of class II histone deacetylases by myocyte enhancer factor-2. J Mol Biol 345:91–102

Hideshima T, Qi J, Paranal RM, Tang W, Greenberg E, West N, Colling ME, Estiu G, Mazitschek R, Perry JA, Ohguchi H, Cottini F, Mimura N, Görgün G, Tai Y-T, Richardson PG, Carrasco RD, Wiest O, Schreiber SL, Anderson KC, Bradner JE (2016) Discovery of selective small-molecule HDAC6 inhibitor for overcoming proteasome inhibitor resistance in multiple myeloma. Proc Natl Acad Sci 113:13162–13167

Hou X, Du J, Liu R, Zhou Y, Li M, Xu W, Fang H (2015) Enhancing the sensitivity of pharmacophore-based virtual screening by incorporating customized ZBG features: a case study using histone deacetylase 8. J Chem Inf Model 55:861–871

Hu E, Dul E, Sung C-M, Chen Z, Kirkpatrick R, Zhang G-F, Johanson K, Liu R, Lago A, Hofmann G, Macarron R, De Los Frailes M, Perez P, Krawiec J, Winkler J, Jaye M (2003) Identification of novel isoform-selective inhibitors within class I histone deacetylases. J Pharmacol Exp Ther 307:720–728

Islam MM, Banerjee T, Packard CZ, Kotian S, Selvendiran K, Cohn DE, Parvin JD (2017) HDAC10 as a potential therapeutic target in ovarian cancer. Gynecol Oncol 144:613–620

Jochems J, Boulden J, Lee B, Blendy J, Jarpe M, Mazitschek R, Duzer V, Berton O (2013) Antidepressant-like properties of novel HDAC6 selective inhibitors with improved brain bioavailability. Neuropsychopharmacology

Kalyaanamoorthy S, Chen Y-PP (2013) Energy based pharmacophore mapping of HDAC inhibitors against class I HDAC enzymes. Biochim Biophys Acta Proteins Proteomics 1834:317–328

Kazantsev AG, Thompson LM (2008) Therapeutic application of histone deacetylase inhibitors for central nervous system disorders. Nat Rev Drug Discov 7:854–868

Krennhrubec K, Marshall BL, Hedglin M, Verdin E, Ulrich SM (2007) Design and evaluation of 'Linkerless' hydroxamic acids as selective HDAC8 inhibitors. Bioorg Med Chem Lett 17:2874–2878

Liu C, Ding H, Li X, Pallasch CP, Hong L, Guo D, Chen Y, Wang D, Wang W, Wang Y, Hemann MT, Jiang H (2015) A DNA/HDAC dual-targeting drug CY190602 with significantly enhanced anticancer potency. EMBO Mol Med 7:438–449

Lu J, McKinsey TA, Zhang CL, Olson EN (2000) Regulation of skeletal myogenesis by association of the MEF2 transcription factor with class II histone deacetylases. Mol Cell 6:233–244

Marek M, Shaik TB, Heimburg T, Chakrabarti A, Lancelot J, Ramos-Morales E, Da Veiga C, Kalinin D, Melesina J, Robaa D, Schmidtkunz K, Suzuki T, Holl R, Ennifar E, Pierce RJ, Jung M, Sippl W, Romier C (2018) Characterization of histone deacetylase 8 (HDAC8) selective inhibition reveals specific active site structural and functional determinants. J Med Chem 61:10000–10016

Martin DT, Hoimes CJ, Kaimakliotis HZ, Cheng CJ, Zhang K, Liu J, Wheeler MA, Kelly WK, Tew GN, Saltzman WM, Weiss RM (2013) Nanoparticles for urothelium penetration and delivery of the histone deacetylase inhibitor belinostat for treatment of bladder cancer. Nanomedicine 9:1124–1134

Milde T, Oehme I, Korshunov A, Kopp-Schneider A, Remke M, Northcott P, Deubzer HE, Lodrini M, Taylor MD, von Deimling A, Pfister S, Witt O (2010) HDAC5 and HDAC9 in medulloblastoma: novel markers for risk stratification and role in tumor cell growth. Clin Cancer Res 16:3240–3252

Morphy R, Kay C, Rankovic Z (2004) From magic bullets to designed multiple ligands. Drug Discov Today 9:641–651

Negmeldin AT, Pflum MKH (2019) Abstract 19: design and synthesis of biaryl indolyl benzamides as HDAC1-selective inhibitors via a fragment-based lead generation approach. Cancer Res 79:19–19

Oehme I, Linke J-P, Böck BC, Milde T, Lodrini M, Hartenstein B, Wiegand I, Eckert C, Roth W, Kool M, Kaden S, Gröne H-J, Schulte JH, Lindner S, Hamacher-Brady A, Brady NR, Deubzer HE, Witt O (2013) Histone deacetylase 10 promotes autophagy-mediated cell survival. Proc Natl Acad Sci 110:E2592–E2601

Ouaïssi M, Sielezneff I, Silvestre R, Sastre B, Bernard JP, Lafontaine JS, Payan MJ, Dahan L, Pirrò N, Seitz JF, Mas E, Lombardo D, Ouaissi A (2008) High histone deacetylase 7 (HDAC7) expression is significantly associated with adenocarcinomas of the pancreas. Ann Surg Oncol 15:2318–2328

Rosini M (2014) Polypharmacology: the rise of multitarget drugs over combination therapies. Future Med Chem 6:485–487

Sakuma T, Uzawa K, Onda T, Shiiba M, Yokoe H, Shibahara T, Tanzawa H (2006) Aberrant expression of histone deacetylase 6 in oral squamous cell carcinoma. Int J Oncol 29:117–124

Schweipert M, Jänsch N, Sugiarto WO, Meyer-Almes FJ (2019) Kinetically selective and potent inhibitors of HDAC8. Biol Chem 400:733–743

Seto E, Yoshida M (2014) Erasers of histone acetylation: the histone deacetylase enzymes. Cold Spring Harb Perspect Biol 6:a018713

Shen J, Woodward R, Kedenburg JP, Liu X, Chen M, Fang L, Sun D, Wang PG (2008) Histone deacetylase inhibitors through click chemistry. J Med Chem 51:7417–7427

Siliphaivanh P, Harrington P, Witter DJ, Otte K, Tempest P, Kattar S, Kral AM, Fleming JC, Deshmukh SV, Harsch A, Secrist PJ, Miller TA (2007) Design of novel histone deacetylase inhibitors. Bioorg Med Chem Lett 17:4619–4624

Singleton WG, Collins AM, Bienemann AS, Killick-Cole CL, Haynes HR, Asby DJ, Butts CP, Wyatt MJ, Barua NU, Gill SS (2017) Convection enhanced delivery of panobinostat (LBH589)-loaded pluronic nano-micelles prolongs survival in the F98 rat glioma model. Int J Nanomedicine 12:1385–1399

Skok Ž, Zidar N, Kikelj D, Ilaš J (2020) Dual inhibitors of human DNA topoisomerase II and other cancer-related targets. J Med Chem 63:884–904

Somoza JR, Skene RJ, Katz BA, Mol C, Ho JD, Jennings AJ, Luong C, Arvai A, Buggy JJ, Chi E, Tang J, Sang BC, Verner E, Wynands R, Leahy EM, Dougan DR, Snell G, Navre M, Knuth MW, Swanson RV, McRee DE, Tari LW (2004) Structural snapshots of human HDAC8 provide insights into the class I histone deacetylases. Structure 12:1325–1334

Son SI, Cao J, Zhu C-L, Miller SP, Lin H (2019) Activity-guided design of HDAC11-specific inhibitors. ACS Chem Biol 14:1393–1397

Sun Q, Yao Y, Liu C, Li H, Yao H, Xue X, Liu J, Tu Z, Jiang S (2013) Design, synthesis, and biological evaluation of novel histone deacetylase 1 inhibitors through click chemistry. Bioorg Med Chem Lett 23:3295–3299

Suzuki T, Kouketsu A, Itoh Y, Hisakawa S, Maeda S, Yoshida M, Nakagawa H, Miyata N (2006) Highly potent and selective histone deacetylase 6 inhibitors designed based on a small-molecular substrate. J Med Chem 49:4809–4812

Suzuki T, Ota Y, Ri M, Bando M, Gotoh A, Itoh Y, Tsumoto H, Tatum PR, Mizukami T, Nakagawa H, Iida S, Ueda R, Shirahige K, Miyata N (2012) Rapid discovery of highly potent and selective inhibitors of histone deacetylase 8 using click chemistry to generate candidate libraries. J Med Chem 55:9562–9575

Suzuki T, Kasuya Y, Itoh Y, Ota Y, Zhan P, Asamitsu K, Nakagawa H, Okamoto T, Miyata N (2013) Identification of highly selective and potent histone deacetylase 3 inhibitors using click chemistry-based combinatorial fragment assembly. PLoS One 8:e68669

Talevi A (2015) Multi-target pharmacology: possibilities and limitations of the "skeleton key approach" from a medicinal chemist perspective. Front Pharmacol 6:205

Trivedi P, Adhikari N, Amin SA, Jha T, Ghosh B (2018) Design, synthesis and biological screening of 2-aminobenzamides as selective HDAC3 inhibitors with promising anticancer effects. Eur J Pharm Sci 124:165–181

Tsukamoto S, Sakae Y, Itoh Y, Suzuki T, Okamoto Y (2018) Computational analysis for selectivity of histone deacetylase inhibitor by replica-exchange umbrella sampling molecular dynamics simulations. J Chem Phys 148:125102

Tu B, Zhang M, Liu T, Huang Y (2020) Nanotechnology-based histone deacetylase inhibitors for cancer therapy. Front Cell Dev Biol 8

Vansteenkiste J, Van Cutsem E, Dumez H, Chen C, Ricker JL, Randolph SS, Schöffski P (2008) Early phase II trial of oral vorinostat in relapsed or refractory breast, colorectal, or non-small cell lung cancer. Investig New Drugs 26:483–488

Vergani B, Sandrone G, Marchini M, Ripamonti C, Cellupica E, Galbiati E, Caprini G, Pavich G, Porro G, Rocchio I, Lattanzio M, Pezzuto M, Skorupska M, Cordella P, Pagani P, Pozzi P, Pomarico R, Modena D, Leoni F, Perego R, Fossati G, Steinkühler C, Stevenazzi A (2019) Novel benzohydroxamate-based potent and selective histone deacetylase 6 (HDAC6) inhibitors bearing a pentaheterocyclic scaffold: design, synthesis, and biological evaluation. J Med Chem 62:10711–10739

Vögerl K, Ong N, Senger J, Herp D, Schmidtkunz K, Marek M, Müller M, Bartel K, Shaik TB, Porter NJ, Robaa D, Christianson DW, Romier C, Sippl W, Jung M, Bracher F (2019) Synthesis and biological investigation of phenothiazine-based benzhydroxamic acids as selective histone deacetylase 6 inhibitors. J Med Chem 62:1138–1166

Wagner FF, Olson DE, Gale JP, Kaya T, Weïwer M, Aidoud N, Thomas M, Davoine EL, Lemercier BC, Zhang Y-L, Holson EB (2013) Potent and selective inhibition of histone deacetylase 6 (HDAC6) does not require a surface-binding motif. J Med Chem 56:1772–1776

Wagner FF, Zhang YL, Fass DM, Joseph N, Gale JP, Weïwer M, McCarren P, Fisher SL, Kaya T, Zhao WN, Reis SA, Hennig KM, Thomas M, Lemercier BC, Lewis MC, Guan JS, Moyer MP, Scolnick E, Haggarty SJ, Tsai LH, Holson EB (2015) Kinetically selective inhibitors of histone deacetylase 2 (HDAC2) as cognition enhancers. Chem Sci 6:804–815

Wang X-Q, Bai H-M, Li S-T, Sun H, Min L-Z, Tao B-B, Zhong J, Li B (2017) Knockdown of HDAC1 expression suppresses invasion and induces apoptosis in glioma cells. Oncotarget 8

Wang D, Li W, Zhao R, Chen L, Liu N, Tian Y, Zhao H, Xie M, Lu F, Fang Q, Liang W, Yin F, Li Z (2019) Stabilized peptide HDAC inhibitors derived from HDAC1 substrate H3K56 for the treatment of cancer stem-like cells in vivo. Cancer Res 79(8):canres.1421.2018

Watson PJ, Millard CJ, Riley AM, Robertson NS, Wright LC, Godage HY, Cowley SM, Jamieson AG, Potter BVL, Schwabe JWR (2016) Insights into the activation mechanism of class I HDAC complexes by inositol phosphates. Nat Commun 7:11262

Woyach JA, Kloos RT, Ringel MD, Arbogast D, Collamore M, Zwiebel JA, Grever M, Villalona-Calero M, Shah MH (2009) Lack of therapeutic effect of the histone deacetylase inhibitor vorinostat in patients with metastatic radioiodine-refractory thyroid carcinoma. J Clin Endocrinol Metab 94:164–170

Yang F, Zhao N, Ge D, Chen Y (2019) Next-generation of selective histone deacetylase inhibitors. RSC Adv 9:19571–19583

Yang F, Zhao N, Hu Y, Jiang C-S, Zhang H (2020) The development process: from SAHA to hydroxamate HDAC inhibitors with branched CAP region and linear linker. Chem Biodivers 17:e1900427

Zhang M, Liu E, Cui Y, Huang Y (2017) Nanotechnology-based combination therapy for overcoming multidrug-resistant cancer. Cancer Biol Med 14:212–227

Zhang L, Zhang J, Jiang Q, Zhang L, Song W (2018) Zinc binding groups for histone deacetylase inhibitors. J Enzyme Inhib Med Chem 33:714–721

Zhang Y, Ying JB, Hong JJ, Li FC, Fu TT, Yang FY, Zheng GX, Yao XJ, Lou Y, Qiu Y, Xue WW, Zhu F (2019) How does chirality determine the selective inhibition of histone deacetylase 6? A lesson from trichostatin A enantiomers based on molecular dynamics. ACS Chem Neurosci 10:2467–2480

Zhang M, Ying JB, Wang SS, He D, Zhu H, Zhang C, Tang L, Lin R, Zhang Y (2020) Exploring the binding mechanism of HDAC8 selective inhibitors: lessons from the modification of Cap group. J Cell Biochem 121:3162–3172

Zhou J, Li M, Chen N, Wang S, Luo H-B, Zhang Y, Wu R (2015) Computational design of a time-dependent histone deacetylase 2 selective inhibitor. ACS Chem Biol 10:687–692

Zimmermann GR, Lehár J, Keith CT (2007) Multi-target therapeutics: when the whole is greater than the sum of the parts. Drug Discov Today 12:34–42

The manufacturer's authorised representative in the EU is Springer Nature Customer Service Centre GmbH, Europaplatz 3, 69115 Heidelberg, Germany. If you have any concerns regarding our products, please contact ProductSafety@springernature.com

Printed and bound by CPI Group (UK) Ltd, Croydon, CR0 4YY
23/09/2026
02221542-0005